# Designing Hydrogels and Hydrogel-Derived Materials for Agriculture and Water Sustainability

# Designing Hydrogels and Hydrogel-Derived Materials for Agriculture and Water Sustainability

Guest Editors

**Daniel A. Palacio**
**Manuel Francisco Melendrez Castro**
**Gustavo Cabrera Barjas**

MDPI

Basel • Beijing • Wuhan • Barcelona • Belgrade • Novi Sad • Cluj • Manchester

*Guest Editors*
Daniel A. Palacio
Departamento de Polímeros
Universidad de Concepción
Concepción
Chile

Manuel Francisco Melendrez Castro
Faculty of Sciences for Healthcare Nutrition and Dietetica School
Universidad San Sebastian
Concepción
Chile

Gustavo Cabrera Barjas
Faculty of Sciences for Healthcare Nutrition and Dietetica School
Universidad San Sebastián
Concepción
Chile

*Editorial Office*
MDPI AG
Grosspeteranlage 5
4052 Basel, Switzerland

This is a reprint of the Special Issue, published open access by the journal *Gels* (ISSN 2310-2861), freely accessible at: https://www.mdpi.com/journal/gels/special_issues/XOZ3307K1I.

For citation purposes, cite each article independently as indicated on the article page online and as indicated below:

Lastname, A.A.; Lastname, B.B. Article Title. *Journal Name* **Year**, *Volume Number*, Page Range.

**ISBN 978-3-7258-6282-5 (Hbk)**
**ISBN 978-3-7258-6283-2 (PDF)**
**https://doi.org/10.3390/books978-3-7258-6283-2**

# Contents

# About the Editors

**Daniel A. Palacio**

Daniel A. Palacio is a chemist at the University of Córdoba and holds a PhD in *Chemical Sciences* with a specialization in polymer science and applications from the University of Concepcion in Chile, where he is currently an academic in the Department of Polymers. His academic trajectory includes two postdoctoral appointments funded by the Vice-Rectory of Research and Development of the University of Concepcion and by the National Agency for Research and Development of Chile. His work has been recognized through awards such as the timely completion of his doctoral program, the Best Doctoral Thesis in Sciences at the University of Concepcion, and institutional awards for scientific impact, knowledge protection, and patents. Between 2015 and 2017, he was selected as a researcher for the Frontier Science Program of the Chilean Academy of Sciences, an honor granted to young scientists under 40 for excellence and scientific projection. His scientific productivity includes more than fifty Web of Science-indexed publications, five book chapters, five granted patents, and participation in over fifty national and international conferences. He has served as Guest Editor of three Special Issues in WoS journals and is currently part of six postgraduate academic committees, where he supervises doctoral and master's students. He has led research projects as the Principal Investigator in national funding schemes, including FONDEF I D and Fondecyt Regular, and participates as a co-investigator in multiple institutional and national initiatives. He leads the Laboratory of Functional Polymers and Environment and the Research Group on Bio-Based and Bioinspired Biomaterials, focusing on the development of bio-based and functional polymeric materials for the removal, detection, and degradation of emerging organic contaminants, and on the design of biomaterials for biomedical and agricultural applications within a one health framework.

**Manuel Francisco Melendrez Castro**

Manuel Francisco Melendrez Castro is a chemist at the University of Córdoba and holds a PhD in *Chemical Sciences* from the University of Concepcion in Chile, where he is currently an professor in the Faculty of Sciences for Healthcare Nutrition, along with in the Dietetica School at San Sebastián University. His research interests focus on the design, synthesis, and characterization of advanced functional materials, including nanostructured polymers, composites, and smart materials with applications in environmental remediation, energy, and biomedical technologies. He has led and participated in national and international research projects supported by competitive funding programs, contributing to technological innovation and scientific development in the field of advanced materials. His academic trajectory includes significant achievements such as peer-reviewed publications, patents, technology transfer initiatives, and collaborations with industry and research institutions. He has supervised graduate students in master's and doctoral programs and actively contributes to academic committees, postgraduate training, and scientific dissemination. His work aims to bridge fundamental material science with applied engineering, promoting sustainable and high-impact solutions for current technological challenges.

**Gustavo Cabrera Barjas**

Gustavo Cabrera Barjas graduated in Chemistry from the University of Havana, Cuba, where he also worked as a researcher at the National Institute of Agricultural Sciences. He later completed a master's degree in Physics and Chemistry of Polymers at the same institution, followed by a PhD in *Chemical Sciences* at the University of Concepcion in Chile. His research interests include biotechnology, biopolymers, natural products, and the development of materials and technologies for food, pharmaceutical, and agricultural applications. He has extensive experience in research and development management, participating in multiple national and international collaborative projects funded by agencies such as FONDECYT, INNOVA, FONDEF, ANID, CYTED, and IBEROEKA. He has served as a professor and researcher at several universities and research institutes in Chile and abroad, and has contributed to the development of agrochemical products currently available on the market. His scientific production includes more than sixty international conference presentations, seventy-two ISI-indexed articles, seven SciELO papers, six book chapters, and seven granted patents in the United States and other countries. He has supervised twenty-one undergraduate theses and two doctoral dissertations, contributing to advanced training and scientific capacity building in the areas of biotechnology and polymer science.

# Preface

This Reprint gathers current advances in the design, development, and application of hydrogels and hydrogel-derived materials that help to improve the sustainability of agriculture and water use. The subject of this work focuses on innovative gel-based solutions that contribute to environmental protection, sustainable resource management, and the mitigation of pollution in soil and water systems. The scope encompasses fundamental research, material synthesis, structural characterization, mechanistic studies, and field-oriented applications using natural, synthetic, or hybrid networks capable of responding to complex environmental challenges.

The aim of this Reprint is to bring together scientific contributions that demonstrate how hydrogels can provide functional, economic, and ecologically responsible alternatives for water remediation, soil conditioning, contaminant sequestration, the controlled release of nutrients, or the improvement of agricultural productivity. The purpose of assembling these contributions is to highlight the interdisciplinary advances that connect polymer chemistry, biotechnology, engineering, and environmental sciences, and to stimulate further research in order to establish sustainable technological solutions.

The motivation for preparing this scientific work stems from the growing awareness that water scarcity, soil degradation, and the presence of emerging contaminants are critical global issues with direct impacts on food security, ecosystem resilience, and public health. Hydrogels represent a versatile platform for addressing these problems due to their tunable physicochemical properties, ability to interact with a wide range of molecules, and compatibility with renewable feedstocks and circular economy principles.

This Reprint is aimed at researchers, practitioners, industry professionals, and policymakers interested in advanced biomaterials, agroenvironmental technologies, water treatment systems, and sustainable development strategies. It aims to serve as both a scientific reference and a source of inspiration for those seeking to develop innovative approaches to mitigating environmental pressures through material science.

**Daniel A. Palacio, Manuel Francisco Melendrez Castro, and Gustavo Cabrera Barjas**
*Guest Editors*

*gels*

MDPI

*Article*

# Polyampholytic Hydrogels from Chitosan Macromonomers with Aryl-Mono and Di-Sulfonated Groups: An Approach to the Removal of Copper Ions and Ciprofloxacin in Aqueous Solutions

**Diana Montoya-Rodríguez [1], Alexis Salas [2], Manuel F. Meléndrez [3], Elizabeth R. Gillies [4,5] and Daniel A. Palacio [1,*]**

[1] Departamento de Polímeros, Facultad de Ciencias Químicas, Universidad de Concepción, Edmundo Larenas 129, Concepcion 4070371, Chile; dianamontoya@udec.cl
[2] Department of Mechanical Engineering (DIM), Faculty of Engineering, University of Concepción, 219 Edmundo Larenas, Concepcion 4070409, Chile; alesalas@udec.cl
[3] Facultad de Ciencias para el Cuidado de la Salud, Universidad San Sebastián, Campus Las Tres Pascuales, Lientur 1457, Concepcion 4060000, Chile; manuel.melendrez@uss.cl
[4] Department of Chemistry, The University of Western Ontario, London, ON N6A 5B7, Canada; egillie@uwo.ca
[5] Department of Chemical and Biochemical Engineering, The University of Western Ontario, London, ON N6A 5B7, Canada
* Correspondence: dapalacio@udec.cl

**Abstract**

Functional hydrogels have significant potential for applications in the pharmaceutical, agricultural, and environmental sectors. This study focuses on the synthesis of polyampholytic hydrogels through free radical polymerization using functionalized chitosans. The chitosan was modified with mono and disulfonic groups at different temperatures (25 °C and 60 °C) and reaction times (1, 8, 24 h), followed by further modification with glycidyl methacrylate to introduce vinyl groups into the polymers structure. The modified polymers were analyzed using proton nuclear magnetic resonance, Fourier transform infrared, scanning electron spectroscopy, thermogravimetric analysis, and solubility tests. Specifically, 0.74 mmol/g and 1.58 mmol/g of the primary amine groups available in the chitosan chain (out of a total of 4.93 mmol/g) were substituted with mono- and disulfonic groups, respectively. Following treatment with glycidyl methacrylate, 3.39 mmol/g and 2.21 mmol/g of the remaining primary amine groups in the mono- and disulfonic polymers, respectively, were substituted. The hydrogels obtained by the modified polymers at optimal conditions of 1 h and 25 °C, were characterized by the techniques already mentioned in addition to rheological tests, and water absorption studies across different pHs. The hydrogels demonstrated potential for environmental remediation, particularly in adsorptions of ciprofloxacin (CPX) and copper ($Cu^{2+}$) from aqueous solutions at pH 7, achieving adsorption efficiencies of 24–25% for CPX and 83% for $Cu^{2+}$. The results suggest that the synthesized hydrogels could provide an eco-friendly and efficient solution to challenges in wastewater treatment.

**Keywords:** chitosan; sulfonation; polyampholytic hydrogels; copper removal; ciprofloxacin removal

## 1. Introduction

Hydrogels (HGs) are three-dimensional structures composed of high molecular weight hydrophilic polymers that are either physically or chemically cross-linked, providing them with a high capacity for water absorption. This property makes them desirable for

 https://doi.org/10.3390/gels11080622

applications across various fields, including environmental remediation, tissue engineering, regenerative medicine, drug administration, and absorbent materials, among others [1,2]. In particular, hydrogels with polyampholytic properties have aroused special interest in environmental remediation. This is because polyampholytic HGs combine intrinsic characteristics of this type of materials such as porosity, with the ability to participate in interactions such as hydrogen bonding, van der Waals and electrostatic forces, cation exchange and complex formation, which improves their potential for the adsorption of pollutants [3]. Among the polymers used to obtain this type of hydrogels, the following stand out: polyacrylamide, cyclodextrin, chitosan, polyacrylate, and cellulose, because their combination has polyampholytic properties conferred by the combination of anionic and cationic polymers. Others are obtained from the functionalization of these polymers with compounds such as 1,3-diaminopropane, 1,4-diaminobutane, graphene oxide, among others [4–9].

Chitin stands out as the second most abundant natural polymer and one of the high molecular weight polymers used to obtain functional HGs. It has a linear structure composed of poly[β-(1-4)-2-acetamide-2-deoxy-D-glucopyranose] [10]. Through the partial deacetylation of chitin, chitosan is obtained, a linear cationic polysaccharide consisting of β-(1-4)-2-deoxy-2-amino-D-glucopyranose (D-glucosamine) and β-(1-4)-2-deoxy-2-acetamido-D-glucopyranose (N-acetyl-D-glucosamine) units [11]. Chitosan has been widely studied in various fields due to its unique properties, including biodegradability, hemocompatibility, biocompatibility, and antimicrobial activity, which make it suitable for numerous biological, chemical, and medical applications. The distinctive structure of chitosan, containing reactive amino and hydroxyl groups, allows for extensive chemical modification that enhances its solubility and reactivity in different environments [12,13].

Extensive research has been conducted to develop chemical, physical or enzymatic modifications to chitosan to enhance or introduce new properties that can be applied across diverse fields, including food, pharmaceutical, biotechnological, medical, textile, paper, agricultural, and environmental applications [11,14,15]. Several chemical modification strategies have been explored, depending on the desired applications and properties to be improved. For instance, modifications such as alkylation, acylation, hydroxyalkylation, carboxyalkylation, phosphorylation, azidation, and thiolation have been explored [16–18].

The research described here used reductive amination to enhance the absorption of different contaminants through chitosan functionalization. Specifically, our aim was to modify chitosan with sulfonic acid groups to provide anionic charge, supported by previous studies showing the effectiveness of these functional groups in the adsorption of pollutants [19–21]. The objective was to improve the solubility of chitosan at basic pH and allow the formation of polyampholytic HGs with characteristics that favor their adsorbent properties and interactions with contaminants. Although the sulfonation approach has been widely explored for biomedical applications and the development of ion exchange filters, most research has focused on the synthesis and characterization of these chitosan derivatives, with less emphasis on their use. Therefore, in this research we propose to go beyond what has already been reported in the literature on the chitosan functionalization, obtaining hydrogels with polyampholytic characteristics conferred by the incorporation of anionic functional groups, for the removal of contaminants in aqueous matrices [19,21,22].

The hypothesis of the present work is that the modification of chitosan with anionic groups, to obtain polyampholytic hydrogels, favors its ability to absorb different types of contaminants in aqueous media. This strategy represents a significant advance by combining the versatility of sulfonic groups with the water retention properties and biocompatibility of chitosan, thus offering an innovative and effective solution to environmental challenges related to water purification. This is particularly relevant in Chile, a country that faces increasing pressure on its freshwater resources due to mining, agriculture, and population growth, all of which contribute to the presence of persistent pollutants in surface and groundwater systems. In order to explore the first application of the polyampholytic hydrogel, two representative contaminants of concern in Chile were selected, $Cu^{2+}$ and ciprofloxacin (CPX), whose cationic and ampholytic nature, respectively, could favor their electrostatic and hydrogen bonding interaction with the hydrogel. CPX is a broad-spectrum antibiotic, which is used in both humans and animals, ranking among the best-selling therapeutics in Chile [23]. Its metabolism is incomplete, so generally a large percentage of the dose is excreted through the urine and feces [24,25], thereby contaminating aquatic ecosystems. This contamination can result in the generation of resistant bacteria, insertion into the food chain and adverse effects such as an impact on the trophic abundance of some species after only one month of exposure [26–28].

Regarding metal ions, Chile is one of the countries with the highest copper production worldwide, with the potential for copper ion accumulation in the environment and its contamination of the food chain. Copper can bioaccumulate, leading to negative health effects such as an increased rates of cancer [29–31]. Consequently, researchers have designed various HG-type materials based on polymers such as lignin, cellulose, graphene or chitosan to adsorb copper ions [32–37]. In Chile, copper-related pollution is especially critical in regions such as Antofagasta and Atacama, where mining wastewater has led to elevated concentrations of heavy metals in rivers and aquifers, posing significant risks to both environmental and human health [38]. The development of hydrogels capable of selectively capturing such metals from contaminated waters aligns with the country's need for accessible and efficient water treatment technologies.

To evaluate the hypothesis, this work begins by describing the preparation of sulfonated polymers through reductive amination, applying different reaction times and temperatures ranging between 1 and 24 h and 25 to 60 °C, to choose those with higher degrees of functionalization. The best conditions were found to be one hour of reaction at 25 °C, reaching a DS of 15 and 32%. The resulting polymers were then modified with glycidyl methacrylate (GMA), to incorporate vinyl bonds that would participate in the subsequent free radical polymerization to form the hydrogels. Free radical polymerization is a relatively simple, versatile, and low-cost crosslinking method [39]. These polyampholytic hydrogels were characterized by FTIR, TGA, and SEM, demonstrating 20 to 21 times the initial mass increase in swelling processes. Finally, the introduction of sulfonic groups into chitosan significantly enhances its water solubility and swelling properties, which makes it particularly suitable for environmental applications such as heavy metal and pharmaceutical removal from water. Our findings suggest that these materials have the potential to address challenges in wastewater treatment, offering an eco-friendly and efficient solution.

## 2. Results and Discussion

### 2.1. Functionalization of CTS with Sodium Sulfonates

#### 2.1.1. Characterizations by FTIR, $^1$H-NMR, and TGA

The chitosan derivates modified with 2-formylbenzylsulfonic (FB1S) and 4-formyl-1,3-benzenedisulfonic (FB2S) under different time and temperature conditions were initially characterized by FTIR spectroscopy to determine whether the evaluated parameters, such as temperature and reaction time, led to the effective functionalization of the biopolymer. The FTIR spectra are shown in Figure 1 following the nomenclature proposed in Table 2, according to the modification conditions. The results reveal two characteristic bands for both FB1S and FB2S, indicating the successful functionalization of the polymer. The first notable signal was observed at 1450 cm$^{-1}$, corresponding to the bending vibrations of the substituted amine groups (Figure 2c,d). Additionally, a series of peaks attributed to the stretching of the S=O bond from the sulfonic groups appeared between 1170 and 1190 $cm^{-1}$ in the spectra of the functionalized derivatives [40–42].

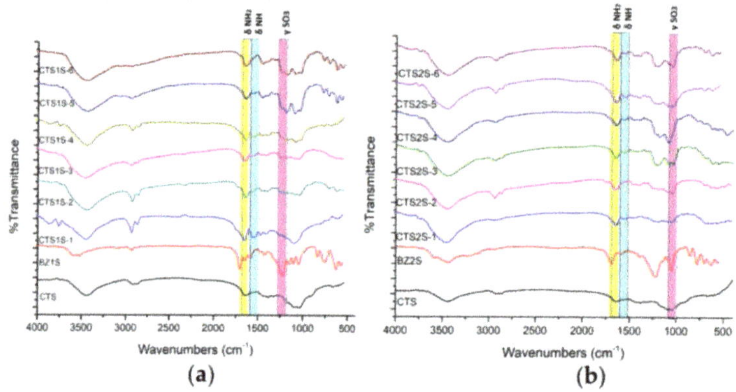


**Figure 1.** FTIR spectra for CTS and CTS modified with sulfonic groups. (**a**) Polymers obtained by modifying CTS with FB1S. (**b**) Polymers obtained by modifying CTS with FB2S.

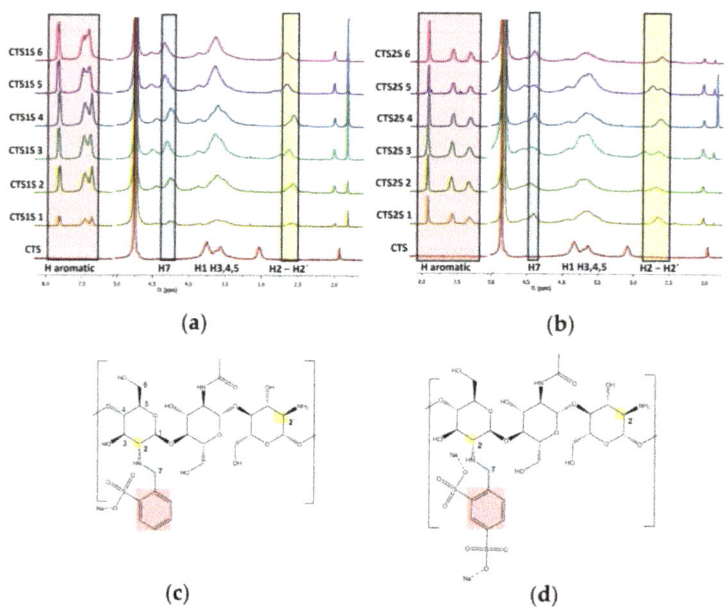


**Figure 2.** $^{1}$H-NMR spectra of polymers modified with sulfonic groups. (**a**) Polymers obtained by modifying chitosan with FB1S. (**b**) Polymers obtained by modifying chitosan with FB2S. (**c**) Expected structures of the polymers modified with FB1S. (**d**) Expected structures of the polymers modified with FB2S.

$^{1}$H-NMR spectra of the biopolymers are shown in Figure 2, where the analysis of two regions of the spectra supports the functionalization of the CTS. First, new peaks appeared between 7.1 and 7.5 ppm corresponding to the protons on the aromatic group. In addition, functionalization resulted in peaks between 4.2 and 4.5 ppm, corresponding to the benzylic protons [43,44]. These results qualitatively indicated that all the conditions evaluated led to a modification of the CTS. To quantify the degree of substitution (DS), Equation (1) was used where the integrals of the peaks corresponding to the aromatic protons (Ar), the proton from the alpha ring adjacent to the secondary amine ($H_2$) or primary amine ($H_2'$), as well as the degree of deacetylation of the starting chitosan (DDA) and the number of aromatic protons, being 3 and 4 for FB1S and FB2S, respectively, were considered [44]. The results obtained are presented in Table 1. It was determined that the conditions that lead to higher degrees of modification were one hour of reaction at 25 °C, reaching a DS of 15 and 32% for CTS1S and CTS2S, respectively. Thus, CTS1S remains structurally with 4.19 mmol of residual primary amino groups, while CTS2S—with 3.35 mmol/g of primary amine groups.

$$\%DS = (Ar/n)/((H_2 + H_2')/DDA) \times 100 \quad (1)$$

**Table 1.** Results obtained by $^{1}$H-NMR spectroscopy, SEM-EDS, and TGA for biopolymers modified with mono- and disulfonic groups under different time and temperature conditions.

| | Samples | 1 h—25 °C | 1 h—60 °C | 8 h—25 °C | 8 h—60 °C | 24 h—25 °C | 24 h—60 °C |
|---|---|---|---|---|---|---|---|
| $^{1}$H-NMR | %DS [a] CTS1S | 15 | 11 | 13 | 11 | 11 | 8 |
| | %DS [a] CTS2S | 32 | 21 | 26 | 22 | 39 | 16 |
| SEM-EDS | %S [b] CTS1S | 12.29 | 9.76 | 2.77 | 8.73 | 4.74 | 11.11 |
| | %S [b] CTS2S | 11.15 | 13.13 | 7.86 | 10.98 | 12.21 | 14.83 |
| TGA | %Wt [c] CTS1S | 21 | 20 | 18 | 13 | 16 | 17 |
| | %Wt [c] CTS2S | 22 | 11 | 13 | 19 | 14 | 16 |

[a] Degree of substitution. [b] Percentage of sulfur. [c] Percent of weight loss of mono or disulfonic aryl.

The degrees of substitution determined by $^{1}$H-NMR spectroscopy were complemented by energy-dispersive X-ray scanning electron microscopy (SEM-EDS) analyses. Specifically, the percentages of the elements of interest such as carbon, nitrogen, oxygen, and sulfur were quantified (Table 1), where the average results of random points in the samples of the polymers are presented. Sulfur was of special interest, as a quantitative indicator of the degree of functionalization. The heterogeneity of the sample makes it difficult to establish comparisons between the results of both techniques, since the modification with sulfonic groups depends on the distribution and availability of the amino groups. However, it was observed that the polymers with higher percentages of sulfur (%S) were prepared with a reaction time of 1 h at 25 °C.

In TGA (See Figure 3), the curves obtained allowed the identification of the stages in which the polymers degraded, where the gradients in weight percentage represented abrupt changes in weight loss over specific temperature intervals. Figure 2 shows the results obtained for the modified polymers, in comparison with the commercial CTS sample. For CTS, the decomposition was evident in two steps—the first, with a maximum rate at 54 °C, corresponding to the loss of water, and the second at 297 °C associated with the decomposition and carbonization of the sugar ring of the CTS. On the other hand, all modified CTS samples decomposed in three steps.

The first step, corresponding to the loss of water was observed at temperatures less than 70 °C. The second step had a maximum rate of mass loss between 260 and 290 °C, and was attributed to the decomposition of the CTS sugar ring as well as the amine and sulfonic functional groups. Finally, the decomposition step corresponding to the organic residues of chitosan and the aryl groups occurred between 340 and 440 °C [12,17]. The results obtained show a thermal stability associated with the electrostatic interactions produced by the modification of sulfonic groups [12,17].

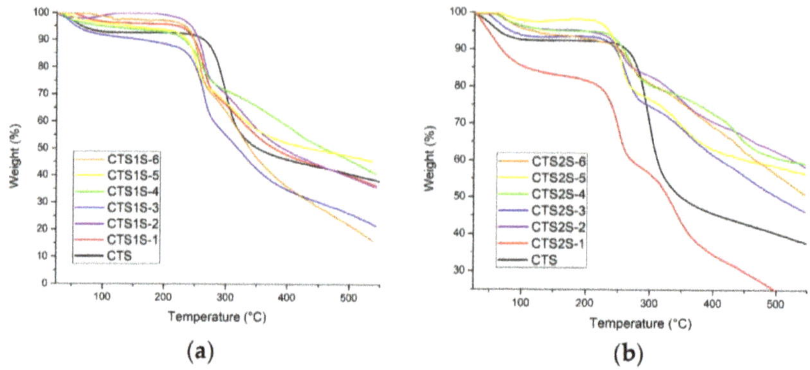


**Figure 3.** Results obtained by thermogravimetric analysis of the modified polymers at different time and temperature conditions. (**a**) CTS1S polymers. (**b**) CTS2S polymers.

Table 1 shows the percentages of weight loss obtained from the analysis of the TGA graphs, corresponding to the third step, attributed to the modification by the aryl-sulfonated groups for each of the samples, with the objective of contrasting the weight losses for products obtained using different reaction conditions. In both cases, the greatest weight losses associated with the aryl groups were obtained for polymers prepared using the conditions of 1 h and 25 °C.

2.1.2. Solubility

Solubility is one of the properties of interest in chemical processes, particularly solubility in water, since aqueous solubility helps minimize the use of organic solvents that contribute to environmental pollution [45,46]. The solubility of commercial chitosan has been reported as a function of the degree of deacetylation of the polymer, as it is related to the number of deacetylated units, going from insoluble chitin (deacetylation < 50%) to a chitosan soluble in acidic aqueous media (pH < 6.5), which is attributed to the protonation of the amino groups [47]. In this research, we sought to determine the influence of the degrees of modification arising from the different synthesis conditions on the solubility of the CTS.

The absorbance results obtained as a function of pH are shown in Figure 4, and are complemented by Figure S3, constructed from the transmittance results. In Figure 4, a significant difference can be seen in the absorbance between the two CTS derivatives, with the polymer modified with FB1S (CTS1S) being soluble at basic pH and in most cases insoluble at acidic pH, contrary to CTS, which at pH > 6 is insoluble, due to the high formation of hydrogen bonds and generating curling. For the polymers obtained from the modification with FB2S (CTS2S), solubility over a wide range of pH values is evident, resulting from introduction of the disulfonated aryls.

The solubility behavior can be associated with the presence of sulfonic and amino groups simultaneously, giving the polymer polyampholytic behavior. The solubility at basic pH is justified by the presence of sulfonic groups (pKa = ~2.5), while at a pH below 5.0, the amino groups of chitosan are protonated [18,47]. However, the differences between one group of modifications and the other can be attributed both to the degree of modification and consequently the availability of amino groups that favor solubility at acidic pH, the effect of the $\pi$-$\pi$ interactions of the aryl groups, and the repulsions between the sulfonic groups, which allows the solubility to be preserved in the case of CTS2S in all pH ranges [21].

Overall, it was possible to achieve different degrees of modification to CTS under different conditions of temperature and reaction time. The qualitative and semiquantitative analyses lead to the conclusion that 1 h and 25 °C are the conditions that give rise to polymers with a higher degree of modification, which means lower economic expenses in terms of energy and time, thus favoring a friendlier process.

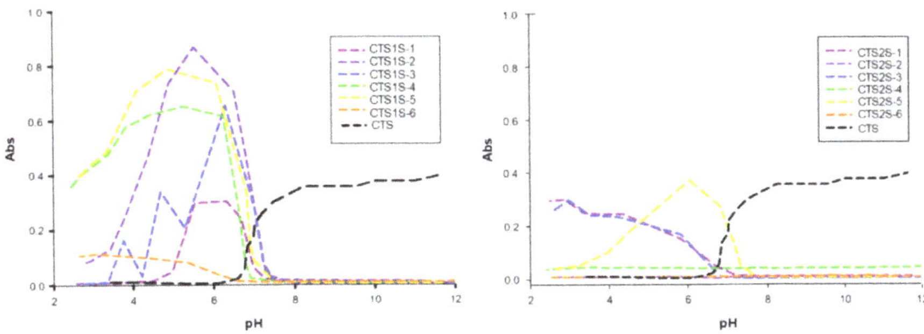


(a) (b)

**Figure 4.** Absorbance results for polymers modified with sulfonic groups. (**a**) Polymers obtained by modifying chitosan with FB1S. (**b**) Polymers obtained by modifying chitosan with FB2S.

## 2.2. *Preparation of Polyampholytic Hydrogels*

Starting from the chitosan modified using the conditions of 1 h and 25 °C, the modification was continued to introduce methacrylate groups for radical polymerization. CTS1S and CTS2S were reacted with GMA at pH 6 and 50 °C for 24 h. The 1H-NMR spectroscopic analysis of the resulting polymers CTSV1S (monosulfonic) and CTSV2S (disulfonic) polymers showed the appearance of new peaks between 5.0 and 5.5 ppm corresponding to the methacrylate groups (Figure 5a).

The NMR analysis agreed with the FTIR analysis (Figure 5b), where three bands of interest were identified, the first two corresponding to the C=C stretching signals between 560 and 630 $cm^{-1}$ and 1325–1396 $cm^{-1}$ [48,49]; and the third, associated with the stretching of OH between 3200 and 3700 $cm^{-1}$. Methacrylation with GMA can result from two different mechanisms, transesterification or opening of the epoxide ring, the latter being the preferred one in acidic media. The reaction results allow us to conclude that the mechanism is the opening of the epoxide rings by the amino groups (still available) and the alkoxy groups. Nucleophilic attack by the amino groups is preferred, with 81 and 66% of the remaining primary amino groups of CTS1S and CTS2S, respectively, functionalized with the GMA. Following treatment with glycidyl methacrylate, 3.39 mmol/g and 2.21 mmol/g of the remaining primary amine groups in the mono- and disulfonic polymers, respectively, were substituted. It evidences an occupation of the GMA by the available amino groups of 3.39 mmol/g of amine for CTS1S and 2.21 mmol/g of amine for CTS2S, which leads to a greater modification of the GMA in the CTS1S.The degree of modification with glycidyl methacrylate (GMA) was calculated using the relationship between the integrals of the NMR signals, and the protons of GMA and CTS [20,49].

After confirming the modifications with GMA, given the presence of alkene carbon signals, the CTSV1S and CTSV2S were polymerized under free radical conditions to obtain the HGCTS1S and HGCTS2S hydrogels, respectively, (Figure 5d). Figure 5b shows the decrease in the signals associated with the vibrations of the C=C bonds for the HG spectra compared to the spectra of the CTSV1S and CTSV2S polymers, arising from their consumption in the radical polymerization [50]. The process of obtaining the polyampholytic hydrogel is summarized in Figure 6.

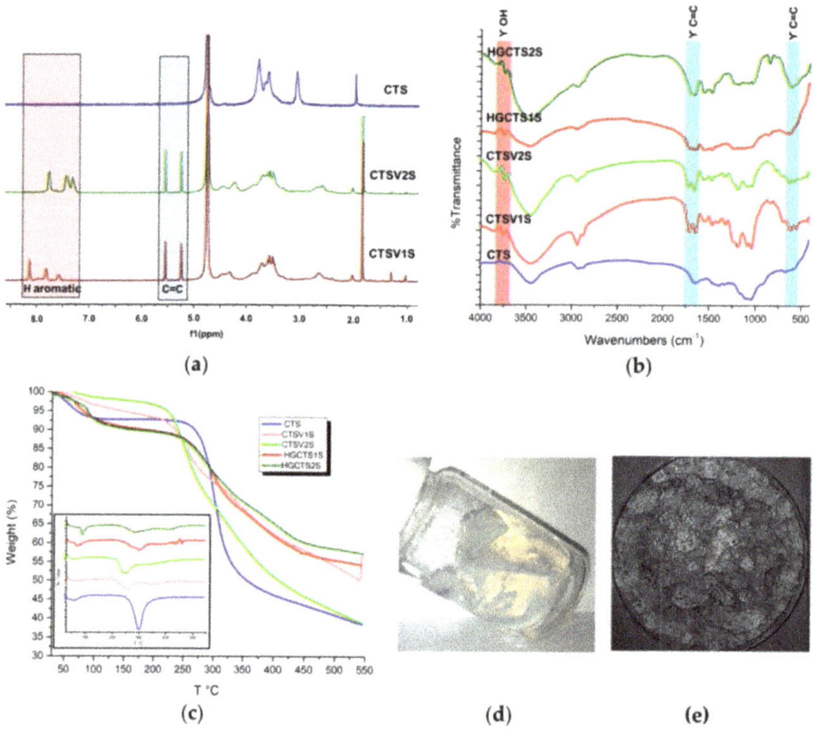


**Figure 5.** (**a**) $^1$H-NMR spectra of CTS and GMA modified polymers. (**b**) FTIR spectra for CTS, GMA-modified polymer and obtained hydrogels. (**c**) TGA and DTG polymer modified with GMA and hydrogels obtained. (**d**,**e**) Image of the hydrogel at the end of the reaction and after drying, respectively.

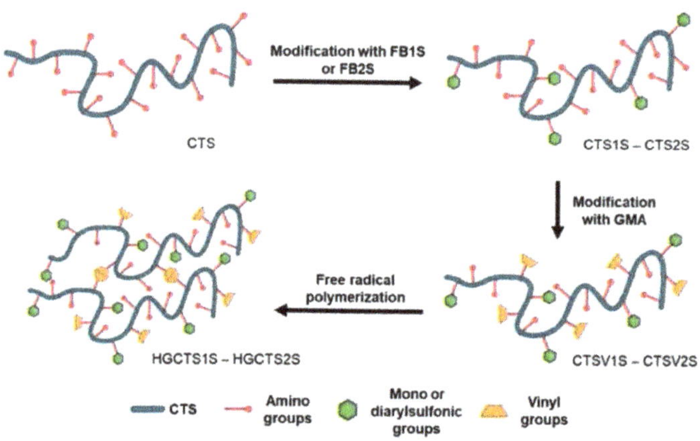


**Figure 6.** Preparation of HGCTS1S and HGCTS2S. CTS: 4.93, CTS1S: 1.13, CTS2S: 3.35, CTSV1S: 0.79, and CTSV2S: 1.13 mmol/g of primary amine, respectively.

The results of the TGA carried out for the sulfonated macromonomers and the hydrogels after several days of drying are shown in Figure 5c. In general, the modified polymers (CTSV1S and CTSV2S) and the hydrogels present similar behaviors to each other with three stages of thermal decomposition and differ from what was observed for the commercial CTS sample, which only shows two stages as mentioned in the previous section. The first stage corresponds to the loss of water, which in the case of the HGs, is larger in magnitude based on the derived thermogravimetry (DTG) (Figure 5c), as a result of the increased hydrophilic capacity of the materials. The second loss for the polymers occurs at 240 and 260 °C, and for the HGs at 305 and 284 °C for HGCTS1S and HGCTS2S, respectively. The differences in behavior for the polymers versus HGs can be attributed to variations in intermolecular interactions. In the case of the polymers, sulfonation can interfere with the formation of hydrogen bonds, whereas in the HGs, these units are now covalently intertwined, favoring thermal stability, which is evident in the residual mass obtained for the HG [50,51].

To evaluate the water absorption capacity, swelling tests for the hydrogels were carried out at four different pH values. Figure 7a shows the degrees of swelling (W) for each hydrogel at different pH values after 24 h, with the HGCTS1S being the one that shows a greater water absorption capacity (20 to 21-fold the initial mass), reaching in almost all cases double those achieved by the HGCTS2S. These results can be associated with both the degree of sulfonate functionalization, which affects the hydrophilicity of the material, and therefore its water absorption capacity, and the degree of cross-linking of the materials. This is related to the degree of modification with GMA, associated with a greater availability of vinyl bonds that participate in the radical addition reaction.

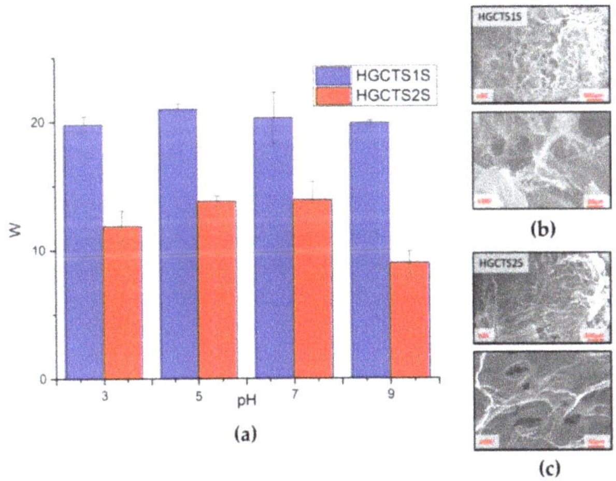


**Figure 7.** (**a**) Swelling degrees of hydrogels as a function of pH (3, 5, 7, and 9). Images achieved with SEM (30× and 300×) for the (**b**) HGCTS1S and (**c**) HGCTS2S. 2.3. Application in Pollutant Removal.

The swelling was also corroborated by images obtained by high-resolution scanning electron microscopy (SEM) (Figure 7b,c), with which the surface morphology of the HG was characterized. Greater roughness was evident in the case of HGCTS1S, associated with a greater degree of modification as has already been reported in other investigations [52–54], in this case, corresponding to the addition of benzyl monosulfonate and disulfonate groups to the main column of chitosan. This is consistent with the results obtained from BET and BJH analyses, which reveal significant differences in the porous structure of the HGCTS1S and HGCTS2S hydrogels, despite their synthesis. HGCTS1S has a surface area of 2.88 $m^2/g$ and an average pore size of 8.79 nm, placing it within the mesoporous range according to the IUPAC classification. In contrast, HGCTS2S displays a very low surface area of only 0.04 $m^2/g$ and no detectable pore distribution, suggesting a compact or non-porous structure from the perspective of nitrogen adsorption analysis.

Rheological studies of the complex modulus account for the total resistance against the applied deformation (Figure S4), revealing better stability of HGCTS1S, compared to HGCTS2S. This is related to the functionalization of amino groups after modification with GMA, so the availability of vinyl groups that participate in radical polymerization favors cross-linking, reflected in a greater storage modulus ($G'$). A higher $G'$ implies enhanced elasticity and mechanical integrity, which are essential for the hydrogel to maintain its three-dimensional network during practical use, especially under repeated compression or handling. Although it is known that greater cross-linking generates a lower water absorption capacity, in this case, the low stability of HGCTS2S could impact an absorption-desorption balance, which reduces the degree of swelling.

Specifically, HGCTS1S, with its higher crosslinking density and elastic modulus, is expected to show better regeneration efficiency due to its ability to retain its structure over multiple cycles. Furthermore, its increased swelling favors interactions such as hydrogen bonding and hydrophobic interactions. In contrast, the denser and stiffer network of HGCTS2S could limit the accessibility of active sites, affecting its adsorption efficiency. Taking into account that both materials were developed under the same conditions, from the methacrylation process to obtaining the hydrogel, the variations in stability and water absorption capacity can be attributed to steric impediments given the presence of more sulfate groups. This is attributed to steric hindrance resulting from the higher number of disulfonate groups, which can restrict the mobility of the polymer network [55–57].

A first approach to the possible applications of these materials for the removal of different types of contaminants in aqueous matrices was carried out by performing removal tests on two types of contaminants of interest present in aqueous systems: the antibiotic CPX and $Cu^{2+}$. These tests were carried out at pH 7 in order to emulate natural conditions and based on previous studies where this pH is reported as optimal for the removal of $Cu^{2+}$ and CPX species in aqueous media [58–61]. The removal percentages achieved after three hours of contact for these contaminants are presented in Figure 8b. The results showed significant differences between CPX versus $Cu^{2+}$, with the adsorption percentages achieved for $Cu^{2+}$ being considerably higher ($61.8 \pm 1.8$–$83.7 \pm 0.3\%$) than for CPX ($25.0 \pm 0.01$–$24.4 \pm 0.01\%$). To further explore the influence of pH on CPX adsorption and evaluate the applicability of the hydrogels under variable environmental conditions, additional adsorption tests were carried out at pH 3, 7, and 9. The results, shown in Figure 8c, indicate that pH has a moderate but significant effect on CPX removal efficiency. For HGCTS1S, the highest adsorption was observed at pH 3 ($28.9 \pm 1.06$ %), with a slight decrease at pH 7 ($25.0 \pm 1.59$ %), and further reduction at pH 9 ($20.5 \pm 1.59$ %). In contrast, HGCTS2S showed $19.9 \pm 0.79$ % at pH 3, a maximum at pH 7 ($24.4 \pm 2.23$ %), and $20.1 \pm 2.83$ % at pH 9.

The slightly better performance of HGCTS1S at this pH may be due to lower steric hindrance compared to HGCTS2S. At pH 7, CPX adopts a zwitterionic form where electrostatic interactions are minimized and other mechanisms, such as hydrogen bonding or hydrophobic interactions, could dominate. At pH 9, the antibiotic becomes negatively charged, which may result in repulsion from the negatively charged hydrogel surfaces, explaining the drop in removal efficiency.

In order to understand the removal mechanism, experiments were carried out for the determination of the zero charge point ($pH_{ZCP}$) for both the chitosan hydrogel (HGCTS) and for each of the sulfonated chitosan-based hydrogels as presented in Figure 8a. The results obtained suggest that the modification with sulfonic groups considerably modifies the $pH_{ZCP}$, being 6.4, 5.0, and 3.9 for HGCTS, HGCTS1S, and HGCTS2S, respectively. Since $pH_{ZCP}$ is the pH at which the value of the net surface charge of the adsorbent is zero, it is understood that below this point, positive charges predominate, and above these, negative charges predominate. The results confirm the significant increase in the arrangement of negative charges on the surface of the material associated with the presence of mono and disulfonic groups, being greater in the case of the latter.

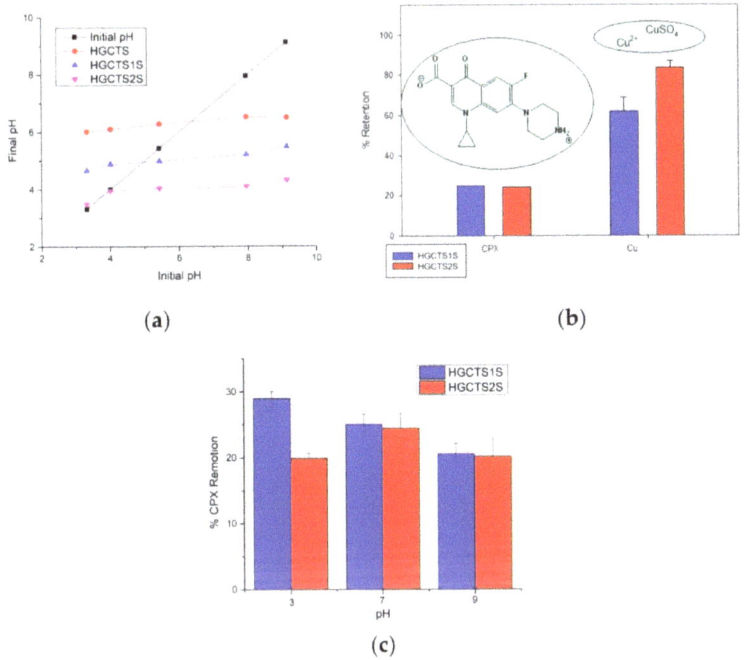


**Figure 8.** (**a**) Point of zero charge curve of hydrogels. (**b**) Retention percentages and chemical structures of the aqueous contaminants studied. (**c**) Ciprofloxacin removal percentage at different pH values.

The $pH_{ZCP}$ results correlate with the results of the percentage of removal achieved by each of these materials, because the differences in the retention of the two contaminants by the hydrogels can be attributed to their different chemical structures and properties of each contaminant. On the one hand, CPX has pKa values of 6 and 8.7, with an isoelectric point of ~7.1 so at pH 7 the antibiotic molecule is largely in the zwitterionic state [41,62], so repulsive or neutral electrostatic interactions may dominate. In the case of the copper ion, according to its speciation, at pH 6 it is present mainly as a free $Cu^{2+}$ ion in solution and in a lower percentage as $CuSO_4$ [63]. Taking into account the $pH_{ZCP}$, it can be concluded that the density of anionic charges in the material given the modification with sulfonic groups provides the material with anionic characteristics at pH 7, so that copper ions ($Cu^{2+}$) benefit by achieving greater retention. Additionally, the differences between both hydrogels are attributed to the fact that HGCTS2S has a greater availability of anionic groups than HGCTS1S (also evidenced in $pH_{ZCP}$), so the adsorption of contaminants in the cationic state such as $Cu^{2+}$ is favored over that of zwitterionic species such as CPX in which they could be impeded by electrostatic repulsions.

These results allow us to know that the modification of chitosan with sulfonic groups favored not only the solubility and water absorption capacity and therefore its swelling, but also when applied preliminarily for the removal of contaminants, it favors the adsorption of contaminants of a cationic nature such as $Cu^{2+}$. As a result of this research, it is also suggested to continue exploring the applications of this type of materials, evaluating different functionalization relationships to enhance the polyampholytic behavior of the hydrogel, such as evaluating different modifications to obtain higher percentages in terms of the elimination of fluoroquinolone antibiotics.

## 3. Conclusions

This study successfully demonstrated the synthesis of polyampholytic hydrogels from chitosan derivatives functionalized with mono- and disulfonic groups. The hydrogels displayed excellent swelling properties and high adsorption efficiencies for copper ions, making them promising candidates for the removal of heavy metals from wastewater. The best degrees of modification with mono- and disulfonic aryl groups were achieved in 1 h at room temperature, compared to longer time periods or higher temperatures. The hydrogel that demonstrated greater water absorption capability was the hydrogel with aryl-monosulfonate groups, which underwent twice as much swelling as the disulfonic one. This swelling may be associated with differing degrees of functionalization with the sulfonic acids and the cross-linkable methacrylate moieties, as corroborated by $^{1}$H-NMR spectroscopy, EDS, and TGA studies. For the CPX and $Cu^{2+}$ removal studies, the aryl-disulfonic chitosan hydrogel favored the adsorption of copper with 83% removal compared to 62% in the case of the hydrogel with aryl-monosulfonate groups. For the CPX at pH 7.0, a removal degree of 25% was achieved, with its lower removal efficiency compared to $Cu^{2+}$ attributed to its zwitterionic rather than cationic state. In conclusion, the modification with anionic groups allowed the preparation of polyampholyte hydrogels, which showed that they can be materials with potential use in the processes of eliminating mainly cationic contaminants of organic and inorganic origin in aqueous systems. It is important to note that this study represents an initial approximation of the hydrogel's potential as an adsorbent for ciprofloxacin and copper ions. Further investigations, including detailed kinetic and isotherm analyses, will be essential to optimize the adsorption performance and fully elucidate the adsorption mechanisms. These studies will be conducted in future work to provide a comprehensive understanding of the hydrogel's capabilities under various conditions.

## 4. Materials and Methods

### *4.1. General Materials and Procedures*

To obtain the zwitterionic polyelectrolyte, chitosan 98% *w*/*w* (CTS) was used with a deacetylation degree of 79.7% (4.93 mmol/g of amino groups) calculated conductometrically, ash percentage of 0.2–0.9%, and molar mass of 150–300 kg/mol, supplied by Quitoquimica (Chile). Sodium cyanoborohydride 95% ($NaBH_3CN$), 2-formylbenzylsulfonic acid sodium salt 95% (FB1S), 4-formyl-1,3-benzenedisulfonic acid 97% (FB2S), glycidyl methacrylate (GMA) 97%, and ammonium persulfate (APS) were purchased from Sigma-Aldrich, Chile. Also, 1 M NaOH, 0.1 M HCl, N,N,N′,N′-tetramethyl ethylenediamine (TEMED) 99%, and methanol (99%) were purchased from Merck, Chile. Glacial acetic acid (99.8%) was obtained from Winkler Ltd., Chile.

### *4.2. CTS Characterization: Potentiometric Measurement of Deacetylation Degree (DDA) Determination, and $^{13}$C-NMR*

The polymer (0.2 g) was dissolved under stirring in 5 mL of HCl (1.0 M). Then, 450 mL of 0.001 M NaCl was added and stirred to achieve a uniform solution. This solution was titrated conductometrically with 60 mL of standardized 0.1 M NaOH. The titration was carried out by discharging 0.5 mL of 0.1 M NaOH for each measurement. Measurements were performed by using Thermo Scientific Orion 5-Star Plus (Thermo Scientific, Waltham, WA, USA). The DDA of CTS was calculated using Equation (2) [64,65]. (Figure S1, Equation (2)). The $^{13}$C NMR spectra were detected by 0.1 M $CF_3COOD$ as solvents, and the NMR spectra was analyzed by MestReNova software v14.2.3-29241 (Figure S2) [43,66,67].

$$DDA = (C_{NaOH} \times \Delta V_{NaOH} \times 16)/(m_{CTS} \times 0.0994) \quad (2)$$

DDA = degree of deacetylation
$C_{NaOH}$ = NaOH concentration (M)
$\Delta V_{NaOH}$ = volume difference between the two inflection points (L)
$m_{CTS}$ = mass of the analyzed chitosan (g)

### 4.3. Functionalization of CTS with Sodium Sulfonates

The modification of chitosan (CTS) with sulfonic groups (Figure 9) was carried out by reductive amination [24,48]; varying the time (1.0, 8.0 and 24.0 h) and temperature (25 and 60 °C) at a fixed molar ratio of 1.5 equivalents of sulfonic acid/available amino group, 100 and 80 mg of CTS were dissolved in 10 mL of a 1% (*v*/*v*) aqueous acetic acid solution and subsequently 9 mL of methanol were added for modification with FB1S and FB2S, respectively. After obtaining a homogeneous mixture, FB1S (0.74 mmol) and FB2S (0.59 mmol), dissolved in 1 mL of water, were added to separate flasks. The reaction mixture was stirred for 3 min and then finally 60 mg of $NaBH_3CN$ are added, followed by stirring for 1, 8 or 24 h, at temperatures of 25 or 60 °C. The product was then dialyzed for four days using a cellulose membrane regenerated (Sigma-Aldrich, St. Louis, MI, USA) with a molecular weight cut-off of 12–14 kDa. Finally, the modified polymers were frozen and then lyophilized for 24 h.

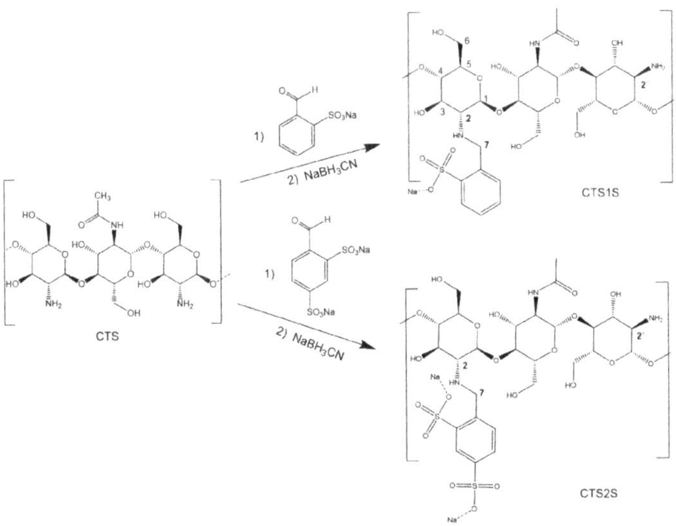


**Figure 9.** Synthesis of benzyl monosulfonate and disulfonate derivatives CTS1S and CTS2S.

Table 2 shows the codes that refer to the different conditions explored.

**Table 2.** Codes referring to the different time and temperature conditions used to prepare polymers modified with mono- and disulfonic groups.

| | 1 h 25 °C | 1 h 60 °C | 8 h 25 °C | 8 h 60 °C | 24 h 25 °C | 24 h 60 °C |
|---|---|---|---|---|---|---|
| FB1S | CTS1S-1 | CTS1S-2 | CTS1S-3 | CTS1S-4 | CTS1S-5 | CTS1S-6 |
| FB2S | CTS2S-1 | CTS2S-2 | CTS2S-3 | CTS2S-4 | CTS2S-5 | CTS2S-6 |

The modified CTSs obtained were characterized by $^1$H-NMR spectroscopy, dissolving 10 mg of polymer in 600 μL of 0.01 M NaOD. Through EDS analysis, the percentage composition of carbon, oxygen, nitrogen, and sulfur was determined by scanning electron microscopy (JEOL JSM 6010 Plus/LV SEM, JEOL Ltd., Tokyo, Japan) after the samples were coated in gold with LUXOR Gold Coater equipment. Thermogravimetric analysis (TGA) was carried out using a Netzsch TG 209F1 iris 220-12-0045-L instrument (NETZSCH-Gerätebau GmbH, Selb, Germany) with nitrogen flow, in a range of 25 to 550 °C, with a heating rate of 10 °C $min^{-1}$. Fourier transform infrared (FTIR) spectroscopy was performed using a Perkin Elmer 1760-X spectrophotometer (PerkinElmer, Inc., Waltham, MA, USA) with a range 4000–400 $cm^{-1}$ and using a KBr pellet of the material.

To determine the solubility as a function of pH, a turbidimetric technique with an ultraviolet-visible (UV-Vis) spectrophotometer (UVmini-1240, Shimadzu, Tokyo, Japan) operating at a wavelength of 660 nm was carried out. Solutions (2 mg·$mL^{-1}$) of the chitosan sulfonate derivatives were prepared in 0.01 mol $L^{-1}$ NaOH, and then the pH was adjusted from 11.0 to 3.0 with the addition of 0.05 mol $L^{-1}$ HCl.

### *4.4. Preparation of Polyampholytic Hydrogels*

Preparation of the hydrogels was carried out in two stages: (*i*) synthesis of the methacrylate-functionalized macromonomer modified with sulfonic groups and (*ii*) preparation of the hydrogels via free radical polymerization. To synthesize the macromonomer, 0.5 g of either the chitosan with mono (CTS1S) or disulfonic (CTS2S) groups was dissolved in 50 mL of water and the pH was adjusted to 6 using HCl (0.1M). The reaction was heated to reflux at 50 °C and then 1.5 mL of glycidyl methacrylate (GMA) was added slowly dropwise under vigorous stirring until a homogeneous solution was obtained. Stirring (200 rpm) was maintained for 24 h [50,68], and then the polymer was purified by dialysis, using a regenerated cellulose membrane with a molecular weight cut-off of 12–14 kDa for 3 days with water exchange every ~4 h. Then, the sample was frozen and lyophilized. The products were characterized by TGA, FTIR spectroscopy, and $^1$H-NMR spectroscopy.

The hydrogels were prepared by dissolving 1.0 g of the mono- or disulfonic macromonomer (CTSV1S or CTSV2S, respectively) in 25 mL of 1 M NaOH in a Schlenk tube. This solution was stirred until a homogeneous mixture was obtained, and then 200 mg of the N,N-methylenebisacrylamide (BIS) cross-linker was added while stirring vigorously. Once homogenized, nitrogen was bubbled through the solution for 10 min and then 100 mg of ammonium persulfate (APS) as an initiator was added with constant stirring, followed by 100 mg of TEMED as a catalyst. The solution was heated at 50 °C resulting in gelation after ~30 min and then the heating was continued for 4 h to allow all the reagents to [50]. The resulting gel was washed extensively with water until a neutral pH was achieved, and then it was dried at 30 °C for 24 h, and finally pulverized and sieved, to produce particles with diameters between 100 and 280 μm.

### *4.5. Characterization of Polyampholytic Hydrogels*

The materials obtained were characterized by TGA and FTIR. Additionally, the morphologies were observed by scanning electron microscopy (SEM-mapping; JEOL JSM 6010 Plus/LV SEM, JEOL Ltd., Tokyo, Japan), freezing and lyophilizing the fully swollen hydrogels, to subsequently coat them in gold using the low-vacuum sputter coating technique. Finally, swelling tests were carried out in water at 4 different pH values (3.0, 5.0, 7.0, and 9.0), weighing dry hydrogel pieces and placing them in water, to determine swelling percentages after 24 h (Equation (3)) [69,70]. The swelling was then calculated using Equation (3).

$$W = (W_t - W_0)/W_0 \quad (3)$$

W = Degree of swelling
$W_0$ = Dry weight
$W_t$ = Weight over time

Additionally, rheological tests were carried out using parallel plates (DHR-3/TA instruments rheometer, TA Instruments, New Castle, DE, USA) ETC/20 mm standard, using oscillatory scanning measurements at a temperature of 25 °C, frequency of 1 Hz and deformation of 0.001 to 500% in triplicate to determine the Storage modulus, Loss modulus, Complex viscosity, and Complex modulus.

Finally, the pH of the zero charge point ($pH_{ZCP}$) of the hydrogels obtained was determined. For this, 0.1 M NaCl solutions with pH between 3 and 9 (adjusted with HCl or NaOH) were prepared. Then, 10 mL of these solutions were taken, and 0.08 g of each hydrogel was added. The suspensions obtained were stirred for 24 h. Then, the dispersions were filtered and the final pH of the solutions was determined. The $pH_{ZCP}$ value of the hydrogels was found from the intersection of the final pH vs. initial pH curve with the abscissa.

### *4.6. Application of the Hydrogels to Adsorption of Ciprofloxacin and Copper Ions in Aqueous Systems*

The application of the polyampholytic hydrogels for the adsorption of contaminants in aqueous systems was carried out at pH 7 from a 20 mg $L^{-1}$ solution of each contaminant: ciprofloxacin (CPX) and $Cu^{2+}$ (from the salt of copper sulfate pentahydrate). The removal calculations for $Cu^{2+}$ were carried out from the $Cu^{2+}$ in solution. A total of 20 mg of hydrogel in triplicate was placed in contact with 10 mL of each solution for 3 h under constant stirring at a temperature of 25 °C. To monitor both contaminants, the respective calibration curves were constructed that varied from 1 to 20 ppm. CPX monitoring was carried out by UV spectroscopy at 271 nm using an Evolution OnePlus UV-Vis Spectophotometer (Thermo Scientific, Waltham, MA, USA), while for $Cu^{2+}$ atomic absorption spectroscopy was performed using a PinAAcle 900F instrument (Perkin Elmer, Waltham, MA, USA) monitoring the removal as a function of the coverage concentration in solution.

**Supplementary Materials:** The following supporting information can be downloaded at: https://www.mdpi.com/article/10.3390/gels11080622/s1, Figure S1. Conductivity for the determination of the degree of deacetylation (DDA). Figure S2. 13C-NMR spectra of chitosan samples. Figure S3. Transmittance percentage results for polymers modified with sulfonic groups, (a) Polymers obtained by modifying chitosan with FB1S. (b) Polymers obtained by modifying chitosan with FB2S. Figure S4. Rheological characterization, (a) complex modulus by Oscillation strain from 0.001% to 500% (b) Comparison of the complex modulus at 0.01% of Oscillation strain.

**Author Contributions:** D.M.-R.: Methodology, Research, Writing—original draft, Writing—review and editing. A.S.: Methodology. M.F.M.: Formal analysis, Writing—review and editing. E.R.G.: Methodology, Formal analysis, Writing—review and editing. D.A.P.: Conceptualization, Methodology, Validation, Data curation, Formal analysis, Resources, Visualization, Project administration, Writing—original draft, Writing—review and editing. All authors have read and agreed to the published version of the manuscript.

**Funding:** This research was funded by FONDECYT (Grants No 3220108 and 1251578) of the Chilean Agency for Research and Development—Chile, for its Spanish acronym ANID.

**Institutional Review Board Statement:** Not applicable.

**Informed Consent Statement:** Not applicable.

**Data Availability Statement:** The data presented in this study are available on request from the corresponding author.

**Acknowledgments:** The authors thank FONDECYT Grant N° 3220108 and 1251578 of the National Research and development Agency of Chile by its acronym in Spanish (ANID). DMM thanks the Graduate Direction (University of Concepcion) for the Ph.D. Scholarship. ERG thanks the Canada Research Chairs Program (CRC-2020-00101).

**Conflicts of Interest:** The authors declare no conflicts of interest.

## Abbreviations

The following abbreviations are used in this manuscript:

| | |
|---|---|
| GMA | Glycidyl methacrylate |
| CTS | Chitosan |
| BIS | N,N-methylenebisacrylamide |
| APS | Ammonium persulfate |
| CPX | Ciprofloxacin |
| HG | Hydrogel |
| FB1S | 2-formylbenzylsulfonic acid sodium salt |
| FB2S | 4-formyl-1,3-benzenedisulfonic acid |
| TEMED | N,N,N′,N′-tetramethyl ethylenediamine |
| $pH_{ZCP}$ | pH of zero charge point |
| DDA | Degree of deacetylation of the starting chitosan |
| DS | Degree of substitution |
| CTS1S | Chitosan modified with FB1S |
| CTS2S | Chitosan modified with FB2S |
| CTSV | Chitosan modified with GMA |
| CTSV1S | CTS1S modified with GMA |
| CTSV2S | CTS2S modified with GMA |
| HGCTS | Hydrogel obtained from CTS |
| HGCTS1S | Hydrogel obtained from CTSV1S |
| HGCTS2S | Hydrogel obtained from CTSV2S |
| $^1$H-NMR | Proton nuclear magnetic resonance |
| FTIR | Fourier transform infrared |
| SEM | Scanning electron microscopy |
| TGA | Thermogravimetric analysis |

## References

1. Zhao, J.; Xing, T.; Li, Q.; Chen, Y.; Yao, W.; Jin, S.; Chen, S. Preparation of chitosan and carboxymethylcellulose-based polyelectrolyte complex hydrogel via SD-A-SGT method and its adsorption of anionic and cationic dye. *J. Appl. Polym. Sci.* **2020**, *137*, 34. [CrossRef]
2. Chen, Y.; Yan, X.; Zhao, J.; Feng, H.; Li, P.; Tong, Z.; Yang, Z.; Li, S.; Yang, J.; Jin, S. Preparation of the chitosan/poly(glutamic acid)/alginate polyelectrolyte complexing hydrogel and study on its drug releasing property. *Carbohydr. Polym.* **2018**, *191*, 8–16. [CrossRef] [PubMed]
3. Lv, S.; Zhang, S.; Zuo, J.; Liang, S.; Yang, J.; Wang, J.; Wei, D. Progress in preparation and properties of chitosan-based hydrogels. *Int. J. Biol. Macromol.* **2023**, *242*, 124915. [CrossRef]
4. Ceper, T.; Mohotti, S.W.; Lange, L.X.; Schacher, F.H. Polyampholyte Hydrogels with pH-Dependent Swelling for Controlled Catch & amp; Release of Model Dyes. *Org. Mater.* **2024**, *6*, 1–11. [CrossRef]
5. Chang, Z.; Chen, Y.; Tang, S.; Yang, J.; Chen, Y.; Chen, S.; Li, P.; Yang, Z. Construction of chitosan/polyacrylate/graphene oxide composite physical hydrogel by semi-dissolution/acidification/sol-gel transition method and its simultaneous cationic and anionic dye adsorption properties. *Carbohydr. Polym.* **2020**, *229*, 115431. [CrossRef]
6. Lipin, V.A.; Poshvina, T.A.; Petrova, Y.A. Interaction of polyampholytic hydrogels based on partially hydrolyzed polyacrylamide with divalent metals. *Russ. Chem. Bull.* **2023**, *72*, 1299–1306. [CrossRef]
7. Lipin, V.A.; Evdokimov, A.N.; Poshvina, T.A.; Petrova, Y.A.; Garcia, D.D.H. Sorption Capacity of Polyampholytic Hydrogels with Respect to Dyes of Different Origins. *Russ. J. Phys. Chem. A* **2024**, *98*, 1285–1292. [CrossRef]

8. Lipin, V.A.; Evdokimov, A.N.; Alekseev, V.G.; Sustavova, T.A.; Petrova, Y.A. Sorption of Anionic Dyes by Polyampholyte Hydrogels Based on Hydrolized Polyacrylamide Modified with Aliphatic Diamines. *Russ. J. Phys. Chem. A* **2022**, *96*, 387–390. [CrossRef]
9. Yan, J.; Li, K. A magnetically recyclable polyampholyte hydrogel adsorbent functionalized with β-cyclodextrin and graphene oxide for cationic/anionic dyes and heavy metal ion wastewater remediation. *Sep. Purif. Technol.* **2021**, *277*, 119469. [CrossRef]
10. Kato, Y.; Onishi, H.; Machida, Y. Application of Chitin and Chitosan Derivatives in the Pharmaceutical Field. *Curr. Pharm. Biotechno.* **2003**, *4*, 303–309. [CrossRef] [PubMed]
11. Bakshi, P.S.; Selvakumar, D.; Kadirvelu, K.; Kumar, N.S. Chitosan as an environment friendly biomaterial—A review on recent modifications and applications. *Int. J. Biol. Macromol.* **2020**, *150*, 1072–1083. [CrossRef]
12. Masteri-Farahani, M.; Shahsavarifar, S. Chemical functionalization of chitosan biopolymer and chitosan-magnetite nanocomposite with sulfonic acid for acid-catalyzed reactions. *Chin. J. Chem. Eng.* **2021**, *39*, 154–161. [CrossRef]
13. Shirdast, A.; Sharif, A.; Abdollahi, M. Effect of the incorporation of sulfonated chitosan/sulfonated graphene oxide on the proton conductivity of chitosan membranes. *J. Power Sources* **2016**, *306*, 541–551. [CrossRef]
14. Balan, V.; Verestiuc, L. Strategies to improve chitosan hemocompatibility: A review. *Eur. Polym. J.* **2014**, *53*, 171–188. [CrossRef]
15. Nunes, Y.L.; de Menezes, F.L.; de Sousa, I.G.; Cavalcante, A.L.G.; Cavalcante, F.T.T.; da Silva Moreira, K.; de Oliveira, A.L.B.; Mota, G.F.; da Silva Souza, J.E.; de Aguiar Falcão, I.R.; et al. Chemical and physical Chitosan modification for designing enzymatic industrial biocatalysts: How to choose the best strategy? *Int. J. Biol. Macromol.* **2021**, *181*, 1124–1170. [CrossRef] [PubMed]
16. Grande, C.D.; Alí, F.; Zuluaga, F. Síntesis de quitosano modificado con poli (ácido acrílico) vía polimerización por transferencia atómica iniciada desde la superficie(SIP-ATRP). *Iberoam. Polímeros* **2010**, *11*, 505–519.
17. Liu, P.; Yang, X.; Chen, W.; Hao, Y. Preparation of the modified chitosan flocculant introduced acryloyloxyethyl trimethyl ammonium chloride and 2-acrylamido-2-methyl propane sulfonic acid for the treatment of papermaking wastewater. *Colloids Surf. A Physicochem. Eng. Asp.* **2024**, *682*, 132934. [CrossRef]
18. Montenegro, M.H.E.; Vega, M.D.; Hernández, M.A. Preparation and characterization of some derivatives of chitosan. *Sci. Rev. Investig. Univ. Panamá* **2019**, *29*, 10–33.
19. Dimassi, S.; Tabary, N.; Chai, F.; Blanchemain, N.; Martel, B. Sulfonated and sulfated chitosan derivatives for biomedical applications: A review. *Carbohydr. Polym.* **2018**, *202*, 382–396. [CrossRef]
20. Wang, J.; Zhuang, S. Chitosan-based materials: Preparation, modification and application. *J. Clean. Prod.* **2022**, *355*, 131825. [CrossRef]
21. Ouerghemmi, S.; Dimassi, S.; Tabary, N.; Leclercq, L.; Degoutin, S.; Chai, F.; Pierlot, C.; Cazaux, F.; Ung, A.; Staelens, J.-N.; et al. Synthesis and characterization of polyampholytic aryl-sulfonated chitosans and their in vitro anticoagulant activity. *Carbohydr. Polym.* **2018**, *196*, 8–17. [CrossRef]
22. Han, X.; Zheng, Z.; Yu, C.; Deng, Y.; Ye, Q.; Niu, F.; Chen, Q.; Pan, W.; Wang, Y. Preparation, characterization and antibacterial activity of new ionized chitosan. *Carbohydr. Polym.* **2022**, *290*, 119490. [CrossRef]
23. *Institute of Public Health*; Ministry of Health Chile ISP Informa Sobre la Resistencia a los Antimicrobianos y los Antibióticos más Vendidos en Chile. Ministerio de Salud Pública: Santiago, Chile, 2019. Available online: https://www.ispch.gob.cl/noticia/isp-informa-sobre-la-resistencia-a-los-antimicrobianos-y-los-antibioticos-mas-vendidos-en-chile/#:~:text=5 (accessed on 28 November 2023).
24. Serna-Galvis, E.A.; Vélez-Peña, E.; Osorio-Vargas, P.; Jiménez, J.N.; Salazar-Ospina, L.; Guaca-González, Y.M.; Torres-Palma, R.A. Inactivation of carbapenem-resistant *Klebsiella pneumoniae* by photo-Fenton: Residual effect, gene evolution and modifications with citric acid and persulfate. *Water Res.* **2019**, *161*, 354–363. [CrossRef]
25. Xiao, R.; He, Z.; Diaz-Rivera, D.; Pee, G.Y.; Weavers, L.K. Sonochemical degradation of ciprofloxacin and ibuprofen in the presence of matrix organic compounds. *Ultrason. Sonochem.* **2014**, *21*, 428–435. [CrossRef]
26. World Health Organization. *Vigilancia de la Resistencia a Los Antimicrobianos*; WHO: Geneva, Switzerland, 2018; Volume 4, pp. 20–33.
27. Nasri, A.; Allouche, M.; Hannachi, A.; Harrath, A.H.; Aldahmash, W.; Alwasel, S.; Mahmoudi, E.; Beyrem, H.; Boufahja, F. Restructuring of a meiobenthic assemblage after sediment contamination with an antibacterial compound: Case study of ciprofloxacin. *Ecotoxicol. Environ. Saf.* **2020**, *205*, 111084. [CrossRef]
28. Gomes, M.C.; Mano, J.F. Chemical modification strategies to prepare advanced protein-based biomaterials. *Biomater. Biosyst.* **2021**, *1*, 100010. [CrossRef] [PubMed]
29. Punia, A. Role of temperature, wind, and precipitation in heavy metal contamination at copper mines: A review. *Environ. Sci. Pollut. Res.* **2021**, *28*, 4056–4072. [CrossRef] [PubMed]
30. Zhuang, P.; Zou, B.; Li, N.Y.; Li, Z.A. Heavy metal contamination in soils and food crops around Dabaoshan mine in Guangdong, China: Implication for human health. *Environ. Geochem. Health* **2009**, *31*, 707–715. [CrossRef] [PubMed]
31. Zhuang, P.; McBride, M.B.; Xia, H.; Li, N.; Li, Z. Health risk from heavy metals via consumption of food crops in the vicinity of Dabaoshan mine, South China. *Sci. Total Environ.* **2009**, *407*, 1551–1561. [CrossRef]

32. Shan, S.; Sun, X.-F.; Xie, Y.; Li, W.; Ji, T. High-Performance Hydrogel Adsorbent Based on Cellulose, Hemicellulose, and Lignin for Copper(II) Ion Removal. *Polymers* **2021**, *13*, 3063. [CrossRef]
33. Palacio, D.A.; Becerra, Y.; Urbano, B.F.; Rivas, B.L. Antibiotics removal using a chitosan-based polyelectrolyte in conjunction with ultrafiltration membranes. *Chemosphere* **2020**, *258*, 127416. [CrossRef]
34. Ma, J.; Xiong, Y.; Dai, X.; Yu, F. Coadsorption behavior and mechanism of ciprofloxacin and Cu(II) on graphene hydrogel wetted surface. *Chem. Eng. J.* **2020**, *380*, 122387. [CrossRef]
35. Afzal, M.Z.; Sun, X.-F.; Liu, J.; Song, C.; Wang, S.-G.; Javed, A. Enhancement of ciprofloxacin sorption on chitosan/biochar hydrogel beads. *Sci. Total Environ.* **2018**, *639*, 560–569. [CrossRef]
36. Yu, F.; Sun, Y.; Yang, M.; Ma, J. Adsorption mechanism and effect of moisture contents on ciprofloxacin removal by three-dimensional porous graphene hydrogel. *J. Hazard. Mater.* **2019**, *374*, 195–202. [CrossRef]
37. Ding, J.; Li, Q.; Xu, X.; Zhang, X.; Su, Y.; Yue, Q.; Gao, B. A wheat straw cellulose-based hydrogel for Cu (II) removal and preparation copper nanocomposite for reductive degradation of chloramphenicol. *Carbohydr. Polym.* **2018**, *190*, 12–22. [CrossRef] [PubMed]
38. Brandt, M.; Bieritz, L.; Mönnig, A.; Großmann, A.M. *Development of Sustainable Mining Strategies in Chile with a Regionalized National Model: Project Introduction and Overview Standard-Nutzungsbedingungen*; Gesellschaft für Wirtschaftliche Strukturforschung (GWS): Osnabrück, Germany, 2017.
39. Mavila, S.; Eivgi, O.; Berkovich, I.; Lemcoff, N.G. Intramolecular Cross-Linking Methodologies for the Synthesis of Polymer Nanoparticles. *Chem. Rev.* **2016**, *116*, 878–961. [CrossRef]
40. Hamza, M.F.; Mira, H.; Wei, Y.; Aboelenin, S.M.; Guibal, E.; Salem, W.M. Sulfonation of chitosan for enhanced sorption of Li(I) from acidic solutions—Application to metal recovery from waste Li-ion mobile battery. *Chem. Eng. J.* **2022**, *441*, 135941. [CrossRef]
41. Palacio, D.A.; Urbano, B.F.; Rivas, B.L. Hydrogels based on alkylated chitosan and polyelectrolyte copolymers. *J. Appl. Polym. Sci.* **2018**, *135*, 46556. [CrossRef]
42. Wang, T.; Zhou, Y.; Xie, W.; Chen, L.; Zheng, H.; Fan, L. Preparation and anticoagulant activity of N-succinyl chitosan sulfates. *Int. J. Biol. Macromol.* **2012**, *51*, 808–814. [CrossRef]
43. Van Poucke, C.; Verdegem, E.; Mangelinckx, S.; Stevens, C.V. Synthesis and unambiguous NMR characterization of linear and branched N-alkyl chitosan derivatives. *Carbohydr. Polym.* **2024**, *337*, 122131. [CrossRef]
44. Elomaa, M. Determination of the degree of substitution of acetylated starch by hydrolysis, 1H NMR and TGA/IR. *Carbohydr. Polym.* **2004**, *57*, 261–267. [CrossRef]
45. Echols, K.R.; Meadows, J.C.; Orazio, C.E. Pollution of Aquatic Ecosystems II: Hydrocarbons, Synthetic Organics, Radionuclides, Heavy Metals, Acids, and Thermal Pollution. In *Encyclopedia of Inland Waters*; Elsevier: Amsterdam, The Netherlands, 2009; pp. 120–128. [CrossRef]
46. Khandelwal, S.; Tailor, Y.K.; Kumar, M. Deep eutectic solvents (DESs) as eco-friendly and sustainable solvent/catalyst systems in organic transformations. *J. Mol. Liq.* **2016**, *215*, 345–386. [CrossRef]
47. Gallardo, M.G.C.; Barbosa, R.C.; Fook, M.V.L.; Sabino, M.A. Synthesis and characterization of a novel biomaterial based on chitosan modified with amino acids. *Matéria* **2019**, *24*, e12397. [CrossRef]
48. Fan, L.; Luo, C.; Li, X.; Lu, F.; Qiu, H.; Sun, M. Fabrication of novel magnetic chitosan grafted with graphene oxide to enhance adsorption properties for methyl blue. *J. Hazard. Mater.* **2012**, *215–216*, 272–279. [CrossRef] [PubMed]
49. Dodi, G.; Hritcu, D.; Lisa, G.; Popa, M.I. Core–Shell magnetic chitosan particles functionalized by grafting: Synthesis and characterization. *Chem. Eng. J.* **2012**, *203*, 130–141. [CrossRef]
50. Palacio, D.A.; Urbano, B.F.; Palencia, M.; Rivas, B.L. Preparation of alkylated chitosan-based polyelectrolyte hydrogels: The effect of monomer charge on polymerization. *Eur. Polym. J.* **2019**, *118*, 551–560. [CrossRef]
51. Mohamed, R.R.; Elella, M.H.A.; Sabaa, M.W. Cytotoxicity and metal ions removal using antibacterial biodegradable hydrogels based on N-quaternized chitosan/poly(acrylic acid). *Int. J. Biol. Macromol.* **2017**, *98*, 302–313. [CrossRef]
52. Elzamly, R.A.; Mohamed, H.M.; Mohamed, M.I.; Zaky, H.T.; Harding, D.R.K.; Kandile, N.G. New sustainable chemically modified chitosan derivatives for different applications: Synthesis and characterization. *Arab. J. Chem.* **2021**, *14*, 103255. [CrossRef]
53. Kandile, N.G.; Elzamly, R.A.; Mohamed, M.I.; Zaky, H.T.; Harding, D.R.K.; Mohamed, H.M. New sustainable antimicrobial chitosan hydrogels based on sulfonamides and its nanocomposites: Fabrication and characterization. *Int. J. Biol. Macromol.* **2023**, *239*, 124280. [CrossRef] [PubMed]
54. Abdalla, T.H.; Nasr, A.S.; Bassioni, G.; Harding, D.R.; Kandile, N.G. Fabrication of sustainable hydrogels-based chitosan Schiff base and their potential applications. *Arab. J. Chem.* **2022**, *15*, 103511. [CrossRef]
55. Jang, J.; Seol, Y.-J.; Kim, H.J.; Kundu, J.; Kim, S.W.; Cho, D.-W. Effects of alginate hydrogel cross-linking density on mechanical and biological behaviors for tissue engineering. *J. Mech. Behav. Biomed. Mater.* **2014**, *37*, 69–77. [CrossRef]
56. Hajikarimi, P.; Nejad, F.M. Introduction to viscoelasticity. In *Applications of Viscoelasticity*; Elsevier: Amsterdam, The Netherlands, 2021; pp. 1–13. [CrossRef]

57. Ortega, F.R.; Rodríguez, G.; Aguilar, M.R.; García-Sanmartín, J.; Martínez, A.; Román, J.S. Comportamiento reológico de geles biodegradables para aplicacions en medicina regenerativa. *Biomecánica* **2012**, *20*, 7–19. [CrossRef]
58. Qalyoubi, L.; Al-Othman, A.; Al-Asheh, S. Removal of ciprofloxacin antibiotic pollutants from wastewater using nano-composite adsorptive membranes. *Environ. Res.* **2022**, *215*, 114182. [CrossRef]
59. Chaurasiya, A.; Pande, P.P.; Shankar, R.; Chaurasia, S.K.; Kumar, A.; Kumar, P.; Kashaudhan, K. Synthesis and characterization of allyl mannitol cross-linker based novel hydrogel for capturing of toxic metal ions from aqueous solution. *Sep. Sci. Technol.* **2024**, *60*, 1–20. [CrossRef]
60. Akpor, O.B.; Otohinoyi, D.A.; Olaolu, T.D.; Aderiye, B.I. Pollutants in wastewater effluents: Impacts and remediation processes. *Int. J. Environ. Res. Earth Sci.* **2014**, *3*, 50–59.
61. Shi, H.; Ni, J.; Zheng, T.; Wang, X.; Wu, C.; Wang, Q. Remediation of wastewater contaminated by antibiotics. A review. *Environ. Chem. Lett.* **2020**, *18*, 345–360. [CrossRef]
62. Palacio, D.A.; Rivas, B.L.; Urbano, B.F. Ultrafiltration membranes with three water-soluble polyelectrolyte copolymers to remove ciprofloxacin from aqueous systems. *Chem. Eng. J.* **2018**, *351*, 85–93. [CrossRef]
63. Moes, S.; van Gestel, K.; van Beek, G. Metal Speciation. In *Environmental Toxicology*; Libretexts: Davis, CA, USA, 2023; Chapter 3.5, p. 2.
64. Wang, Q.Z.; Chen, X.G.; Liu, N.; Wang, S.X.; Liu, C.S.; Meng, X.H.; Liu, C.G. Protonation constants of chitosan with different molecular weight and degree of deacetylation. *Carbohydr. Polym.* **2006**, *65*, 194–201. [CrossRef]
65. Dinculescu, D.D.; Apetroaei, M.R.; Gîjiu, C.L.; Anton, M.; Enache, L.; Schröder, V.; Isopescu, R.; Rău, I. Simultaneous Optimization of Deacetylation Degree and Molar Mass of Chitosan from Shrimp Waste. *Polymers* **2024**, *16*, 170. [CrossRef]
66. Kasaai, M.R. Determination of the degree of N-acetylation for chitin and chitosan by various NMR spectroscopy techniques: A review. *Carbohydr. Polym.* **2010**, *79*, 801–810. [CrossRef]
67. Wang, J.; Jiang, J.-Z.; Chen, W.; Bai, Z.-W. Data of 1H/13C NMR spectra and degree of substitution for chitosan alkyl urea. *Data Brief* **2016**, *7*, 1228–1236. [CrossRef]
68. Palencia, M.S.; Mora, M.A.; Palencia, S.L. Biodegradable Polymer Hydrogels Based in Sorbitol and Citric Acid for Controlled Release of Bioactive Substances from Plants (Polyphenols). *Curr. Chem. Biol.* **2017**, *11*, 36–43. [CrossRef]
69. Cao, J.; Wang, Y.; He, C.; Kang, Y.; Zhou, J. Ionically crosslinked chitosan/poly(acrylic acid) hydrogels with high strength, toughness and antifreezing capability. *Carbohydr. Polym.* **2020**, *242*, 116420. [CrossRef]
70. Saghandali, F.; Salehi, M.B.; Taghikhani, V. Hydrogel nanocomposite network elasticity parameters as a function of swelling ratio: From micro to macro flooding. *J. Ind. Eng. Chem.* **2023**, *125*, 163–177. [CrossRef]

gels

MDPI

Article

# Development of Eco-Friendly Date Palm Biomass-Based Hydrogels for Enhanced Water Retention in Soil

**Faisal S. Alsubaie *, Mouyed Srdar, Osama Fayraa, Faris M. Alsulami, Feras Omran and Khalid A. Alamry ***

Chemistry Department, Faculty of Science, King Abdulaziz University, Jeddah 21589, Saudi Arabia; mabdoalhaqserdar@stu.kau.edu.sa (M.S.); oifayraa@stu.kau.edu.sa (O.F.); fbaderalsulami@stu.kau.edu.sa (F.M.A.); fmohammedomran@stu.kau.edu.sa (F.O.)
* Correspondence: falsubaie0068@stu.kau.edu.sa (F.S.A.); kaalamri@kau.edu.sa (K.A.A.)

**Abstract:** The growth of plants highly depends on the soil's water availability and properties. Hydrogels (HGs) have been used for decades to enhance soil water retention, whereas developing eco-friendly and sustainable HGs for agricultural applications is still necessary to ensure water and food security. In this study, renewable and cost-effective HGs were prepared from all-lignocellulose fibers of date palm biomass after carboxymethylation followed by citric acid (CA) crosslinking. HGs showed high equilibrium swelling capacity (EWC%), even in salty media, whereas purified HGs showed about 700–400 EWC% in deionized water. Further, HGs' effect on germination was studied on Chico III tomato, mint, Basilico red, and chia seeds. The results revealed that HGs enhanced the soil properties, with taller and healthier plants observed in HG-amended soil. FTIR, thermal analysis, and microscope imaging were utilized to evaluate HGs' and raw materials' characteristics. The findings in this study support the idea that all-lignocellulose could be used for HG production without separation.

**Keywords:** hydrogel; lignocellulosic biopolymers; all-lignocellulose; water retention; soil conditioning; germination; plant growth

## 1. Introduction

Sustainable agriculture systems, water availability, and proper soil are key to food security. The agriculture sector in arid and semi-arid regions faces multiple challenges due to drought, desertification, climate change, sandy soil, and harsh weather. In arid conditions, irrigational water tends to evaporate from the soil due to high temperatures. Also, the soil type stimulates water infiltration, increasing irrigational water demand and irrigation cycles. Further, arid and semi-arid regions suffer from a shortage of freshwater resources [1–4]. On the other hand, the demand for freshwater is increasing daily due to the growing population, urbanization, and industrialization. By 2030, the expected water demand will be 50% higher than today's demand. This may result in water scarcity if efficient water management systems are still underdeveloped [5]. It is worth mentioning that about 70% of global freshwater was estimated to be used for the irrigation of only 25% of the world's croplands which supply only 45% of international food [6]. A helpful solution may lie in inventing sustainable irrigation systems and soil conditioners for water preservation to reduce the total water demand from the farming sector [5]. In 2021, Saudi Arabia initiated the "Saudi & Middle East Green Initiatives" to plant more than 10 billion trees to face climate change, desertification, and air pollution [7]. This number of trees will require vast irrigational water, arable soil, and rehabilitated land, which are mandatory for these initiatives to succeed, considering that water resources are scarce in regions like the Middle East and North Africa (MENA). Hence, local, modern solutions to enhance planting should be explored and utilized.

Hydrogels (HGs) and superabsorbent hydrogels (SHs), also known as superabsorbent polymers (SAPs), have been an attractive solution for soil conditioning since the 80s owing to their moisture and water preservation ability [8]. HGs are formulated from different types of crosslinked hydrophilic polymers. Different physical and chemical methods could be used for crosslinking polymer chains. The physical crosslinking methods lean on physical interactions like hydrogen bonding and Van der Waals interactions, while chemical crosslinking relies on covalent bonds formation. Epichlorohydrin, glutaraldehyde, and citric acid (CA) are some of the commonly used crosslinking agents [3]. CA is considered an eco-friendly and cost-effective crosslinker in comparison with other common crosslinkers like epichlorohydrin and glutaraldehyde [9]. After crosslinking, HG materials can hold water, organic matter, or inorganic materials within their 3D network matrix and release them upon some stimulus, such as pH, temperature, and ionic strength [3,10,11]. Investigations have found that HGs have huge water capacities comprised of their original weight, which could be higher than 1000% of their weights, enhancing soil moisture and oxygenation [12]. HGs have been well-utilized in the industrial field in various applications, such as in the biomedical and agricultural industries. Many studies have explored HGs and their effects as a soil conditioner, and they have been considered an adequate solution for enhancing planting and soil [13,14]. Demitri et al. studied the impact of HGs prepared from hydroxyethylcellulose (HEC)/carboxymethylcellulose sodium salt (CMCNa) that was citric acid-crosslinked on cherry tomato cultivation in greenhouses and claimed that HG-amended soil showed three times higher humidity with respect to the control soil. Also, longer survival times were observed in hydrogel-containing soil [15]. Chitosan (CS) is a chitin-derived polymer that is highly abundant in living organisms and has also been utilized for hydrogel production and for controlled fertilizer release. Jamnongkan et al. developed chitosan/poly(vinyl alcohol)-based hydrogel for the controlled release of potassium. Their results revealed that the incorporation of hydrogel in the soil enhanced the water retention period [16]. In addition, starch-based HGs have been applied in many studies for agricultural applications. Starch is one of the most abundant polysaccharides, and it is biodegradable and biocompatible. Chamorro et al. developed citric acid-crosslinked cassava starch hydrogels for the slow release of ammonium. They claimed that the prepared hydrogel reduced the fertilizer release rate in the soil [9]. Both synthetic polymers and natural polymers have been used for HG preparation by physical or chemical crosslinking. To some extent, synthetic HGs have some drawbacks regarding their environmental footprint. Therefore, natural polymers are a good candidate due to their biodegradability, biocompatibility, and natural abundance [12,17]. Natural polymers or biopolymers are the types of polymers that can be found in nature within biological systems, whether in plant-based or animal-based systems.

Agricultural biomass is one of the wealthiest sources of biopolymers. It is a plant-based agrarian waste that is biodegradable, biocompatible, and highly available. Biomass is an underutilized mine of biopolymers, and it mostly becomes landfill waste or is burned, which is considered polluting and a waste of valuable resources. The lignocellulosic matrix is the main constituent of biomass, comprising about 65% of biomass weight. This matrix comprises polysaccharides, such as cellulose and hemicellulose, alongside lignin, a polyphenol biopolymer [18,19]. These biopolymers have been utilized for decades directly or after modification toward different applications and uses. Cellulose is the main component of the lignocellulose matrix, and it is the most abundant natural polymer on earth. Cellulose is a homopolymer of anhydroglucoses ($C_6H_{10}O_5$) linked by β-1,4 glycosidic bond, and the result is a water-insoluble and highly crystalline long-chain polymer. The second major component is hemicellulose, which is amorphous in structure and constituted of different penta- and hexa-sugars. The third component by percentage is lignin, which is composed of three different phenols with alternative ratios, which are p-coumaroyl alcohol, coniferyl alcohol, and sinapyl alcohol, forming three-dimensional polyphenol [20,21]. Many studies have used those biopolymers to prepare hydrogels and show their capability in different applications [22–24]. Several pretreatments are required to reduce the biomass recalcitrance and separate the biopolymers from the matrix to produce HGs. Common pretreatments for biomass recalcitrance reduction are costly and energy extensive [25]. The most efficient and common pretreatments are chemical pretreatments, but unfortunately, they include the use of harsh chemicals that are accompanied by environmental pollution [26,27]. Also, it is worth mentioning that lignocellulosic biopolymers are naturally insoluble in water, so chemical modification reactions should be performed to yield water-soluble polymers to form HGs. Many chemical modifications, including acetylation [28] and etherification [29], have been applied to biopolymers to enhance their solubility and reactivity. To increase water solubility, carboxymethylation is considered one of the most common modifications for polysaccharides, where a carboxymethyl group is introduced to the biopolymer molecular chain in an etherification reaction with monochloroacetic acid (MCA), which is a well-studied reaction [30].

The date palm tree (*Phoenix dactylifera* L.) is one of the most common crops cultivated in the MENA and is one of the oldest tree species. Worldwide, there are approximately 100 million date palm trees [31]. In Saudi Arabia alone, there are about 30 million date palm trees, where the present research was conducted [32]. In addition, there are about 40 million trees in the United Arab Emirates (UAE) [28], 18.7 million in Algeria [29], and 4.45 million in Morocco [31]. In Saudi Arabia, the amount of date palm biomass released is about 1 million metric tons annually, and it is primarily underutilized and disposed of in landfills or burned [32]. In addition, date palm biomass, specifically date palm rachis (DPR), is one of the most common non-edible date byproducts and is full of a lignocellulosic matrix, which can be fractionated and used as biopolymers or used according to the desired application [33]. It is worth mentioning that, like other biomass, date palm biomass contains 60–95% biopolymers [34].

This study aims to utilize DPR biomass to prepare HGs for water retention in soil using a facile and eco-friendly method. The suggested approach is to directly convert DPR to carboxymethyl date palm rachis (CMDPR) by the MCA etherification reaction without chemically pretreating the biomass before the modification reaction, reducing the production cost and footprint. The prepared water-soluble product was crosslinked by an eco-friendly agent, i.e., CA, using different ratios to yield HGs. Different tests and characterization were applied to the resulting CMDPR to confirm the modification occurrence and to the prepared HGs to confirm their suitability in soil water retention. The prepared HGs showed suitable water swelling, thermal stability, and a positive impact on plant germination.

## 2. Results and Discussion

### 2.1. Chemical Composition of DPR and Pretreatment

The chemical composition of DPR is shown in Table 1. The DPR used in this study contained 13.74%, 29.82%, and 21.69% of lignin, α-cellulose, and hemicellulose, respectively, which is similar to the ratios that are found in the literature and presented in Table 1 [31,35–37]. Based on Table 1, cellulose content varied between 47–29% of DPR weight. The hemicellulose and lignin also varied in different collected DPR from different places in a lower range. These variations were mostly due to differences between soil conditions, geographic location, and climate [38]. As a result, the total amount of lignocellulosic biopolymer (lignin, cellulose, and hemicellulose) within the DPR used in this study was 65.25% of the total weight of the DPR. Total lignocellulosic biopolymers indicate the total of the most modifiable polymers (that could be carboxymethylated). Two types of pretreatments were performed on DPR to ensure efficient carboxymethylation: physical pretreatment, which includes grinding and milling, followed by screening. This pretreatment was performed to increase the modification degree, where the particle size tremendously impacts the degree of substitution [39]. Subsequently, hot water pretreatment was performed to reduce the amount of hot-water-soluble (HWS) matter to avoid the possibility of reaction reagents consumption, i.e., NaOH and MCA, during the modification reaction [40].

**Table 1.** Chemical composition of date palm rachis from different countries and locations.

| Biomass Type | Cellulose | Hemicellulose | Lignin | Ash | Extractives | Hot Water Soluble | Ref. |
|---|---|---|---|---|---|---|---|
| Date palm rachis from Jeddah City in Saudi Arabia | 29.82 | 21.69 | 13.74 | 5.89 | 9.5 | 10.84 | This work |
| Date palm rachis from Marrakesh City in Morocco | 39.8 | 31.4 | 14.0 | 9.2 | NA * | 16.8 | [31] |
| Date palm rachis from southern Algeria | 35.87 | NA * | 16.94 | NA * | 8.66 | NA * | [37] |
| Date palm rachis from Biskra City in Algeria | 47.31 | 25.72 | 15.67 | 5.47 | 5.80 | NA * | [35] |
| Date palm rachis from Khoozestan City in Iran | 38.26 | 28.17 | 22.53 | 5.96 | 5.08 | NA * | [36] |

* NA: not available or mentioned.

### 2.2. Characterization of DPR and WTDPR

The pretreated DPR was named water-treated DPR (WTDPR), which was then characterized by comparison with DPR. Fourier-transformed infrared spectrometry (FTIR) was employed to confirm the structural differences and crystalline indexes, while morphological characteristics were evaluated by a digital microscope and dynamic light scattering (DLS). The FTIR spectrum DPR and WTDPR are shown in Figure 1A. Both samples presented typical vibration bands of lignocellulosic biopolymers, i.e., cellulose, hemicellulose, and lignin chemical functional groups, as presented in Table 2. A broad absorption band was observed around 3400 $cm^{-1}$, which is attributed to the −OH group stretching of lignocellulose biopolymers and water, whereas the absorbance band at 2919 $cm^{-1}$ was assigned to the −CH group stretching vibrations from cellulose and hemicellulose [41]. A peak observed at 1731 $cm^{-1}$ was attributed to C=O of ester or carboxylic acid group in hemicellulose and lignin [42,43], and a vibration band was observed at 1637 $cm^{-1}$, which was attributed to conjugated carbonyl group [43]. In the following, three sequential peaks were attributed to the lignin aromatic ring at 1618 $cm^{-1}$, which was assigned aromatic skeletal vibration of lignin, and −C=O stretching of the lignin ring plus flavonoids [44,45]; a second peak at 1507 $cm^{-1}$, which was assigned to −C=C stretching of the lignin ring; and lastly at 1458 $cm^{-1}$, which was assigned to −CH group deformation within the lignin ring [46,47]. The absorbance at 1425 $cm^{-1}$ was mainly attributed to the $-CH_2$ symmetric bending in cellulose and −CH deformation in the lignin ring [47]. The absorption band at 1374 $cm^{-1}$ was assigned to the bending vibration of the −CH group of hemicellulose and cellulose, while the band at 1321 $cm^{-1}$ was attributed to the $-CH_2$ rocking vibration in cellulose [47]. The peak at 1250 $cm^{-1}$ was assigned to the C−O stretching of the acetyl group in the lignin [48]. The peak at 1158 cm$^{-1}$ was assigned to C−O−C asymmetric stretching in hemicellulose and cellulose or the ester group in lignin, while the peak at 1049 cm$^{-1}$ was attributed to ether group C−O stretching vibration within the cellulose, hemicellulose, and lignin [46,47]. The band at 1108 cm$^{-1}$ was attributed to C−C/C−O stretching within the aliphatic skeletal of cellulose and hemicellulose [49]. The band at 897 cm$^{-1}$ was assigned to the β-glycosidic bonds between the monosaccharides [50]. It is noteworthy that there were no notable peak absences since the hot water treatment does not affect cellulose and lignin structure while negligibly affecting hemicellulose [51].

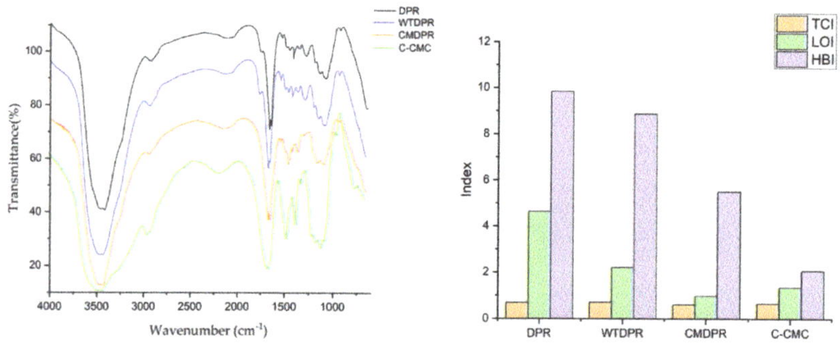


(A) (B)

**Figure 1.** FTIR spectrum (**A**) and crystallinity indices (**B**) of DPR, WTDPR, CMDPR, and C-CMC.

**Table 2.** Comparison of FTIR spectra of samples.

| Peak Assignment | Peak Wavenumber (cm$^{-1}$) | | | | | | Ref. |
|---|---|---|---|---|---|---|---|
| | **DPR** | **WTDPR** | **CMDPR** | **C-CMC** | **CM15** | **CM20** | |
| $-\mathrm{OH}_{\text{stretching}}$ | 3400 | 3400 | 3400 | 3400 | 3384 | 3402 | [41] |
| $-\mathrm{CH}_{\text{stretching}}$ | 2919 | 2919 | 2919 | 2919 | 2913 | 2923 | [41] |
| $\mathrm{C{=}O}_{\text{ester or }-\mathrm{COOH}}$ | 1731 | 1731 | - | - | 1726 | 1726 | [42,43] |
| $\mathrm{C{=}O}_{\text{conjugated}}$ | 1637 | 1637 | 1637 | - | - | - | [43] |
| $\mathrm{C{=}C/C{=}O}_{\text{aromatic or carboxylic acid}}$ | 1618 | 1618 | 1618 | 1618 | 1587 | 1586 | [44,45] |
| $\mathrm{C{=}C}_{\text{aromatic}}$ | 1507 | 1507 | 1507 | - | - | - | [46,47] |
| $-\mathrm{CH}_{\text{aromatic}}$ | 1458 | 1458 | 1458 | - | - | - | [46,47] |
| $-\mathrm{CH}_{2\,\text{aliphatic}}/-\mathrm{CH}_{\text{aromatic}}$ | 1425 | 1425 | 1422 | 1420 | 1414 | 1414 | [47] |
| $-\mathrm{CH}_{\text{deformation}}$ | 1374 | 1374 | 1374 | - | - | - | [47] |
| $-\mathrm{CH}_{\text{rocking}}$ | 1321 | 1321 | 1330 | 1330 | 1318 | 1316 | [47] |
| $\mathrm{C{-}O}_{\text{aromatic}}$ | 1250 | 1250 | 1270 | - | 1233 | 1235 | [48] |
| $\mathrm{C{-}O{-}C}_{\text{stretching}}$ | 1158 | 1158 | 1158 | 1158 | - | - | [46,47] |
| C−C/C−O | 1108 | 1108 | 1120 | 1120 | - | - | [49] |
| $\mathrm{C{-}O}_{\text{stretching}}$ | 1049 | 1049 | 1049 | 1059 | 1042 | 1042 | [46,47] |
| $\mathrm{C{-}O{-}C}_{\beta\text{-glycosidic}}$ | 897 | 897 | 897 | 901 | 898 | 898 | [50] |
| $-\mathrm{OH}_{\text{stretching}}$ | | | | | | | |

Different indices were calculated utilizing the ratio of certain bands' intensities from the FTIR spectra of samples to evaluate the crystallinity changes. The lateral order index (LOI) is an empirical crystallinity index of cellulose correlated to the overall order degree of cellulose. LOI was proposed by Nelson and O'Connor, alongside the total crystallinity index (TCI), which is proportional to the crystallinity degree of cellulose [52,53]. In addition, the hydrogen bond intensity (HBI) index was also used to evaluate the crystallinity of samples, which was proposed by Nada et al. [54]. HBI also indicates the degree of crystallinity and bound water content, where it increases as crystallinity decreases. By calculating the ratio between the absorbance intensities of $CH_2$ flexion at 1430–1420 cm$^{-1}$ of the crystalline cellulose and the intensity of the amorphous cellulose at 900–890 $cm^{-1}$ that is assigned to the β-glycosidic bonds between the monosaccharides, LOI could be obtained [55,56]. TCI is the ratio between the absorbance intensities at ~2900 $cm^{-1}$, which corresponds to stretching vibrations of –CH, and ~1374 $cm^{-1}$, which is assigned to the deformation vibration of CH in cellulose. In contrast, TCI is proportional to the degree of crystallinity of cellulose [55,57]. On the other hand, HBI is the ratio between the absorbance intensities of OH stretching and hydrogen bonding between molecules around 3330–3400 $cm^{-1}$ and the absorbance of at ~1320 $cm^{-1}$ from CH rocking vibration of glucose ring [58,59]. The TCI was obtained from the intensity ratio of A1374/A2919 and LOI from the ratio A1425/A897. The ratio A3336/A1321 was used for HBI. The TCI of DPR was found to be 0.697 with 3.245 LOI and 3.647 HBI, as presented in Figure 1B. After the hot water treatment, both TCI and LOI were noticed to increase to 0.707 and 3.635, respectively. The resulting increase was mostly attributed to the reduction in HWS and extractives, where samples with higher extractive content showed less TCI [59]. Furthermore, the HBI was found to decrease to 3.005 [59]. These results are in harmony with findings in the literature regarding the hot water treatment effect on biomass, where researchers have found that both LOI and TCI increased with liquid hot water treatment in different ranges of temperature and pretreatment time [60]. Also, another study found that LOI and TCI increased after thermal pretreatment from 1.346 to 1.384 and from 0.282 to 0.295, respectively, when the temperature increased from 20 °C to 160 °C [61]. In addition, researchers have found that TCI increases by increasing the sterilization temperature with a reduction in both LOI and HBI; this may be due to the degradation of hemicellulose and amorphous fraction of cellulose, which results in cellulose with better crystallinity [62].

Figure 2 shows the morphology of DPR and WTDPR fibers, which was recorded by a digital microscope. The DPR fiber shows a smooth surface with a compact structure and aligned ordered microfibril lines. Morphological changes were observed after the hot water treatment, where the WTDPR fiber showed a swelled structure with a rough surface and some cracks. Additionally, small holes are observed after the hot water treatment, which may attributed to the leaching of silica bodies from WTDPR. These results are similar to biomass after different pretreatments, where cracks and silica cavities were observed [50,63].

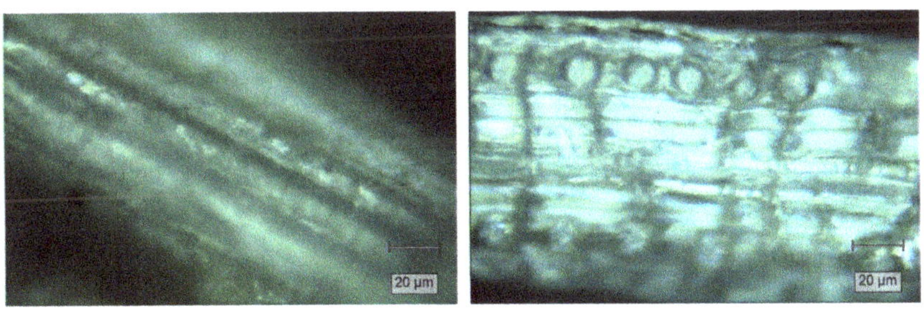


**Figure 2.** Digital microscope imaging of DPR (**left**) and WTDPR (**right**).

The results of the DLS analysis of DPR and WTDPR are shown in Figure 3. The analysis was performed to assess the particle size distribution change after the milling and hot water treatment. The volume-weighted mean diameter $D_{[4,3]}$ of DPR was 157 µm. The median volume distribution of the $Dv_{(50)}$ particles was 120 µm, meaning that 50% of the particles were smaller while 50% were larger. After the hot water treatment, an increase was recorded where $D_{[4,3]}$ was 179 µm while recorded $Dv_{(50)}$ was 148 µm; the uniformity was found to be 0.805 and 0.657 for DPR and WTDPR, respectively.

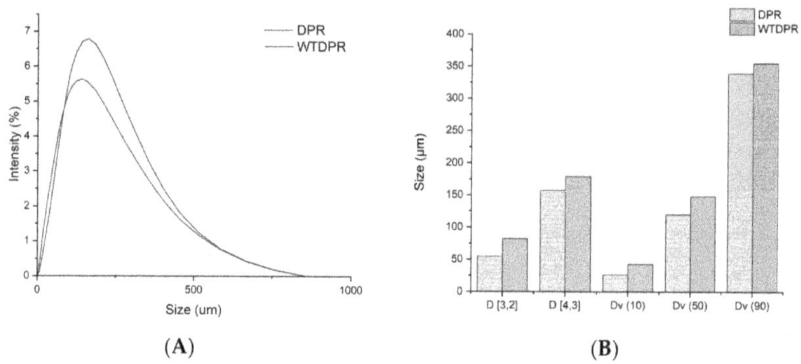


**Figure 3.** Size distribution of particles by intensities (**A**) diameter and distribution by size distribution (**B**).

This may have been due to the removal of hot water solubles and silica, which increased the accessible surface area of the lignocellulosic biopolymers. This explanation is in harmony with findings in the literature, where it was reported that the degradation of hemicellulose or lignin increases the surface area of cellulose, which promotes aggregation [64,65]. Besides, the aqueous media and drying process were reported to promote the self-aggregation of lignocellulose biopolymers [66,67].

### *2.3. Preparation of CMDPR, Degree of Substitution, and Yield*

Carboxymethyl date palm rachis (CMDPR) was prepared from WTDPR with 179 µm $D_{[4,3]}$. The carboxymethylation was carried out according to the known two-step carboxymethylation reaction. In the first step, NaOH activated lignocellulose biopolymers' hydroxyl groups, and in the second step, MCA reacted with the activated groups to yield CMDPR, as shown in Figure 4. The resulting CMDPR mainly comprised carboxymethyl lignin, carboxymethyl cellulose, and carboxymethyl hemicellulose. The estimated degree of substitution of the resulting CMDPR was found to be 1.14, with high solubility and fine low insoluble particles, as shown in Figure 5, compared with DPR and WTDPR visually. The yield was 137.29%, which is close to the findings in other literature (Table 3). A study reported that using date palm rachis with a size range of 200 µm–1000 µm resulted in carboxymethylated biomass with 1.17 DS [68]. Another study showed that holocellulose with a particle size of 100 resulted in CMC with 1.83 DS and 182.55% yield [39]. Besides the other reaction conditions, particle size has an important effect on DS.

**Table 3.** Comparison of different studies on carboxymethylation of lignocellulosic matrix.

| Starting Material | Conditions | DS | Ref. |
|---|---|---|---|
| Date palm rachis | 30% NaOH, 1 g substrate/1.2 g MCA, 55 °C, 3.5 h, Isopropanol | 1.14 | This work |
| Date palm rachis | 40% NaOH, 1 g substrate/1.749 g MCA, 80 °C, 8 h in n-Butanol | 1.17 | [68] |
| Posidonia oceanica | 40% NaOH, 1 g substrate/1.749 g MCA, 80 °C, 8 h in n-Butanol | 1 | [68] |
| Cunninghamia lanceolata | 30% NaOH, 1 g substrate/1.4 g MCA, 80 °C, 2 h, Isopropanol | 1.36 | [69] |
| Lepidium Polyuronide | 45% NaOH, 1 g substrate/15 g MCA, 70 °C, 2 h | 1.75 | [70] |

### *2.4. Characterization of CMDPR and C-CMC*

The prepared CMDPR was characterized using FTIR, a digital microscope, thermalgravimetric analysis (TGA) in an oxygen-rich atmosphere, and differential scanning calorimetry (DSC) and compared with commercial carboxymethyl cellulose (C-CMC) with 0.8 DS.

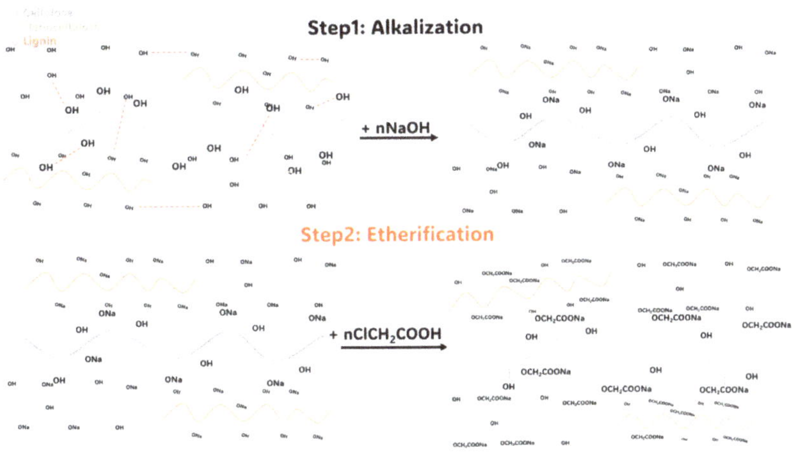


**Figure 4.** Carboxymethylation reaction of lignocellulosic matrix.

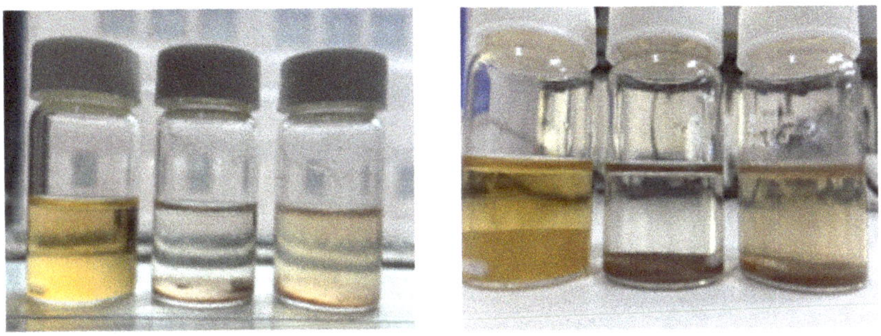

**Figure 5.** Visual photos of 1% (*w*/*v*) mixtures of CMDPR (**left**), WTDPR (**mid**), and DPR (**right**) after 1 h of stirring.

FTIR spectroscopy was used to confirm the carboxymethylation of WTDPR and other structural changes regarding the functional groups and crystallinity indices. Significant changes were observed in the FTIR spectrum between WTDPR and CMDPR, whereas a noticeable reduction of the following peaks around 1158 $cm^{-1}$, 1250 $cm^{-1}$, and 1731 $cm^{-1}$ were noticed in CMDPR, as shown in Figure 1A. These peaks were assigned to the ether or esters groups of lignocellulosic polymers, C−O of acetyl of lignin, and ester or carboxylic groups in hemicellulose and lignin, respectively [43,46]. The absence of these peaks indicates the possibility of hemicellulose and lignin solubilization or the hydrolysis of the mentioned functional groups due to the alkalization step, which tends to hydrolyze branched groups on lignin and hemicellulose [71–74]. An observed reduction in the band at 1374 $cm^{-1}$ also supports the possibility of hemicellulose solubilization, which is assigned to the bending vibration of the −CH group in hemicellulose and cellulose. Intensity increases were noted in three bands associated with carboxymethylated polysaccharides in prepared CMDPR at 1618 $cm^{-1}$, 1425 $cm^{-1}$, 1326 $cm^{-1}$, which were assigned to −C=O stretching in the carboxyl group, the vibration of the carboxyl groups salts, and −OH bending vibration [39]. Those peaks were also observable in the C-CMC with 0.8 DS. It is noteworthy that commercial carboxymethyl cellulose is often prepared from pure cellulose; for this reason, aromatic bands of lignin around 1637 $cm^{-1}$, 1507 $cm^{-1}$, 1458 $cm^{-1}$, and 1374 $cm^{-1}$ were not detected in comparison with CMDPR, which was prepared from a lignocellulosic matrix that contains lignin and hemicellulose.

Furthermore, crystallinity changes occurred within CMDPR (Figure 1B), where TCI decreased from 0.707 in WTDPR to 0.606 in CMDPR. Also, a drastic reduction in LOI was observed from 3.635 in WTDPR to 0.977 in CMDPR. As expected, an increase in HBI was noticed from 3.005 to 3.413, where HBI increased with a decrease in crystallinity. C-CMC showed better TCI at 0.662 and 1.351 LOI and lower HBI at 2.088. These results match with the results from other studies using different techniques, where it was found that the carboxymethylation reaction highly reduced the crystallinity of cellulose, which was most likely due to the substitution of hydroxyl groups with carboxymethyl groups and sodium hydroxide ions, which cleave and restrict the formation of hydrogen bonding, hindering the adoption of order arrangements and increasing the distance between the chains [75–78]. This may explain the reason that C-CMC with a lower DS of 0.8 had higher crystallinity than CMDPR with a higher DS of 1.14.

Figure 6 shows the micrographs of CMDPR, which also confirmed the separation of fibers. Meanwhile, much smaller fibers appeared after carboxymethylation with smooth surfaces, reduced diameters, and ribbon-like shapes compared to WTDPR and raw DPR, which showed relatively compact structures. These results are in alignment with other studies [76,79,80]. In addition, a change in fiber color was observed, where the fiber color turned to golden yellow-like from brown in WTDPR.

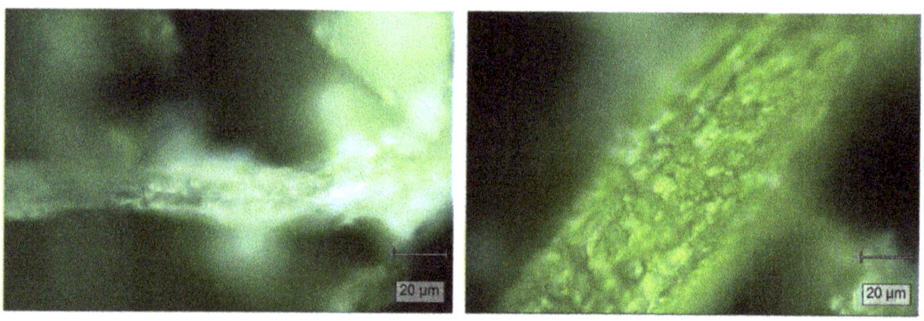


**Figure 6.** Digital microscope imaging of carboxymethylated date palm rachis (CMDPR).

TGA and DSC established CMDPR with 1.14 DS and C-CMC with 0.8 DS thermal characteristics and stability. Figure 7A presents TG curves and the derivative DTG curves of both CMDPR and C-CMC, showing decomposition peaks, which are illustrated in Table 4. An initial mass loss started at 50 °C and ended at 180 °C in both samples, which was most likely due to the evaporation of volatile organic and absorbed moisture attributed to the hygroscopic nature of the samples. The initial mass loss percentages were 12.16% for CMDPR and 15.26% for C-CMC. Further, several decomposition peaks were observed in CMDPR between 230 °C and 530 °C. The decompositions in this range mainly comprised the degradation of carboxymethyl groups alongside the degradation of the lignocellulosic matrix. Cellulose, hemicellulose, and lignin degradation were exhibited at 360–400 °C, 200–315 °C, and 160–900 °C [39,81]. In both samples, a typical decomposition curve of carboxylic and hydroxyl groups was observed; the decomposition of CMDPR was between 230 °C and 324 °C, with a weight loss of about 33.5%. For C-CMC, the decomposition started at 246 °C and went to 314 °C with a weight loss of about 39.7% [82]. In addition, a degradation peak was observed between 324 and 373 °C in CMDPR and between 360 and 400 °C in C-CMC, mostly representing the degradation of cellulose [47]. The decomposition at 382–421 °C represents the methyl-aryl ether bonds of lignin, whereas the decomposition between 421 °C and 460 °C was due to the degradation of lignin polymeric backbone chains [83]. The last decomposition peak in both CMDPR and C-CMC was attributed to the oxidation of residual carbonaceous materials [82].

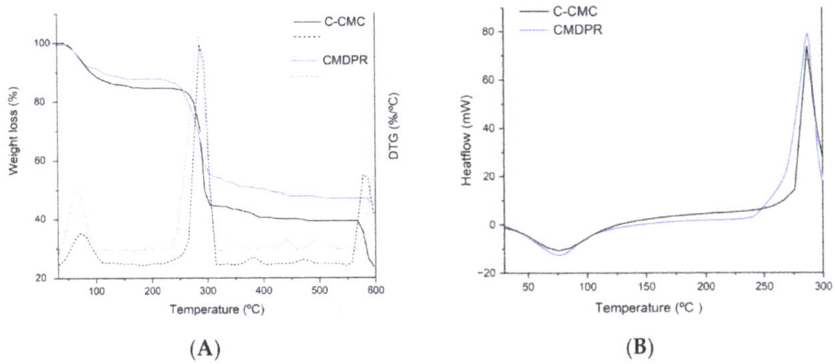


**Figure 7.** TGA, DTG (**A**), and DSC analyses (**B**) of CMDPR and C-CMC.

**Table 4.** Summary of TGA analysis results.

| Sample | Stage | Range | Decomposition Assignment | Residue (%) |
|---|---|---|---|---|
| CMDPR | 1st | 50–180 °C | Volatile compounds and moisture | 45.0% |
| | 2nd | 230–324 °C | Carboxyl and hydroxyl groups | |
| | 3rd | 324–373 °C | Cellulose chain | |
| | 4th | 382–421 °C | Methyl-aryl ether bonds of lignin | |
| | 5th | 421–460 °C | Lignin chain | |
| | 6th | 590–600 °C | Oxidation of residual carbonaceous | |
| C-CMC | 1st | 50–180 °C | Volatile compounds and moisture | 23.8% |
| | 2nd | 246–314 °C | Carboxyl and hydroxyl groups | |
| | 3rd | 360–400 °C | Cellulose chain | |
| | 4th | 587.5–600 °C | Oxidation of residual carbonaceous | |
| CM15 | 1st | 158.8–207.3 °C | Non-crosslinked CA | 67.0% |
| | 2nd | 207.3–343.7 °C | Crosslinked CA and CMDPR chain | |
| CM20 | 1st | 207–343.3 °C | Crosslinked CA and CMDPR chain | 74.1% |

Figure 7B presents the DSC analysis for C-CMC and CMDPR. For both samples, a single endothermic peak attributed to the glass transition ($T_g$) and a single exothermic peak attributed to the polymer destruction were observed, which were confirmed by TGA. For C-CMC, the $T_g$ was found at 71 °C, while for CMDPR, it was at 75 °C. The polymer destruction for C-CMC was about 286 °C and 288 °C for CMDPR. These results prove that CMDPR has good thermal stability in comparison with C-CMC.

### 2.5. *Hydrogels Fabrication and Characterization*

Three different concentrations of CA were chosen to assess the HG preparation potential of prepared CMDPR 10%, 15%, and 20% and the resulting HG-abbreviated CM10, CM15, and CM20, respectively. The gelation mechanism is illustrated in Figure 8. The gelation reaction consisted of two major steps. In contrast, CA underwent a dehydration reaction to yield cyclic anhydride that further reacted with OH groups within the polymer chain and formed an ester bond [84]. The properties of prepared HGs were demonstrated by performing different tests for applicability reasons.

EWC% illustrates the hydrogels' water absorption capacity, which is a key characteristic of designed hydrogels for agricultural applications [85]. It is mostly influenced by the polymer's nature; crosslinking ratio; and medium conditions like ionic strength, pH, temperature, etc. [86]. Three different mediums (deionized water (DW), tap water (TW), and 0.9% NaCl solution) were chosen to assess the EWC% in different water types. Additionally, the tested samples in deionized water were washed and tested again in deionized water, where it was found that CM10 was semi-soluble and was thus excluded from the following tests. Figure 9A shows the effect of each medium on EWC%.

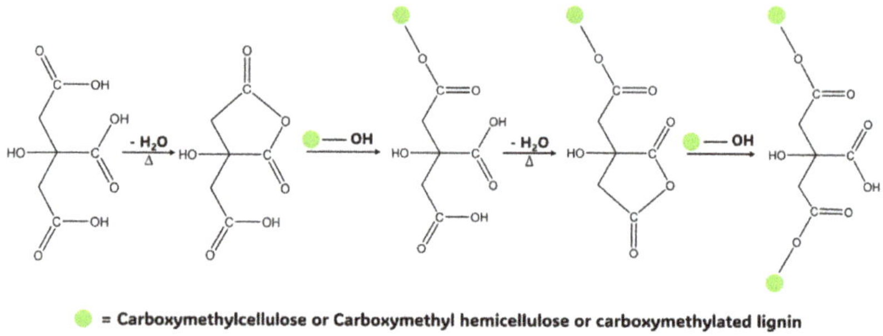


**Figure 8.** Illustration of citric acid crosslinking reaction mechanism with carboxymethylated biopolymers.

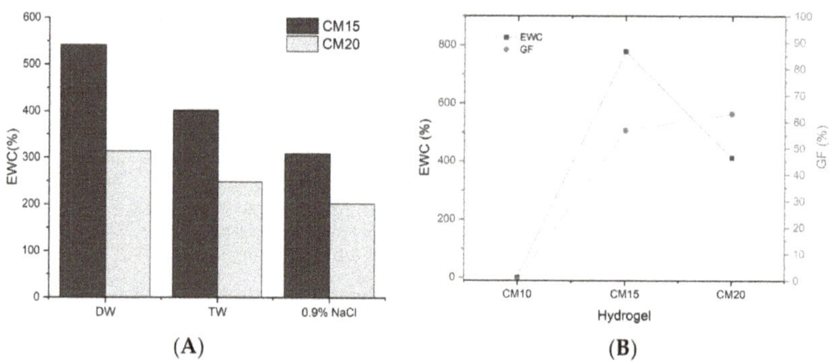


**Figure 9.** EWC% in different mediums before purification (**A**). GF% and EWC% of purified hydrogels in deionized water (**B**).

Ionic strength considerably impacts EWC% where the increment of ionic strength reduces the swelling capability [84]. However, for CM15, the EWC% was found to be 541.7%, 401.7%, and 308.5% in DW, TW, and 0.9% NaCl solution, respectively. For CM20, 313.5%, 247.7%, and 201.56% were found in DW, TW, and 0.9% NaCl solution, respectively. The electrical conductivity (EC) of the three mediums were 5 μS/cm for DW, 230 μS/cm for TW, and 15,400 μS/cm for 0.9% NaCl. Meanwhile, the increase in salt concentration resulted in an increase in EC and ionic strength, with a good correlation [87]. In addition, the purified hydrogels showed an EWC% of about 777.8% and 411.5% in DW for CM15 and CM20, respectively, as presented in Figure 9B. These findings clarify the effect of each medium on the HG swelling capacity with respect to the crosslinker concentration. These findings are relative to the findings in other studies, where Demitri et al. [84] reported that 24 h citric acid-crosslinked CMCNa, with 10% and 20% CA ratios, absorb about 500% and 300% of their weights, respectively. Capanema et al. [88] reported that prepared hydrogels from low and high molecular weight CMC absorbed about 500% when crosslinked by 10% CA and decreased as the CA ratio increased, as found in our study. In addition, different studies have shown that swelling decreased as the ionic strength increased [85,89]. Remarkably, CM20 seems to have a minor sensitivity to ionic strength compared to CM15, where a drastic decrease in EWC% was observed when the ionic strength increased.

The gel fraction (GF%) represents an important characteristic of hydrogels, determining the weight of insoluble (crosslinked) and soluble fractions, the total weight of the used polymers, and the crosslinker; the test was carried out in deionized water. Both CM15 and

CM20 were stable in deionized water, where the GF% averages for CM15 and CM20 were 56.8% and 63.0%, respectively, as shown in Figure 9B. These results are similar to Saputra et al.'s findings, where 15% citric acid-crosslinked CMC had a GF% of 58% [90]. Meanwhile, Capanema et al. claimed that the GF% for CMC hydrogels with Mw 250 and 700 kDa were 60% and 80%, respectively, when 25% citric acid was used for crosslinking [88]. In addition, Durpekova et al. reported that the GF% for CMC hydrogels prepared in acid whey solutions exhibited 67.65% GF% for 15% citric acid-crosslinked hydrogel with a decreasing GF% with a decrease in citric acid concentration [85]. It is well known that there is a proportional relation between GF% and crosslinker concentration, in this case, CA [85]. These results could be attributed to the fact that when CMC is used alone, the electrostatic repulsion between the functional groups will lead to poor crosslinking. In addition, as the DS increases, the crosslinking decreases [84].

FTIR spectrometry, TGA, DTG, and microscope imaging confirmed the formation of hydrogels and their characteristics.

FTIR spectra of hydrogels, CMDPR, and CA are depicted in Figure 10. The spectra of hydrogels showed two new peaks in the ester group compared with CMDPR of around 1235 $cm^{-1}$ and 1726 $cm^{-1}$. Meanwhile, the crosslinking reaction of CA with CMDPR yielded two ester bonds between the polymer chain and CA [84]. The peak intensity around 1726 $cm^{-1}$ at CM15 showed a weak and overlapped peak, which may be attributed to poor crosslinking of CM15; a similar band was also noticed by de Lima et al. for 3% citric acid-crosslinked CMC hydrogels [91]. Also, in both samples, a weaker band at 3400 $cm^{-1}$ was observed, which was also noticed by Capanema et al. [88]. A shift in the carboxylic band from 1618 $cm^{-1}$ to 1587 $cm^{-1}$ was noticed, as reported by Priya et al. [92]. In addition, TCI, LOI, and HBI indices were calculated to assess the crystallinity changes. Both CM15 and CM20 showed increased TCIs of 1.062 and 0.956, respectively. Also, an increase was noticed in LOI from 0.977 in CMDPR to 1.825 and 1.084 in CM20 and CM15, while a reduction in HBI was recorded in both samples from 3.413 to 2.138 and 1.267, respectively. The reduction in HBI is attributed to the fact that −OH groups in crosslinked hydrogel were reduced due to the esterification reaction, while the increase in both TCI and LOI may be related to the overlapping intensities of CH groups from CA and CMDPR in the region of 1425–1300 $cm^{-1}$.

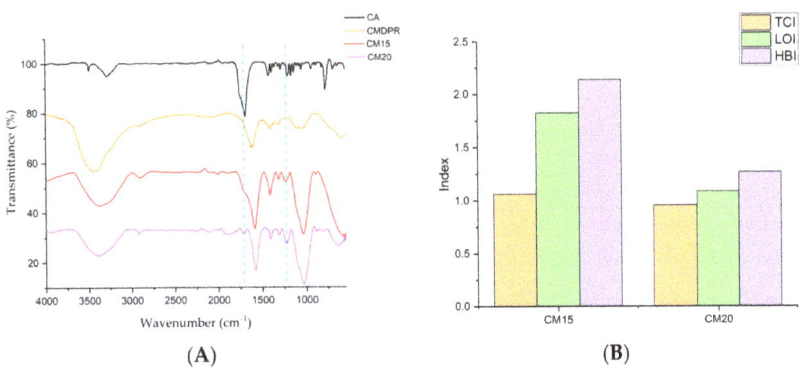


**Figure 10.** FTIR spectrum (**A**) of CA, CMDPR, CM15, and CM20 and crystallinity indices (**B**) of HGs.

The morphological characteristics were recorded using microscope imaging, as shown in Figure 11. The hydrogels showed rough and rigid morphological characteristics. Also, CM20 showed a stiffer structure than CM15, most likely due to the higher crosslinker ratio. Small crystals were noticed in CM15, which were attributed to the unreacted CA, also observed through TGA. Both samples showed gaps and cavities, promoting the absorbance of high amounts of water, whereas CM20, with a higher CA ratio, showed higher porosity.

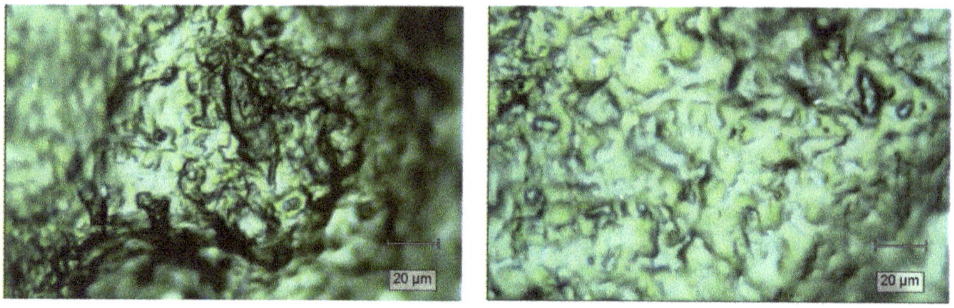


**Figure 11.** Digital microscope imaging of CM20 (**left**) and CM15 (**right**).

This case was in harmony with Ghilan et al.'s study, where they found that increasing the phytic acid ratio in CMC/phytic acid hydrogels increased the porosity [93]. In contrast, Pratinthong et al. reported that in 2–10% crosslinked citric acid-CMC/Poly(vinyl alcohol) hydrogels, an increase in the surface roughness and porosity were noticed from 2–6%; with increasing citric acid concentration, the pores of hydrogels were smaller [94]. In addition, Durpekova et al. mentioned that increasing citric acid concentration from 5% to 15% showed a less sponge-like structure [85]. This may be attributed to the hydrogel compositions, wherein the case of CM15 and CM20 contained residual lignin and hemicellulose, which may justify the porosity increase in respect to the crosslinker.

Figure 12 presents the TG curves and DTG of the HGs. The DTG analysis of CM15 showed three major decomposition peaks in comparison with CMDPR, which showed major decomposition between 230 and 324 °C. On the other hand, CM20 showed only two. In particular, the first decomposition of CM15 was between 158.8 °C and 207.3 °C with a weight loss of about 6%, which was attributed to non-crosslinked CA that starts degrading above 148 °C [95]. This was also observed through microscope imaging. A second peak was exhibited between 207.3 °C and 343.7 °C with three steps; the first step was primarily due to the decomposition of crosslinked CA, while the last two steps represented the synchronous decarboxylation and CMDPR chains degradation which took place in CMDPR between 230 and 324 °C [88,92]. The total weight loss was about 33%. CM20 showed better homogeneity, where no non-crosslinked CA was detected. The thermal decomposition of CM20 started from 207 °C to 343.3 °C, where three decomposition steps were also noticed, with a total weight loss of 25.9%. These results are close to the results found by Priya et al. [92]. Two main decomposition peaks were noticed for 2% citric acid-crosslinked CMC. Also, de Lima et al. reported that 3% citric acid-crosslinked CMC showed two main decompositions [91]. These findings prove the presence of non-crosslinked CA within CM15 that exhibited three decomposition peaks.

### 2.6. *Hydrogels' Effect on Germination*

To assess the applicability of HGs in agricultural applications and soil water retention, the effect of the HGs on different plants' seeds (Chico III tomato, mint, Basilico red, and chia seeds) was studied. The experiment was conducted in a seedling tray and monitored for 20 days (presented in Figure 13), with an average weather temperature of 24 degrees. The soil characteristics are presented in Table 5.

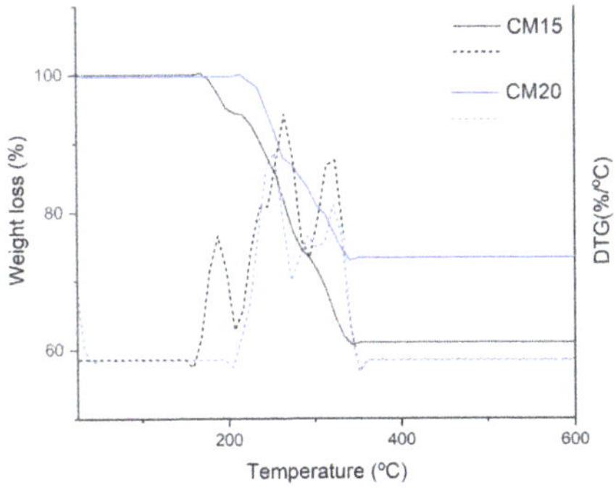


**Figure 12.** TGA and DTG analysis of CM15 and CM20.

**Figure 13.** Visual images of Ch-Ref (**A**), Ch-CM15 (**B**), and Ch-CM20 (**C**) on the fourth day and To-Ref (**D**), To-CM15 (**E**), and To-CM20 (**F**) on the sixth day. Chia is in the upper row and tomato in the lower row on day 13 (**G**), where the two pots in red circles are the ref samples, the orange circle is the CM20 samples, and the blue is soil with CM15.

**Table 5.** Used soil characteristics.

| pH | EC | N | P | K |
|---|---|---|---|---|
| 6–7 | 1.78 µS/cm | 2.08% | 0.015% | 0.22% |

The soil was amended with 0.5% hydrogel of the soil's total weight. Within 12 days, most seeds had already grown on different days. On the first day, the soil was irrigated and then covered with a plastic sheet for two days to prevent water evaporation. The pots were abbreviated as follows. For pots, the first two letters of the seed name were used, where "To" was for tomato, "Ch" was for chia, "Mi" was for mint, and "Ba" was for Basilico red. For example, Ch-CM15 stands for chia pot with CM15 hydrogel. "Ref" was used for soil without HGs.

## 3. Conclusions

HGs have the ability to aid the sustainability of both water and food by enhancing agriculture outcomes while reducing the amount of water needed. An eco-friendly route that reduces the cost and footprint of HG manufacturing was used for HG preparation. Date palm biomass-based carboxymethylated all-lignocellulose was crosslinked by CA to form HGs. The yielded HGs showed a positive impact on tomato, chia, mint, and Basilico red germination in comparison with reference soil. The crosslinking of CM15 and CM20 was confirmed by FTIR analysis and GF%. TGA revealed the thermal stability of HGs prepared below 200 °C. In addition, TGA revealed that in CM15, some unreacted CA was confirmed by microscopic imaging, which explains the low GF%. CM15, after purification, showed an EWC% of about 777.8%, while CM20 showed about 411.5%. Furthermore, a good EWC% was also observed in tap water and saline water (0.9%NaCl) with unpurified hydrogels.

These findings support the utilization of a modified lignocellulosic matrix for low-cost hydrogel fabrication without any fractionation, resulting in HGs with suitable properties for soil water retention and different agricultural applications; however, the effect of the all-lignocellulose composition and molecular weight, alongside HGs' stability and performance in different weather conditions, soil types, and pH mediums, must be investigated for agricultural usage. Further, future studies should consider nutrient and fertilizer loading and release investigations.

## 4. Materials and Methods

### 4.1. Materials

Date palm rachis (DPR) were gathered from the campus of King Abdulaziz University in Jeddah City, Saudi Arabia. Ethanol and methanol were purchased from Sigma Aldrich, St. Louis, MI, USA. Isopropanol (IPA), monochloroacetic acid (MCA), and sulfuric acid ($H_2SO_4$) were purchased from BDH Chemicals Ltd., Poole, UK. Sodium hydroxide (NaOH) was purchased from Merck, Billerica, MA, USA. Citric acid was purchased from Techno Pharmchem, Bahadurgarh, India. Commercial carboxymethyl cellulose (C-CMC) with 0.7–0.8 DS was purchased from BDH Chemicals Ltd., Poole, UK, and was used for comparison purposes.

### 4.2. Biomass Compositional Analysis and Pretreatment

The collected DPR was washed with tap water and wiped with a cloth to remove any physically attached impurities, then left to dry for a week at room temperature. The dried DPR was subjected to physical pretreatment where DPR was chopped by a wood saw until it became stick-like, then grounded by a coffee grinder. In sequence, to get fine powdered biomass, the DPR was sieved by a 1 mm US standard sieve, milled in a Laboratory Mill 3100 (Perten Instruments, Huddinge, Sweden), and then the fine powder was sieved by a 0.21 mm mesh. The compositional analysis for the powdered DPR was performed according to NERL LAPs: NREL/TP-510-42622 for ash content [96], NREL/TP-510-42619 for extractives content [97], NREL/TP-510-42618 for lignin content [98], and TAPPI standard (T 203 cm-99) for cellulose and hemicellulose contents [99].

### 4.3. Biomass Carboxymethylation

In the modification for carboxymethylated date palm rachis (CMDPR) preparation, the DPR was subjected to hot water treatment, where 20 g of DPR was treated with 1 L hot water for 1.5 h at 90–100 °C to eliminate the water-soluble matters to avoid its reaction with MCA that would lead to MCA consumption. The water-treated DPR (WTDPR) was left to dry at room temperature until a constant weight was achieved. The modification reaction was established as follows: 5 g of the air-dried WTDPR was suspended in 105 mL of IPA at room temperature for 15 min. Further, 25 mL of 30% NaOH was added to the suspension portion-wise and kept under stirring for 1 h. Later, a solution of 6 g MCA dissolved in 20 mL of IPA was added to the suspension and heated at 55 °C for 3.5 h under stirring. After the end of the reaction, the mixture was left to reach room temperature and then filtered. The residual was suspended in 100 mL of absolute methanol and neutralized by 90% acetic acid. Then, it was washed four times with 70% ethanol and then with absolute ethanol. The CMDPR was left in a drying oven at 50 °C until a constant weight was achieved.

### 4.4. CMDPR Degree of Substitution and Yield

The yield and DS for the prepared CMDPR was determined using the same method for determining the degree of substitution method based on pure cellulose according to [79] with some modifications. In detail, 0.5 g of CMDPR was suspended in 20 mL of $HNO_3$–methanol (1:1 $v/v$) solution and left undisturbed for 3 h. In sequence, the solution was rinsed with 70% methanol solution ($v/v$) and then dried in an oven until no weight changes were observed. Later, 0.2 g of the product was dissolved in 20 mL of deionized water and 3 mL of 1 N NaOH. This solution was titrated against 1 N HCl using phenolphthalein as an indicator. The titration was considered complete when the color transitioned from red to yellow. The DS was calculated using Equations (1) and (2) [79]. The yield of CMDPR was calculated according to Equation (3) [79].

$$\mathrm{DS} = 0.162\mathrm{A}/(1 - 0.0584\mathrm{A}) \quad (1)$$

$$\mathrm{A} = (\mathrm{BC} - \mathrm{DE})/\mathrm{F} \quad (2)$$

where

A is milliequivalents of acid consumed per gram of sample.
B is NaOH solution added amount in mL.
C is the concentration of the NaOH solution in normality.
D is the amount of HCl consumed in the titration in mL.
E is the concentration of the HCl in normality.
F is acidified CMDPR used (g).
The value 162 is the gram molecular mass of the anhydrous glucose unit of cellulose.
The value 584 is the net increase in molecular mass for each carboxymethyl group substituted.

$$\text{Yield of CMDPR } (\%) = (\mathrm{W_p}/\mathrm{W_0}) \times 100\% \quad (3)$$

where

$W_p$ is the weight of produced CMDPR.
$W_0$ is the weight of WTDPR.

### 4.5. Hydrogels Fabrication

For the fabrication of hydrogels, CMDPR was crosslinked by citric acid in three different ratios, which were 10%, 15%, and 20% of CMDPR weight, which corresponded to 200 mg, 300 mg, and 400 mg, respectively. Three solutions of CMDPR were prepared with 2% consistency. In detail, 2 g of CMDPR was dissolved in 100 mL overnight. Then, an adequate amount of crosslinker was added to each solution and left to dissolve for 1 h. Further, 15 mL of each solution was cast in silicon molds and left to dry at room temperature for 3 days to form hydrogel films. The dried films were cured in an oven at 80 °C for 24 h to get the crosslinked hydrogels.

### 4.6. Hydrogel Properties

#### 4.6.1. Gel Fraction

The gel fraction (GF%) represents an important characteristic of hydrogels, determining the weight of insoluble (crosslinked) and soluble fractions. To determine the gel fraction of the prepared hydrogels, a certain weight ($W_0$, initial mass) of each hydrogel was immersed in 100 mL of deionized water for 24 h at room temperature, then brought out and wiped with tissue paper to remove the adsorbed water on the surface, and then its weight was recorded ($W_s$, swollen mass). Later, the hydrogel was dried in an oven at 50 °C until a constant weight ($W_f$, final mass) was achieved. The GF% was calculated using Equation (4) [88].

$$GF\ (\%) = ((W_0 - W_f)/W_0) \times 100\% \quad (4)$$

#### 4.6.2. Equilibrium Swelling Capacity

The hydrogels' equilibrium swelling capacity (EWC%) was examined in three different media and determined gravimetrically. A known hydrogel weight was taken and then soaked in 100 mL of the examination solution: DW, TW, and 0.9% NaCl, with EC of 5 μS/cm, 230 μS/cm, and 15,400 μS/cm, respectively. The hydrogel was kept in the solution for equilibrium for 24 h at room temperature. After that, the swollen hydrogel was carefully removed from the solution using tweezers; the surface water was wiped with tissue paper; and then the hydrogel was weighed to get the EWC%, which was determined according to Equation (5) [88].

$$EWC\ (\%) = ((W_S - W_0)/W_0) \times 100\% \quad (5)$$

### 4.7. Effects of Hydrogels on Germination

Four different plant seeds were tested in a seedling tray to study the effect of the prepared hydrogels on germination, which were abbreviated using the first two letters of the seed: "To" for tomato, "Ch" for chia, "Mi" for mint, and "Ba" for Basilico red. The soil used in this study was commercial organic soil with a pH range of 6–7, EC 1.78 μS/cm, 2.08% nitrogen content (N), 0.015% phosphorous content (P), and 0.22% potassium content (K). The tray was divided into 4 groups, each group consisting of 4 cavities, for a total of 16 cavities. Two groups were defined as references with unmodified soil, which were abbreviated as "Ref", and two groups with hydrogel-amended soil using CM15 and CM20. For the hydrogel-amended groups, the hydrogels were grounded and mixed thoroughly with the soil, where the percentage of hydrogel was 0.5% of the soil weight. Each cavity was filled with approximately 8 g of soil.

### 4.8. Characterization

#### 4.8.1. Fourier Transform Infrared Spectrometry (FTIR)

The samples were recorded through a Fourier transform infrared spectroscope to investigate the structural changes and crystallinity indices. Powder samples were recorded using the KBr pellet method using Nicolet iS50 FT–IR (Thermo Scientific, Madison, WI, USA) with a resolution of 4 $cm^{-1}$ for 32 scans. CA and prepared HGs were recorded with a Fourier-transform Infrared Spectrometer in attenuated total reflection (ATR) mode using Spectrum 100, (Perkin Elmer, Shelton, CT, USA) with a 4 $cm^{-1}$ resolution for 16 scans. The scanning range was from 500 cm to 4000 $cm^{-1}$ for all samples. Crystallinity indexes of samples were evaluated using the lateral order index (LOI, A1425/A897), which is the ratio between the absorbance of $CH_2$ around 1430–1420 $cm^{-1}$ and the absorbance of β-glycosidic bonds around 900–890 $cm^{-1}$ [55,56]; total crystallinity index (TCI, A1374/A2919), which is the ratio of between the absorbance of –CH stretching vibration around 2900 $cm^{-1}$ and the absorbance of CH deformation vibration around 1374 $cm^{-1}$ [55,57]; and hydrogen bond intensity (HBI, A3336/A1321), which is the ratio between the absorbance of OH stretching and hydrogen bonding between molecules around 3330–3400 $cm^{-1}$ and the absorbance of ~1320 $cm^{-1}$ from CH rocking vibration of the glucose ring [58,59]. To calculate the crystallinity indexes, the transmittance values (T) were converted to absorbance (A) using Equation (6) from Beer's Law [100].

$$A = 2 - \log_{10} T \tag{6}$$

#### 4.8.2. Microscope Imaging

The morphological properties of the sample were recorded using a research-grade digital microscope with a 50× magnification and 50× objective lens on a Leica DM2700 optical microscope (Wetzlar, Germany). The sample images were directly taken in a dry state without any modification.

#### 4.8.3. Dynamic Light Scattering (DLS)

Dynamic light scattering analysis was performed to investigate the particle size distributions using a Mastersizer 3000 equipped with an Aero S dispersion unit (Malvern Instruments Ltd., Malvern, UK). The air pressure was set to 2 bar with a 20% feed rate. For the analysis, 0.25 g of a sample was used.

#### 4.8.4. Thermal Analysis

Thermal behaviors were investigated by Thermogravimetric Analysis (TGA) using a Synchronous Thermal Analyzer STA-1200 (Bioevopeak, Jinan, China). The heating rate was 10 °C/min from 30 °C to 600 °C. Also, the samples' behaviors were studied using differential scanning calorimetry (DSC) using DSC 823 (Mettler Toledo, Nänikon, Switzerland) from 30 °C to 300 °C with a heating rate of 5 °C/min.

**Author Contributions:** Conceptualization, F.S.A. and K.A.A.; methodology, F.S.A. and K.A.A.; validation, F.S.A. and K.A.A.; formal analysis, F.S.A. and M.S.; investigation, F.S.A., M.S., O.F., F.M.A. and F.O.; resources, K.A.A. and F.S.A.; data curation, F.S.A. and M.S.; writing—original draft preparation, F.S.A. and K.A.A.; writing—review and editing, F.S.A. and K.A.A.; visualization, F.S.A.; supervision, K.A.A.; project administration, F.S.A. and K.A.A.; funding acquisition, F.S.A. and K.A.A. All authors have read and agreed to the published version of the manuscript.

**Funding:** This project was funded by KAU Endowment (WAQF) at King Abdulaziz University, Jeddah, under grant number (**WAQF: 295-130-2024**). The authors, therefore, acknowledge with thanks WAQF and the Deanship of Scientific Research (DSR) for technical and financial support.

**Institutional Review Board Statement:** Not applicable.

**Informed Consent Statement:** Not applicable.

**Data Availability Statement:** Data are contained within the article.

**Conflicts of Interest:** The authors declare that there are no conflicts of interest.

## References


1. Adjuik, T.A.; Nokes, S.E.; Montross, M.D.; Wendroth, O. Lignin-Based Hydrogel for Water Retention in Silt Loam Soil. In Proceedings of the American Society of Agricultural and Biological Engineers Annual International Meeting, ASABE 2021, Virtual, 12–16 July 2021; American Society of Agricultural and Biological Engineers: Saint Joseph, MI, USA, 2021; Volume 2, pp. 657–669.
2. Elhag, M. Evaluation of Different Soil Salinity Mapping Using Remote Sensing Techniques in Arid Ecosystems, Saudi Arabia. *J. Sens.* **2016**, *2016*, 7596175. [CrossRef]
3. Vedovello, P.; Sanches, L.V.; da Silva Teodoro, G.; Majaron, V.F.; Bortoletto-Santos, R.; Ribeiro, C.; Putti, F.F. An Overview of Polymeric Hydrogel Applications for Sustainable Agriculture. *Agriculture* **2024**, *14*, 840. [CrossRef]
4. Patra, S.K.; Poddar, R.; Brestic, M.; Acharjee, P.U.; Bhattacharya, P.; Sengupta, S.; Pal, P.; Bam, N.; Biswas, B.; Barek, V.; et al. Prospects of Hydrogels in Agriculture for Enhancing Crop and Water Productivity under Water Deficit Condition. *Int. J. Polym. Sci.* **2022**, *2022*, 4914836. [CrossRef]
5. Cannazza, G.; Cataldo, A.; de Benedetto, E.; Demitri, C.; Madaghiele, M.; Sannino, A. Experimental Assessment of the Use of a Novel Superabsorbent Polymer (SAP) for the Optimization of Water Consumption in Agricultural Irrigation Process. *Water* **2014**, *6*, 2056–2069. [CrossRef]
6. Koech, R.; Langat, P. Improving Irrigation Water Use Efficiency: A Review of Advances, Challenges and Opportunities in the Australian Context. *Water* **2018**, *10*, 1771. [CrossRef]
7. Ghanem, A.M.; Alamri, Y.A. The Impact of the Green Middle East Initiative on Sustainable Development in the Kingdom of Saudi Arabia. *J. Saudi Soc. Agric. Sci.* **2023**, *22*, 35–46. [CrossRef]
8. Sojka, R.E.; Bjorneberg, D.L.; Entry, J.A.; Lentz, R.D.; Orts, W.J. Polyacrylamide in Agriculture and Environmental Land Management. *Adv. Agron.* **2007**, *92*, 75–162.
9. Chamorro, A.F.; Palencia, M.; Arrieta, Á.A. Development of High-Efficiency Fertilizer by Hydrogels Obtained from Cassava Starch and Citric Acid for Slow Release of Ammonium and Potassium. *Gels* **2024**, *10*, 434. [CrossRef]
10. Tariq, Z.; Iqbal, D.N.; Rizwan, M.; Ahmad, M.; Faheem, M.; Ahmed, M. Significance of Biopolymer-Based Hydrogels and Their Applications in Agriculture: A Review in Perspective of Synthesis and Their Degree of Swelling for Water Holding. *RSC Adv.* **2023**, *13*, 24731–24754. [CrossRef]
11. Zhao, B.; Moore, J.S. Fast PH- and Ionic Strength-Responsive Hydrogels in Microchannels. *Langmuir* **2001**, *17*, 4758–4763. [CrossRef]
12. Adjuik, T.A.; Nokes, S.E.; Montross, M.D. Biodegradability of Bio-Based and Synthetic Hydrogels as Sustainable Soil Amendments: A Review. *J. Appl. Polym. Sci.* **2023**, *140*, e53655. [CrossRef]
13. Adjuik, T.A.; Nokes, S.E.; Montross, M.D.; Wendroth, O. The Impacts of Bio-Based and Synthetic Hydrogels on Soil Hydraulic Properties: A Review. *Polymers* **2022**, *14*, 4721. [CrossRef]
14. Agbna, G.H.D.; Zaidi, S.J. Hydrogel Performance in Boosting Plant Resilience to Water Stress—A Review. *Gels* **2025**, *11*, 276. [CrossRef] [PubMed]
15. Demitri, C.; Scalera, F.; Madaghiele, M.; Sannino, A.; Maffezzoli, A. Potential of Cellulose-Based Superabsorbent Hydrogels as Water Reservoir in Agriculture. *Int. J. Polym. Sci.* **2013**, *2013*, 435073. [CrossRef]
16. Jamnongkan, T.; Kaewpirom, S. Potassium Release Kinetics and Water Retention of Controlled-Release Fertilizers Based on Chitosan Hydrogels. *J. Polym. Environ.* **2010**, *18*, 413–421. [CrossRef]
17. Bao, Z.; Xian, C.; Yuan, Q.; Liu, G.; Wu, J. Natural Polymer-Based Hydrogels with Enhanced Mechanical Performances: Preparation, Structure, and Property. *Adv. Healthc. Mater.* **2019**, *8*, 1900670. [CrossRef]
18. Gschwend, F.J.V.; Brandt-Talbot, A.; Chambon, C.L.; Hallett, J.P. Ultra-Low Cost Ionic Liquids for the Delignification of Biomass. In *Ionic Liquids: Current State and Future Directions*; ACS Symposium Series; American Chemical Society: Washington, DC, USA, 2017; Volume 1250, pp. 209–223.
19. Shahi, N.; Wang, P.; Adhikari, S.; Min, B.; Rangari, V.K. Biopolymers Fractionation and Synthesis of Nanocellulose/Silica Nanoparticles from Agricultural Byproducts. *ACS Sustain. Chem. Eng.* **2021**, *9*, 6284–6295. [CrossRef]
20. Isikgor, F.H.; Becer, C.R. Lignocellulosic Biomass: A Sustainable Platform for the Production of Bio-Based Chemicals and Polymers. *Polym. Chem.* **2015**, *6*, 4497–4559. [CrossRef]
21. Shrestha, S.; Kognou, A.L.M.; Zhang, J.; Qin, W. Different Facets of Lignocellulosic Biomass Including Pectin and Its Perspectives. *Waste Biomass Valorization* **2021**, *12*, 4805–4823. [CrossRef]
22. Gabrielii, I.; Gatenholm, P. Preparation and Properties of Hydrogels Based on Hemicellulose. *J. Appl. Polym. Sci.* **1998**, *69*, 1661–1667. [CrossRef]
23. Palantöken, S.; Bethke, K.; Zivanovic, V.; Kalinka, G.; Kneipp, J.; Rademann, K. Cellulose Hydrogels Physically Crosslinked by Glycine: Synthesis, Characterization, Thermal and Mechanical Properties. *J. Appl. Polym. Sci.* **2020**, *137*, 48380. [CrossRef]
24. Wu, L.; Huang, S.; Zheng, J.; Qiu, Z.; Lin, X.; Qin, Y. Synthesis and Characterization of Biomass Lignin-Based PVA Super-Absorbent Hydrogel. *Int. J. Biol. Macromol.* **2019**, *140*, 538–545. [CrossRef] [PubMed]

25. Haq, I.; Qaisar, K.; Nawaz, A.; Akram, F.; Mukhtar, H.; Zohu, X.; Xu, Y.; Mumtaz, M.; Rashid, U.; Ghani, W.; et al. Advances in Valorization of Lignocellulosic Biomass towards Energy Generation. *Catalysts* **2021**, *11*, 309. [CrossRef]
26. Verdía Barbará, P.; Abouelela Rafat, A.; Hallett, J.P.; Brandt-Talbot, A. Purifying Cellulose from Major Waste Streams Using Ionic Liquids and Deep Eutectic Solvents. *Curr. Opin. Green Sustain. Chem.* **2023**, *41*, 100783. [CrossRef]
27. Sosa, F.H.B.; Dias, R.M.; da Costa Lopes, A.M.; Coutinho, J.A.P.; da Costa, M.C. Fast and Efficient Method to Evaluate the Potential of Eutectic Solvents Todissolve Lignocellulosic Components. *Sustainability* **2020**, *12*, 3358. [CrossRef]
28. Shaikh, H.M.; Anis, A.; Poulose, A.M.; Al-Zahrani, S.M.; Madhar, N.A.; Alhamidi, A.; Aldeligan, S.H.; Alsubaie, F.S. Synthesis and Characterization of Cellulose Triacetate Obtained from Date Palm (*Phoenix dactylifera* L.) Trunk Mesh-Derived Cellulose. *Molecules* **2022**, *27*, 1434. [CrossRef]
29. Nasatto, P.; Pignon, F.; Silveira, J.; Duarte, M.; Noseda, M.; Rinaudo, M. Methylcellulose, a Cellulose Derivative with Original Physical Properties and Extended Applications. *Polymers* **2015**, *7*, 777–803. [CrossRef]
30. Chakka, V.P.; Zhou, T. Carboxymethylation of Polysaccharides: Synthesis and Bioactivities. *Int. J. Biol. Macromol.* **2020**, *165*, 2425–2431. [CrossRef]
31. Saadaoui, N.; Rouilly, A.; Fares, K.; Rigal, L. Characterization of Date Palm Lignocellulosic By-Products and Self-Bonded Composite Materials Obtained Thereof. *Mater. Des.* **2013**, *50*, 302–308. [CrossRef]
32. Shaikh, H.M.; Anis, A.; Poulose, A.M.; Al-Zahrani, S.M.; Madhar, N.A.; Alhamidi, A.; Alam, M.A. Isolation and Characterization of Alpha and Nanocrystalline Cellulose from Date Palm (*Phoenix dactylifera* L.) Trunk Mesh. *Polymers* **2021**, *13*, 1893. [CrossRef]
33. Galiwango, E.; Abdel Rahman, N.S.; Al-Marzouqi, A.H.; Abu-Omar, M.M.; Khaleel, A.A. Isolation and Characterization of Cellulose and α-Cellulose from Date Palm Biomass Waste. *Heliyon* **2019**, *5*, e02937. [CrossRef] [PubMed]
34. Ghori, W.; Saba, N.; Jawaid, M.; Asim, M. A Review on Date Palm (*Phoenix dactylifera*) Fibers and Its Polymer Composites. *IOP Conf. Ser. Mater. Sci. Eng.* **2018**, *368*, 012009. [CrossRef]
35. Haddadou, I.; Aliouche, D.; Brosse, N.; Amirou, S. Characterization of Cellulose Prepared from Some Algerian Lignocellulosic Materials (Zeen Oak Wood, Aleppo Pine Wood and Date Palm Rachis). *Eur. J. Wood Wood Prod.* **2015**, *73*, 419–421. [CrossRef]
36. Mirmehdi, S.M.; Zeinaly, F.; Dabbagh, F. Date Palm Wood Flour as Filler of Linear Low-Density Polyethylene. *Compos. B Eng.* **2014**, *56*, 137–141. [CrossRef]
37. Amirou, S.; Zerizer, A.; Pizzi, A.; Haddadou, I.; Zhou, X. Particleboards Production from Date Palm Biomass. *Eur. J. Wood Wood Prod.* **2013**, *71*, 717–723. [CrossRef]
38. Suriyatem, R.; Noikang, N.; Kankam, T.; Jantanasakulwong, K.; Leksawasdi, N.; Phimolsiripol, Y.; Insomphun, C.; Seesuriyachan, P.; Chaiyaso, T.; Jantrawut, P.; et al. Physical Properties of Carboxymethyl Cellulose from Palm Bunch and Bagasse Agricultural Wastes: Effect of Delignification with Hydrogen Peroxide. *Polymers* **2020**, *12*, 1505. [CrossRef]
39. Rahman, M.S.; Mondal, M.I.H.; Yeasmin, M.S.; Abu Sayeed, M.; Hossain, M.A.; Ahmed, M.B. Conversion of Lignocellulosic Corn Agro-Waste into Cellulose Derivative and Its Potential Application as Pharmaceutical Excipient. *Processes* **2020**, *8*, 711. [CrossRef]
40. Moreira, P.H.S.S.; de Oliveira Freitas, J.C.; Braga, R.M.; Araújo, R.M.; de Souza, M.A.F. Production of Carboxymethyl Lignin from Sugar Cane Bagasse: A Cement Retarder Additive for Oilwell Application. *Ind. Crops Prod.* **2018**, *116*, 144–149. [CrossRef]
41. Alsulami, R.A.; El-Sayed, S.A.; Eltaher, M.A.; Mohammad, A.; Almitani, K.H.; Mostafa, M.E. Thermal Decomposition Characterization and Kinetic Parameters Estimation for Date Palm Wastes and Their Blends Using TGA. *Fuel* **2023**, *334*, 126600. [CrossRef]
42. Younis, M.; Farag, H.A.; Alhamdan, A.; Aboelasaad, G.; Zein El-Abedein, A.I.; Kamel, R.M. Utilization of Palm Residues for Biochar Production Using Continuous Flow Pyrolysis Unit. *Food Chem. X* **2023**, *20*, 100903. [CrossRef]
43. Sun; Tomkinson, J.; Zhu, W.; Wang, S.Q. Delignification of Maize Stems by Peroxymonosulfuric Acid, Peroxyformic Acid, Peracetic Acid, and Hydrogen Peroxide. 1. Physicochemical and Structural Characterization of the Solubilized Lignins. *J. Agric. Food Chem.* **2000**, *48*, 1253–1262. [CrossRef]
44. Saleem, H.; Saud, A.; Zaidi, S.J. Sustainable Preparation of Graphene Quantum Dots from Leaves of Date Palm Tree. *ACS Omega* **2023**, *8*, 28098–28108. [CrossRef] [PubMed]
45. Sizirici, B.; Fseha, Y.H.; Yildiz, I.; Delclos, T.; Khaleel, A. The Effect of Pyrolysis Temperature and Feedstock on Date Palm Waste Derived Biochar to Remove Single and Multi-Metals in Aqueous Solutions. *Sustain. Environ. Res.* **2021**, *31*, 9. [CrossRef]
46. Zrelli, A.; Elfalleh, W.; Ghorbal, A.; Chaouachi, B. Valorization of Date Palm Wastes by Lignin Extraction to Be Used for the Improvement of Polymeric Membrane Characteristics. *Period. Polytech. Chem. Eng.* **2022**, *66*, 70–81. [CrossRef]
47. Boumediri, H.; Bezazi, A.; Del Pino, G.G.; Haddad, A.; Scarpa, F.; Dufresne, A. Extraction and Characterization of Vascular Bundle and Fiber Strand from Date Palm Rachis as Potential Bio-Reinforcement in Composite. *Carbohydr. Polym.* **2019**, *222*, 114997. [CrossRef]
48. Bezazi, A.; Belaadi, A.; Bourchak, M.; Scarpa, F.; Boba, K. Novel Extraction Techniques, Chemical and Mechanical Characterisation of *Agave americana* L. Natural Fibres. *Compos. B Eng.* **2014**, *66*, 194–203. [CrossRef]
49. Kasyapi, N.; Chaudhary, V.; Bhowmick, A.K. Bionanowhiskers from Jute: Preparation and Characterization. *Carbohydr. Polym.* **2013**, *92*, 1116–1123. [CrossRef]

50. Bezazi, A.; Boumediri, H.; Garcia del Pino, G.; Bezzazi, B.; Scarpa, F.; Reis, P.N.B.; Dufresne, A. Alkali Treatment Effect on Physicochemical and Tensile Properties of Date Palm Rachis Fibers. *J. Nat. Fibers* **2020**, *19*, 3770–3787. [CrossRef]
51. Wan, C.; Li, Y. Effect of Hot Water Extraction and Liquid Hot Water Pretreatment on the Fungal Degradation of Biomass Feedstocks. *Bioresour. Technol.* **2011**, *102*, 9788–9793. [CrossRef]
52. Nelson, M.L.; O'Connor, R.T. Relation of Certain Infrared Bands to Cellulose Crystallinity and Crystal Lattice Type. Part II. A New Infrared Ratio for Estimation of Crystallinity in Celluloses I and II. *J. Appl. Polym. Sci.* **1964**, *8*, 1325–1341. [CrossRef]
53. Nelson, M.L.; O'Connor, R.T. Relation of Certain Infrared Bands to Cellulose Crystallinity and Crystal Latticed Type. Part I. Spectra of Lattice Types I, II, III and of Amorphous Cellulose. *J. Appl. Polym. Sci.* **1964**, *8*, 1311–1324. [CrossRef]
54. Nada, A.M.A.; Hassan, M.L. Thermal Behavior of Cellulose and Some Cellulose Derivatives. *Polym. Degrad. Stab.* **2000**, *67*, 111–115. [CrossRef]
55. Hrčka, R.; Kučerová, V.; Hönig, V. Dry-Matter Loss and Changes in the Chemical Composition of Spruce Wood after Long-Term Storing in the Form of Roundwood. *Polymers* **2022**, *14*, 3400. [CrossRef]
56. Singh, J.K.; Chaurasia, B.; Dubey, A.; Noguera, A.M.F.; Gupta, A.; Kothari, R.; Upadhyaya, C.P.; Kumar, A.; Hashem, A.; Alqarawi, A.A.; et al. Biological Characterization and Instrumental Analytical Comparison of Two Biorefining Pretreatments for Water Hyacinth (*Eicchornia crassipes*) Biomass Hydrolysis. *Sustainability* **2021**, *13*, 245. [CrossRef]
57. Krutul, D.; Szadkowski, J.; Výbohová, E.; Kučerová, V.; Čabalová, I.; Antczak, A.; Szadkowska, D.; Drożdżek, M.; Zawadzki, J. Effect of Steam Explosion Pretreatment on Chosen Saccharides Yield and Cellulose Structure from Fast-Growing Poplar (*Populus Deltoides* × *Maximowiczii*) Wood. *Wood Sci. Technol.* **2024**, *58*, 441–458. [CrossRef]
58. Dos Santos Dias, D.; Otoni, C.G.; da Silva, R.R.; Meneguin, A.B.; Mattoso, L.H.C.; da Barud, H.S.; Ribeiro, C.A. Large Scale Manufacturing of Puree-Only Edible Films from Onion Bulb (*Allium cepa* L.): Probing Production and Structure–Processing–Property Correlations. *Ind. Crops Prod.* **2020**, *145*, 111847. [CrossRef]
59. Rafidison, B.H.; Ramasawmy, H.; Chummun, J.; Florens, F.B.V. Using Infrared Spectrum Analyses to Predict Tensile Strength of Fibres in a Group of Closely Related Plant Species: Case of Mascarenes *Pandanus* spp. *SN Appl. Sci.* **2020**, *2*, 1922. [CrossRef]
60. Kučerová, V.; Výbohová, E.; Hönig, V.; Čabalová, I. Chemical Changes within Solids during Liquid Hot Water Pretreatment of Wood. *BioResources* **2020**, *15*, 38. [CrossRef]
61. Hrčka, R.; Kučerová, V.; Hýrošová, T.; Hönig, V. Cell Wall Saturation Limit and Selected Properties of Thermally Modified Oak Wood and Cellulose. *Forests* **2020**, *11*, 640. [CrossRef]
62. Kačík, F.; Šmíra, P.; Kačíková, D.; Vel'ková, V.; Nasswettrová, A.; Vacek, V. Chemical Alterations of Pine Wood Saccharides during Heat Sterilisation. *Carbohydr. Polym.* **2015**, *117*, 681–686. [CrossRef]
63. Isroi; Ishola, M.; Millati, R.; Syamsiah, S.; Cahyanto, M.; Niklasson, C.; Taherzadeh, M. Structural Changes of Oil Palm Empty Fruit Bunch (OPEFB) after Fungal and Phosphoric Acid Pretreatment. *Molecules* **2012**, *17*, 14995–15012. [CrossRef] [PubMed]
64. Yang, X.; Berglund, L.A. Recycling without Fiber Degradation—Strong Paper Structures for 3D Forming Based on Nanostructurally Tailored Wood Holocellulose Fibers. *ACS Sustain. Chem. Eng.* **2020**, *8*, 1146–1154. [CrossRef]
65. Huang, C.; Cheng, J.; Zhan, Y.; Liu, X.; Wang, J.; Wang, Y.; Yoo, C.G.; Fang, G.; Meng, X.; Ragauskas, A.J.; et al. Utilization of Guaiacol-Based Deep Eutectic Solvent for Achieving a Sustainable Biorefinery. *Bioresour. Technol.* **2022**, *362*, 127771. [CrossRef]
66. Wahlström, N.; Edlund, U.; Pavia, H.; Toth, G.; Jaworski, A.; Pell, A.J.; Choong, F.X.; Shirani, H.; Nilsson, K.P.R.; Richter-Dahlfors, A. Cellulose from the Green Macroalgae *Ulva lactuca*: Isolation, Characterization, Optotracing, and Production of Cellulose Nanofibrils. *Cellulose* **2020**, *27*, 3707–3725. [CrossRef]
67. Berglund, J.; Mikkelsen, D.; Flanagan, B.M.; Dhital, S.; Gaunitz, S.; Henriksson, G.; Lindström, M.E.; Yakubov, G.E.; Gidley, M.J.; Vilaplana, F. Wood Hemicelluloses Exert Distinct Biomechanical Contributions to Cellulose Fibrillar Networks. *Nat. Commun.* **2020**, *11*, 4692. [CrossRef] [PubMed]
68. Khiari, R.; Mhenni, M.F.; Belgacem, M.N.; Mauret, E. Valorisation of Vegetal Wastes as a Source of Cellulose and Cellulose Derivatives. *J. Polym. Environ.* **2011**, *19*, 80–89. [CrossRef]
69. Jin, Z.; Yu, Y.; Shao, S.; Ye, J.; Lin, L.; Iiyama, K. Lignin as a Cross-Linker of Acrylic Acid-Grafted Carboxymethyl Lignocellulose. *J. Wood Sci.* **2010**, *56*, 470–476. [CrossRef]
70. Rani, D.; Ahuja, M. Carboxymethylation of Lepidium Sativum Polyuronide, Its Characterization and Evaluation as a Nanometric Carrier. *Int. J. Biol. Macromol.* **2017**, *99*, 233–240. [CrossRef]
71. Sankhla, S.; Sardar, H.H.; Neogi, S. Greener Extraction of Highly Crystalline and Thermally Stable Cellulose Micro-Fibers from Sugarcane Bagasse for Cellulose Nano-Fibrils Preparation. *Carbohydr. Polym.* **2021**, *251*, 117030. [CrossRef]
72. Yan, Z.; Li, J.; Chang, S.; Cui, T.; Jiang, Y.; Yu, M.; Zhang, L.; Zhao, G.; Qi, P.; Li, S. Lignin Relocation Contributed to the Alkaline Pretreatment Efficiency of Sweet Sorghum Bagasse. *Fuel* **2015**, *158*, 152–158. [CrossRef]
73. Figueiredo, P.; Lintinen, K.; Hirvonen, J.T.; Kostiainen, M.A.; Santos, H.A. Properties and Chemical Modifications of Lignin: Towards Lignin-Based Nanomaterials for Biomedical Applications. *Prog. Mater. Sci.* **2018**, *93*, 233–269. [CrossRef]
74. Lu, Y.; He, Q.; Fan, G.; Cheng, Q.; Song, G. Extraction and Modification of Hemicellulose from Lignocellulosic Biomass: A Review. *Green Process. Synth.* **2021**, *10*, 779–804. [CrossRef]

75. Rachtanapun, P.; Jantrawut, P.; Klunklin, W.; Jantanasakulwong, K.; Phimolsiripol, Y.; Leksawasdi, N.; Seesuriyachan, P.; Chaiyaso, T.; Insomphun, C.; Phongthai, S.; et al. Carboxymethyl Bacterial Cellulose from Nata de Coco: Effects of NaOH. *Polymers* **2021**, *13*, 348. [CrossRef]
76. Dos Santos, D.M.; de Bukzem, A.L.; Ascheri, D.P.R.; Signini, R.; de Aquino, G.L.B. Microwave-Assisted Carboxymethylation of Cellulose Extracted from Brewer's Spent Grain. *Carbohydr. Polym.* **2015**, *131*, 125–133. [CrossRef] [PubMed]
77. Moussa, I.; Khiari, R.; Moussa, A.; Belgacem, M.N.; Mhenni, M.F. Preparation and Characterization of Carboxymethyl Cellulose with a High Degree of Substitution from Agricultural Wastes. *Fibers Polym.* **2019**, *20*, 933–943. [CrossRef]
78. Yeasmin, M.S.; Mondal, M.I.H. Synthesis of Highly Substituted Carboxymethyl Cellulose Depending on Cellulose Particle Size. *Int. J. Biol. Macromol.* **2015**, *80*, 725–731. [CrossRef]
79. Churam, T.; Usubharatana, P.; Phungrassami, H. Sustainable Production of Carboxymethyl Cellulose: A Biopolymer Alternative from Sugarcane (*Saccharum officinarum* L.) Leaves. *Sustainability* **2024**, *16*, 2352. [CrossRef]
80. Klunklin, W.; Jantanasakulwong, K.; Phimolsiripol, Y.; Leksawasdi, N.; Seesuriyachan, P.; Chaiyaso, T.; Insomphun, C.; Phongthai, S.; Jantrawut, P.; Sommano, S.R.; et al. Synthesis, Characterization, and Application of Carboxymethyl Cellulose from Asparagus Stalk End. *Polymers* **2021**, *13*, 81. [CrossRef]
81. Lin, B.-J.; Chen, W.-H. Sugarcane Bagasse Pyrolysis in a Carbon Dioxide Atmosphere with Conventional and Microwave-Assisted Heating. *Front. Energy Res.* **2015**, *3*, 4. [CrossRef]
82. Lakshmi, D.S.; Trivedi, N.; Reddy, C.R.K. Synthesis and Characterization of Seaweed Cellulose Derived Carboxymethyl Cellulose. *Carbohydr. Polym.* **2017**, *157*, 1604–1610. [CrossRef]
83. Sathawong, S.; Sridach, W.; Techato, K. Lignin: Isolation and Preparing the Lignin Based Hydrogel. *J. Environ. Chem. Eng.* **2018**, *6*, 5879–5888. [CrossRef]
84. Demitri, C.; Del Sole, R.; Scalera, F.; Sannino, A.; Vasapollo, G.; Maffezzoli, A.; Ambrosio, L.; Nicolais, L. Novel Superabsorbent Cellulose-based Hydrogels Crosslinked with Citric Acid. *J. Appl. Polym. Sci.* **2008**, *110*, 2453–2460. [CrossRef]
85. Durpekova, S.; Di Martino, A.; Dusankova, M.; Drohsler, P.; Sedlarik, V. Biopolymer Hydrogel Based on Acid Whey and Cellulose Derivatives for Enhancement Water Retention Capacity of Soil and Slow Release of Fertilizers. *Polymers* **2021**, *13*, 3274. [CrossRef] [PubMed]
86. Mignon, A.; De Belie, N.; Dubruel, P.; Van Vlierberghe, S. Superabsorbent Polymers: A Review on the Characteristics and Applications of Synthetic, Polysaccharide-Based, Semi-Synthetic and 'Smart' Derivatives. *Eur. Polym. J.* **2019**, *117*, 165–178. [CrossRef]
87. Pintro, J.; Inoue, T.T. A Comparative Study of Determined and Calculated Values of Ionic Strength of Different Nutrient Solutions Consisting of an Ionic Pair. *J. Plant. Nutr.* **1999**, *22*, 1223–1231. [CrossRef]
88. Capanema, N.S.V.; Mansur, A.A.P.; Carvalho, I.C.; Carvalho, S.M.; Mansur, H.S. Bioengineered Water-Responsive Carboxymethyl Cellulose/Poly(Vinyl Alcohol) Hydrogel Hybrids for Wound Dressing and Skin Tissue Engineering Applications. *Gels* **2023**, *9*, 166. [CrossRef]
89. Dai, H.; Huang, H. Enhanced Swelling and Responsive Properties of Pineapple Peel Carboxymethyl Cellulose-*g*-Poly(Acrylic Acid-*Co*-Acrylamide) Superabsorbent Hydrogel by the Introduction of Carclazyte. *J. Agric. Food. Chem.* **2017**, *65*, 565–574. [CrossRef]
90. Saputra, A.H.; Hapsari, M.; Pitaloka, A.B.; Wulan, P.P.D.K. Synthesis and Characterization of Hydrogel from Cellulose Derivatives of Water-Hyacinth (*Eichhornia crassipes*) through Chemical Cross-Linking Method by Using Citric Acid. *J. Eng. Sci. Technol.* **2015**, *10*, 75–86.
91. de Lima, G.F.; de Souza, A.G.; dos Rosa, D.S. Nanocellulose as Reinforcement in Carboxymethylcellulose Superabsorbent Nanocomposite Hydrogels. *Macromol. Symp.* **2020**, *394*, 2000126. [CrossRef]
92. Priya, G.; Narendrakumar, U.; Manjubala, I. Thermal Behavior of Carboxymethyl Cellulose in the Presence of Polycarboxylic Acid Crosslinkers. *J. Therm. Anal. Calorim.* **2019**, *138*, 89–95. [CrossRef]
93. Ghilan, A.; Nita, L.E.; Pamfil, D.; Simionescu, N.; Tudorachi, N.; Rusu, D.; Rusu, A.G.; Bercea, M.; Rosca, I.; Ciolacu, D.E.; et al. One-Step Preparation of Carboxymethyl Cellulose—Phytic Acid Hydrogels with Potential for Biomedical Applications. *Gels* **2022**, *8*, 647. [CrossRef] [PubMed]
94. Pratinthong, K.; Punyodom, W.; Jantrawut, P.; Jantanasakulwong, K.; Tongdeesoontorn, W.; Sriyai, M.; Panyathip, R.; Thanakkasaranee, S.; Worajittiphon, P.; Tanadchangsaeng, N.; et al. Modification of a Carboxymethyl Cellulose/Poly(Vinyl Alcohol) Hydrogel Film with Citric Acid and Glutaraldehyde Crosslink Agents to Enhance the Anti-Inflammatory Effectiveness of Triamcinolone Acetonide in Wound Healing. *Polymers* **2024**, *16*, 1798. [CrossRef] [PubMed]
95. Barbooti, M.M.; Al-Sammerrai, D.A. Thermal Decomposition of Citric Acid. *Thermochim. Acta* **1986**, *98*, 119–126. [CrossRef]
96. Sluiter, A.; Hames, B.; Ruiz, R.; Scarlata, C.; Sluiter, J.; Templeton, D. *Determination of Ash in Biomass*; Technical Report NREL/TP-510-42622; National Renewable Energy Laboratory: Golden, CO, USA, 2005.
97. Sluiter, A.; Ruiz, R.; Scarlata, C.; Sluiter, J.; Templeton, D. *Determination of Extractives in Biomass*; Technical Report NREL/TP-510-42619; National Renewable Energy Laboratory: Golden, CO, USA, 2005.

98. Sluiter, A.; Hames, B.; Ruiz, R.; Scarlata, C.; Sluiter, J.; Templeton, D.; Crocker, D. *Determination of Structural Carbohydrates and Lignin in Biomass*; NREL/TP-510-42618; National Renewable Energy Laboratory: Golden, CO, USA, 2012.
99. *TAPPI, Alpha-, Beta- and Gamma-Cellulose in Pulp T 203 cm-99*; Technical Association of the Pulp and Paper Industry: Atlanta, GA, USA, 1999.
100. Azmi, I.S.; Azizan, A.; Mohd Salleh, R. Pretreatment of Oil Palm Frond (OPF) with Ionic Liquid. *IOP Conf. Ser. Mater. Sci. Eng.* **2018**, *358*, 012071. [CrossRef]

*gels*

MDPI

*Article*

# Nano-Enabled Seed Treatment Using Bisepoxide-Polyoxypropylenetriamine Polymeric Gel with Different Embedded Zinc Sources

**Felipe B. Alves [1], Adela S. M. Goñi [1,2], Bruno A. Fico [1], Vanessa S. A. Silva [1], Renato P. Orenha [1], Renato L. T. Parreira [1], Heber E. Andrada [1], Gabriel Sgarbiero Montanha [3], Higor J. F. A. da Silva [3], Eduardo de Almeida [3], Hudson W. P. de Carvalho [3,4], Natália Chittolina [3], Clíssia B. Mastrangelo [3] and Eduardo F. Molina [1,*]**

1 University of Franca, Av. Dr. Armando Salles Oliveira 201, Franca 14404-600, SP, Brazil; felipeb.alves@hotmail.com (F.B.A.); adelasanmartin64@gmail.com (A.S.M.G.); bruno.fico@outlook.com (B.A.F.); vanessasilveira3012@gmail.com (V.S.A.S.); rpo9@hotmail.com (R.P.O.); renato.parreira@unifran.edu.br (R.L.T.P.); heberandrada@hotmail.es (H.E.A.)
2 Public University of Navarre, Arrosadia Campus, Av. Cataluña, 31006 Pamplona, Navarre, Spain
3 Center for Nuclear Energy in Agriculture (CENA), University of São Paulo, Piracicaba 13416-000, SP, Brazil; gabriel.montanha@usp.br (G.S.M.); higorjfas@usp.br (H.J.F.A.d.S.); edualm@cena.usp.br (E.d.A.); hudson.carvalho@um6p.ma (H.W.P.d.C.); nchittolina@usp.br (N.C.); clissia@usp.br (C.B.M.)
4 Mohammed VI Polytechnic University, Lot 660, Ben Guerir 43150, Morocco
* Correspondence: eduardo.molina@unifran.edu.br or molina_ferreira@yahoo.com.br

**Abstract:** In the 21st century, sustainable agriculture is expected to become a major contributor to food security and improved nutrition. Amine–epoxide-based materials have great potential for use in agriculture due to their tunable physicochemical features, which are dependent on the concentration and composition of the monomers. In this work, catalyst-free green synthesis, using only water as a solvent, was performed to obtain a nanocarrier (TGel) capable of transporting nutrients after seed priming. The synthesis was based on the opening of the epoxy ring by nucleophile attack, using an amine-terminated polyether. Transmission electron microscopy (TEM) and dynamic light scattering (DLS) techniques showed the spherical morphology of the particles, which ranged in size from 80 nm (unloaded TGel) to 360 nm (zinc-loaded TGel), respectively. Theoretical bonding analysis revealed that Zn cation species from the $ZnSO_4$ source interact with the polymer via σ-bonds, whereas EDTA forms hydrogen bonds with the polymer, thereby enhancing noncovalent interactions. Micro X-ray fluorescence (μ-XRF) and energy-dispersive X-ray fluorescence spectroscopy (EDXRF) provided details of the distributions of Zn in the seed compartments and shoots of cucumber plants after seed priming and plant growth, respectively. The use of the Zn-loaded TGels did not affect the physiology of the cucumber plants, as indicated by the photosynthetic efficacy, chlorophyll, and anthocyanin indices.

**Keywords:** seed priming; amine–epoxide; zinc-based fertilizers; nutrient distribution

## 1. Introduction

The modern agricultural system encounters numerous challenges originating from the impacts of global climate change, soil degradation, pest and disease, and the inefficient use of agrochemicals, which significantly hinder agricultural productivity. Addressing these challenges requires the development and implementation of innovative agricultural strategies to improve efficiency and promote long-term sustainability [1]. Seed priming with nanomaterials offers a promising approach by improving seed protection during storage, enhancing germination rates and uniformity, and supporting early plant development [2]. In the last decade, there have been significant advances in the development and application of polymeric materials (natural, synthetic, and/or functionalized polymers) in agriculture and the food industry [3,4]. This progress has been primarily driven by the growing demand for food, as well as concerns about food safety. These factors highlight the need for increased efficiency in crop cultivation, food production, and processing methods to achieve high yields, improved quality, and mitigation of potential health risks associated with food consumption [5,6]. The advantageous properties exhibited by polymers (considering their physical form, porosity, permeability, diffusion capacity, chemical reactivity, and stability) make them highly suitable for use in the agricultural area [7,8], where their integration with nano-engineering systems can assist in addressing key issues related to health standards, nutritional enrichment, environmental conservation, pollution mitigation, and economic growth. Consequently, the utilization of polymers in agriculture and the food industry is a dynamic and expanding theme that can contribute to meeting emerging demands and challenges in various sectors essential for human sustenance and well-being [9].

The specific properties of a polymeric material can be tailored based on fundamental considerations, involving two critical stages: (i) the selection of appropriate monomers to achieve the desired arrangement of structural constituents within macromolecules and (ii) the application of optimized polymerization techniques [10,11]. Polymers have diverse applications within the agricultural sector, including (but not limited to) soil conditioning, formulation of planting and transplantation gels, seed coatings to regulate germination, soil aeration, and soil sterilization [12–15].

Micro/nanosized polymeric particles have many uses in the areas of chemicals and materials [16,17]. The development of methods for the fabrication of polymeric particles with adjustable properties is highly desirable for obtaining controlled-release formulations of various agrochemicals, which can reduce the amounts of these chemicals released to the environment [18]. Therefore, this study investigated the formation of an epoxide monomer from diepoxy poly(ethylene glycol) (DPEG) and an amine from a PPG-based polyetheramine of the Jeffamine T-series to produce amine–epoxide particles. Jeffamine T-series is a polyetheramine characterized by a backbone composed of repeating oxypropylene units. The amine groups are positioned on secondary carbon atoms at the end of aliphatic polyether chains. The "click" reaction was used to obtain the polymeric systems (micro- or nanoparticles), using a catalyst-free chemical reaction that formed hydroxyl groups after the reaction of the epoxide with the amine from the polyoxypropylene backbone [19]. Subsequently, the polymeric amine–epoxide formulation was loaded with different Zn sources ($ZnSO_4$ or Zn-EDTA) to study the potential of the materials for use in seed priming applications as carriers for nutrients and other substances essential for plant development. These Zn sources were selected because they are commonly used as fertilizers and are generally in soluble forms. Zn is an essential micronutrient for plant nutrition, playing a critical role in various physiological and biochemical processes. Ethylenediaminetetraacetic acid (EDTA) is a widely used chelating agent in agriculture. This is the first time that Zn-loaded amine–epoxide particles have been investigated as carriers for applications in agriculture as a new seed priming technology.

One of the most effective strategies for enhancing plant development, growth, and yield is seed treatment. This cost-effective and straightforward technique induces a priming effect, could lead to a strengthening seedling vigor, and enhances germination rate, biomass accumulation, crop productivity, and the maintenance of ionic homeostasis [16]. This work demonstrated the potential of the technique by using Zn-loaded amine–epoxide polymeric particles to improve the absorption of micronutrients by the different seed compartments without any adverse effects on plant growth. After initial seed priming with the Zn-loaded polymeric formulations, the physiological parameters of cucumber seedlings were evaluated during plant growth over a period of 18 days.

## 2. Results and Discussion

### 2.1. Characterization of the Zn-Loaded and Unloaded TGels

The synthesized TGels, both unloaded and loaded with Zn, were thoroughly characterized using DLS and TEM techniques (Figure 1). The unloaded TGel particles had a mean diameter of ~267 nm and polydispersity index (PdI) of 0.136 (Figure 1a). At pH 7, the zeta potential ($\zeta$) of the TGel was +34 mV. When the zinc sources were added, the diameters observed by DLS were ~361 nm for TGel-ZnEDTA (with PdI of 0.110) and ~489 nm for TGel-$ZnSO_4$ (with PdI of 0.225), as shown in Figures 1b and 1c, respectively. The zeta potentials of the loaded TGel-ZnEDTA and TGel-$ZnSO_4$ decreased, compared with the unloaded TGel, with $\zeta$ values (at pH 7) of +15 mV and +18 mV, respectively. Transmission electron microscopy showed that the unloaded TGel particles presented spherical-like morphology, with an average size of 80 $\pm$ 5 nm (Figure 1d). The discrepancy between this size and the value obtained by DLS could be explained by the hydration of the TGel in an aqueous environment. Similarly, TGel-ZnEDTA particles exhibited a spherical morphology, comprising both smaller and larger particles, with the latter suggesting an agglomeration (adsorption process) of the former particles by interactions of TGel, EDTA ligand, and $Zn^{2+}$ ions. The average particle size ranged from 180 to 600 $\pm$ 20 nm (Figure 1e). Notably, the TGel-$ZnSO_4$ showed a distinct morphology (Figure 1f), with the apparent formation of polymer clusters. This different morphology could have been due to the presence of ions (such as $Zn^{2+}$ and $SO_4{}^{2-}$) in the solution, which induced the formation of polymer clusters when using $ZnSO_4$ embedded in the polymeric gel. These results indicated that the final morphology of the TGel was determined by the nature of the Zn source embedded in the polymeric network. The loaded TGels containing Zn-EDTA and $ZnSO_4$ were subsequently used as seed priming agents, aiming at agricultural applications.

Theoretical calculations were performed to explore the primary chemical interactions between a molecule based on the TGel framework and either the $Zn^{2+}$ cation or the $[Zn(EDTA)]^{2-}$ complex (Figure 2). The EDA method was used to elucidate the mechanisms of bonding between these structures (Table 1). For the $\text{Polymer}_{\text{Frag}}\cdots[Zn(EDTA)]^{2-}$ complex, two chemical bonding analyses were conducted. The first focused on the interaction between the $[\text{Polymer}_{\text{Frag}}(EDTA)]^{4-}$ complex and the $Zn^{2+}$ cation, while the second examined the chemical interaction between $\text{Polymer}_{\text{Frag}}$ and $[Zn(EDTA)]^{2-}$. In the EDA method [20], the interaction energy, $\Delta E_{\text{int}}$, is decomposed into four main components, as shown below:

$$\Delta E_{\text{int}} = \Delta V_{\text{elstat}} + \Delta E_{\text{Pauli}} + \Delta E_{\text{oi}} + \Delta E_{\text{disp}} \quad (1)$$

In Equation (1), the electrostatic energy term ($\Delta V_{\text{elstat}}$) represents classical electrostatic interactions between the unaltered charge distributions of the interacting fragments. The Pauli repulsion term ($\Delta E_{\text{Pauli}}$) accounts for destabilizing interactions between occupied orbitals, which are linked to steric effects. The orbital interaction energy ($\Delta E_{\text{oi}}$) encompasses charge transfer (interactions between occupied orbitals of one fragment and unoccupied orbitals of another) and polarization (the mixing of occupied and unoccupied orbitals induced by the presence of the other fragment). Lastly, the $\Delta E_{\text{disp}}$ term incorporates dispersion corrections, as introduced by Grimme and colleagues [21].

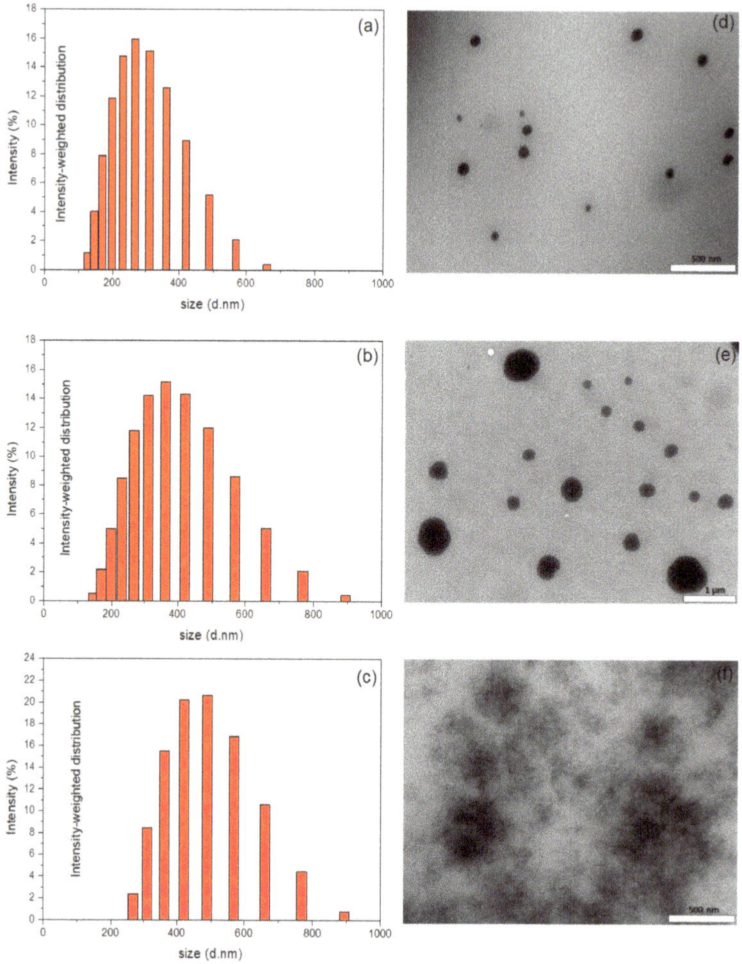


**Figure 1.** Size distributions obtained by DLS for (**a**) pure TGel, (**b**) TGel-ZnEDTA, and (**c**) TGel-$ZnSO_4$. TEM images of (**d**) pure TGel, (**e**) TGel-ZnEDTA, and (**f**) TGel-$ZnSO_4$. The scale bars in the images correspond to 500 nm (**d**,**f**) and 1 μm (**e**).

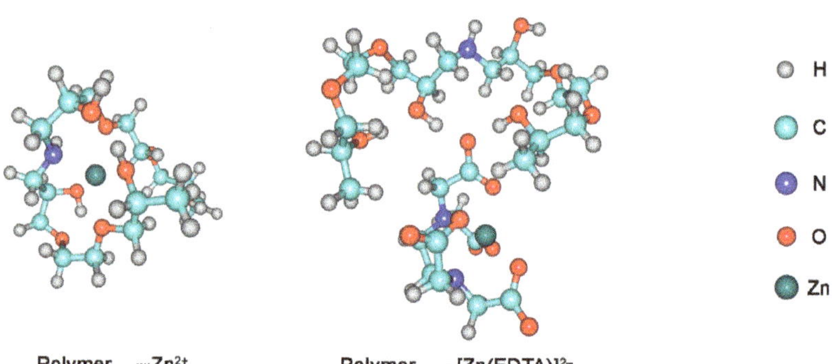


**Figure 2.** Optimized geometries of the analyzed complexes: Polymer$_{Frag}$····$Zn^{2+}$ and Polymer$_{Frag}$····$[Zn(EDTA)]^{2-}$.

**Table 1.** Investigation of the bonding scenarios: (i) $Polymer_{Frag}\cdots Zn^{2+}$, (ii) $[Polymer_{Frag}(EDTA)]^{4-}\cdots Zn^{2+}$, and (iii) $Polymer_{Frag}\cdots[Zn(EDTA)]^{2-}$, employing the EDA-NOCV methodology. The energy values are reported in kcal $mol^{-1}$.

| Interaction | $\Delta E_{int}$ | $\Delta V_{elstat}$ | $\Delta E_{Pauli}$ | $\Delta E_{oi}$ | $\Delta E_{disp}$ | $\Delta E_{oi,1}$ | $\Delta E_{oi,2}$ | $\Delta E_{oi,3}$ | $\Delta E_{oi,4}$ | $\Delta E_{oi,5}$ | $\Delta E_{oi,6}$ |
|---|---|---|---|---|---|---|---|---|---|---|---|
| $Polymer_{Frag}\cdots Zn^{2+}$ | −415.6 | −252.3 | 111.8 | −257.7 | −17.4 | −67.0 | −28.6 | −25.6 | −21.7 | −11.1 | −9.7 |
| $[Polymer_{Frag}(EDTA)]^{4-}\cdots Zn^{2+}$ | −1003.2 | −822.2 | 114.5 | −280.9 | −14.6 | −67.8 | −27.7 | −25.6 | −24.6 | −16.2 | −13.4 |
| $Polymer_{Frag}\cdots[Zn(EDTA)]^{2-}$ | −65.1 | −68.0 | 64.7 | −47.1 | −14.7 | −16.1 | −8.8 | - | - | - | - |

$Polymer_{Frag}$ refers to a fundamental repeating unit within the TGel polymer network. It represents a structural fragment that contains the key functional groups characteristic of the TGel polymer, playing a crucial role in defining its overall architecture and properties. $\Delta E_{int}$ (interaction energy), $\Delta V_{elstat}$ (electrostatic energy), $\Delta E_{Pauli}$ (Pauli repulsion energy), $\Delta E_{oi}$ (orbital interaction energy), and $\Delta E_{disp}$ (dispersion energy) correspond to the bonds analyzed as follows: $\Delta E_{int} = \Delta V_{elstat} + \Delta E_{Pauli} + \Delta E_{oi} + \Delta E_{disp}$, while $\Delta E_{oi1\text{–}6}$ represent the energetic contributions of each deformation density channel ($\Delta\rho_{1\text{–}6}$) associated with $\Delta E_{oi}$.

In the $Polymer_{Frag}\cdots Zn^{2+}$ interaction, the contributions from $\Delta V_{elstat}$ (48%) and $\Delta E_{oi}$ (49%) were comparable, while $\Delta E_{disp}$ accounted for a smaller proportion (3%) of the total stabilization energy ($\Delta V_{elstat} + \Delta E_{oi} + \Delta E_{disp}$). These findings were indicative of a partially covalent character of the $Polymer_{Frag}\cdots Zn^{2+}$ bond. The attractive $\Delta E_{int}$ energy, supporting the interaction between polymer fragment and $Zn^{2+}$ ion, predominantly arose from the favorable $\Delta V_{elstat}$, $\Delta E_{oi}$, and $\Delta E_{disp}$ components, which collectively overcame the destabilizing $\Delta E_{Pauli}$ term (Table 1).

The attractive $[Polymer_{Frag}(EDTA)]^{4-}\cdots Zn^{2+}$ interaction was predominantly driven by $\Delta V_{elstat}$ (74%), with smaller contributions from $\Delta E_{oi}$ (25%) and $\Delta E_{disp}$ (1%) to the total stabilization energy ($\Delta V_{elstat} + \Delta E_{oi} + \Delta E_{disp}$). This suggested that the $[Polymer_{Frag}(EDTA)]^{4-}\cdots Zn^{2+}$ interaction was primarily noncovalent in nature. Similarly, the favorable $Polymer_{Frag}\cdots[Zn(EDTA)]^{2-}$ interaction exhibited a dominant $\Delta V_{elstat}$ contribution (52%), together with a significant role of $\Delta E_{oi}$ (36%) and a smaller influence of $\Delta E_{disp}$ (11%) in the total attractive energy ($\Delta V_{elstat} + \Delta E_{oi} + \Delta E_{disp}$). These results were also indicative of the largely noncovalent character of the $Polymer_{Frag}\cdots[Zn(EDTA)]^{2-}$ bond.

The Natural Orbitals for Chemical Valence (NOCV) method provides valuable insights into important orbital interactions, such as those between $Polymer_{Frag}$ and $Zn^{2+}$, by decomposing the interaction into pairwise contributions from the most significant molecular orbitals. Each pairwise orbital interaction within a given chemical bond can be visualized by means of deformation density channels, $\Delta\rho_k(r)$, where red regions represent electron density loss, while blue regions indicate electron density gain. Additionally, the NOCV method quantifies the energetic contribution ($\Delta E_{oi,k}$) of each deformation density channel ($\Delta\rho_k$) to the total orbital interaction energy ($\Delta E_{oi}$) [22].

The key density deformation channels, $\Delta\rho_{1\text{–}6}$, are associated with the $Polymer_{Frag}\cdots Zn^{2+}$ chemical bonds are shown in Figure 3. These results suggested that the interaction between the polymer fragment and the $Zn^{2+}$ ion occurred primarily through the nitrogen and (especially) the oxygen atoms of the polymer, involving σ-bonds. In the case of the interaction between $[Polymer_{Frag}(EDTA)]^{4-}$ and $Zn^{2+}$, the NOCV methodology revealed that the $Zn^{2+}$ cation interacted with the oxygen atoms in the EDTA structure by means of σ-bonds (Figure 3). The NOCV analysis of the main density deformation channels, $\Delta\rho_1$ and $\Delta\rho_2$, related to the $Polymer_{Frag}\cdots[Zn(EDTA)]^{2-}$ bond showed that the polymer fragment in the $Polymer_{Frag}\cdots[Zn(EDTA)]^{2-}$ complex interacted with the EDTA compound by O–H⋯O hydrogen bonds (Figure 4). The energies associated with the $\Delta\rho_{1\text{–}6}$ channels, $\Delta E_{oi,1\text{–}6}$, are summarized in Table 1.

Theoretical bonding analysis allows us to establish the following correlation. On the one hand, an analysis of the polymer framework and $Zn^{2+}$ cation indicates that the coordination of this ionic species with the polymer framework occurs via σ-bonds. Since these results (optimized geometries) only consider the $Polymer_{Frag}\cdots Zn^{2+}$ interaction, the presence of anionic species $SO_4^{2-}$ (originating from the $ZnSO_4$ source) may also contribute to interactions with the hydrogen from amine (NH) present at the interface of water–polymer gel, leading to a decrease in ζ values from +34 mV (pure TGel) to +15 mV (TGel-$ZnSO_4$). Furthermore, the partial covalent character of the $Polymer_{Frag}\cdots Zn^{2+}$ bond, along with the possible interaction of $SO_4{}^{2-}$ with the NH group of the TGel interface, could hinder the formation of spherical particles, as demonstrated in Figure 1c. On the other hand, due to the hydrogen bonds formed by the EDTA compound (O–H····O) with the TGel polymer—primarily through noncovalent interactions—and the $Zn^{2+}$ cation maintaining interactions with the oxygen atoms of EDTA (see Figure 3), these features suggest that the EDTA ligand cannot diffuse into polymeric nanoparticles but is instead adsorbed onto the surface of TGel spherical particles. This mechanism strongly aligns with the results from DLS, ζ-potential analysis, and TEM image, which show that the incorporation of ZnEDTA into TGel particles leads to an increase in particle size (see Figure 1b), a decrease in ζ-potential, and the presence of larger, deformed spherical TGel-ZnEDTA particles (Figure 1e).

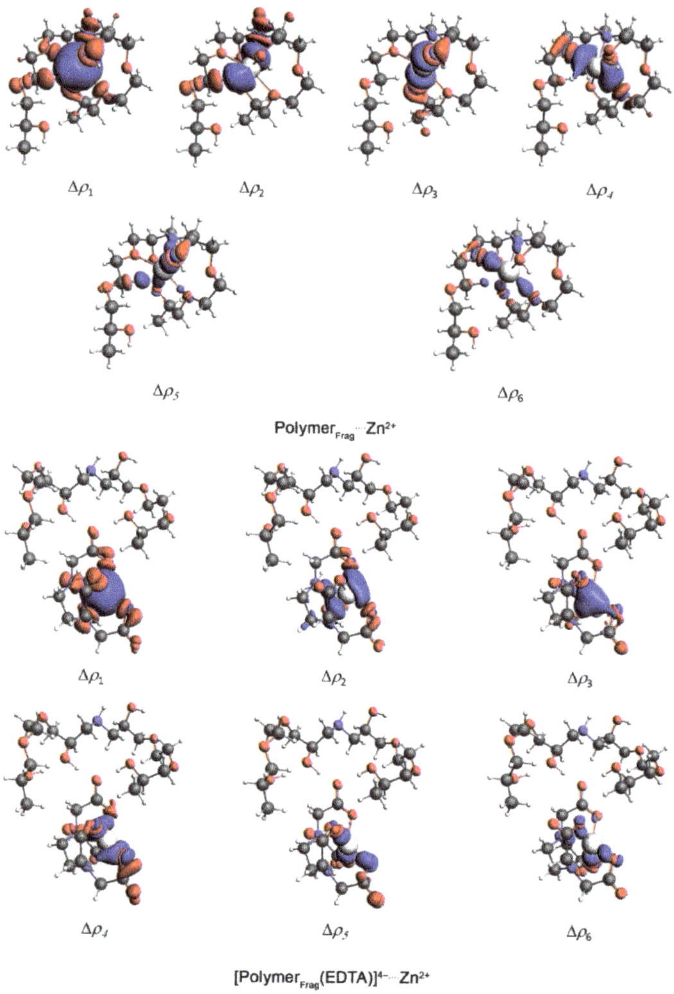


**Figure 3.** Surface plots of the first six density deformation channels, $\Delta_{\rho 1\text{–}6}$, with an isovalue of 0.001 a.u., for the $Polymer_{Frag}\cdots Zn^{2+}$ and $[Polymer_{Frag}(EDTA)]^{4-}\cdots Zn^{2+}$ complexes. Atom color coding: hydrogen (white), carbon (gray), nitrogen (blue), oxygen (red), and zinc (silver).

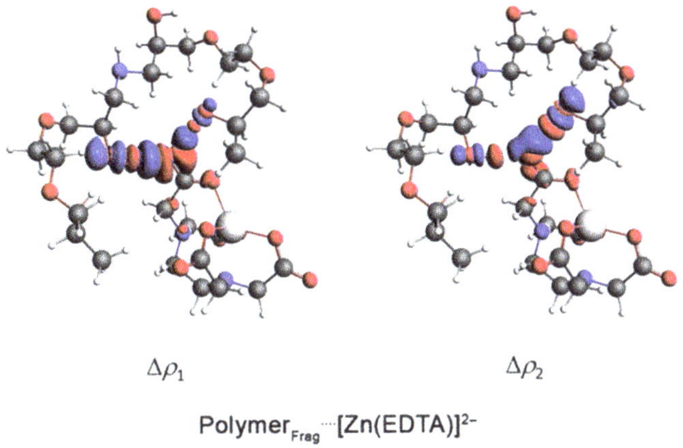


**Figure 4.** Surface plots of the first two density deformation channels, $\Delta\rho_1$ and $\Delta\rho_2$, with an isovalue of 0.001 a.u for the $Polymer_{Frag}\cdots[Zn(EDTA)]^{2-}$ compound. Atom color coding: hydrogen (white), carbon (gray), nitrogen (blue), oxygen (red), and zinc (silver).

*2.2. TGels Containing Different Zn Sources as Safe Seed Nanopriming Agents*

The effect of seed nanopriming on the growth of cucumber plants was evaluated by treating seeds with the TGels containing the different Zn sources (Zn-EDTA and $ZnSO_4$) and measuring the lengths of roots and shoots during a 12-day period following germination. For comparison, Zn-EDTA and $ZnSO_4$ solutions were employed as control seed priming treatments. In these experiments, day 1 was designated as the day that the seeds were placed in Petri dishes after priming. For all the seed treatments, the germination rate exceeded 85%. At a macroscopic level, no phytotoxic effects were observed following the application of the Zn-loaded TGels as nanopriming agents, as evidenced by the germination progression shown in Figure 5, comparing the Zn-EDTA control, TGel-ZnEDTA, $ZnSO_4$ control, and TGel-$ZnSO_4$ treatments.

Figure 6 shows the effects of Zn-loaded TGel nanopriming on cucumber seedling growth after germination, compared with the control Zn treatments, in terms of the evolution of the root and shoot lengths over time. Regardless of the type of polymeric gel, root growth was considerably higher for seed nanopriming with TGel-ZnEDTA and TGel-$ZnSO_4$ compared with the use of the control ZnEDTA and $ZnSO_4$ solutions (Figure 6a,b). After 6 days, the plants from seeds nanoprimed with the Zn-loaded TGels showed substantially higher root growth, which could be attributed to the effective uptake (by diffusion) of the loaded TGel throughout the seed tissues and the influence of subsequent Zn release on the germination process. In contrast, the seeds treated with the Zn controls showed apparent stabilization of root growth after 6 days, which could be explained by the high amounts of Zn available during seed priming using the control solutions, leading to sub-optimal utilization of the nutrients by the seeds. It should be noted that for all the seed treatments (Zn-loaded TGels, ZnEDTA, and $ZnSO_4$ control solutions), the concentration of the zinc source was the same (100 mg $L^{-1}$). Hence, the results for the root lengths indicated that the Zn-loaded TGels provided better accessibility of Zn for plant uptake compared with the control solutions. For the shoot lengths, the results for both TGel-ZnEDTA and TGel-$ZnSO_4$ were close to those for the corresponding Zn control solutions (Figure 6c,d).

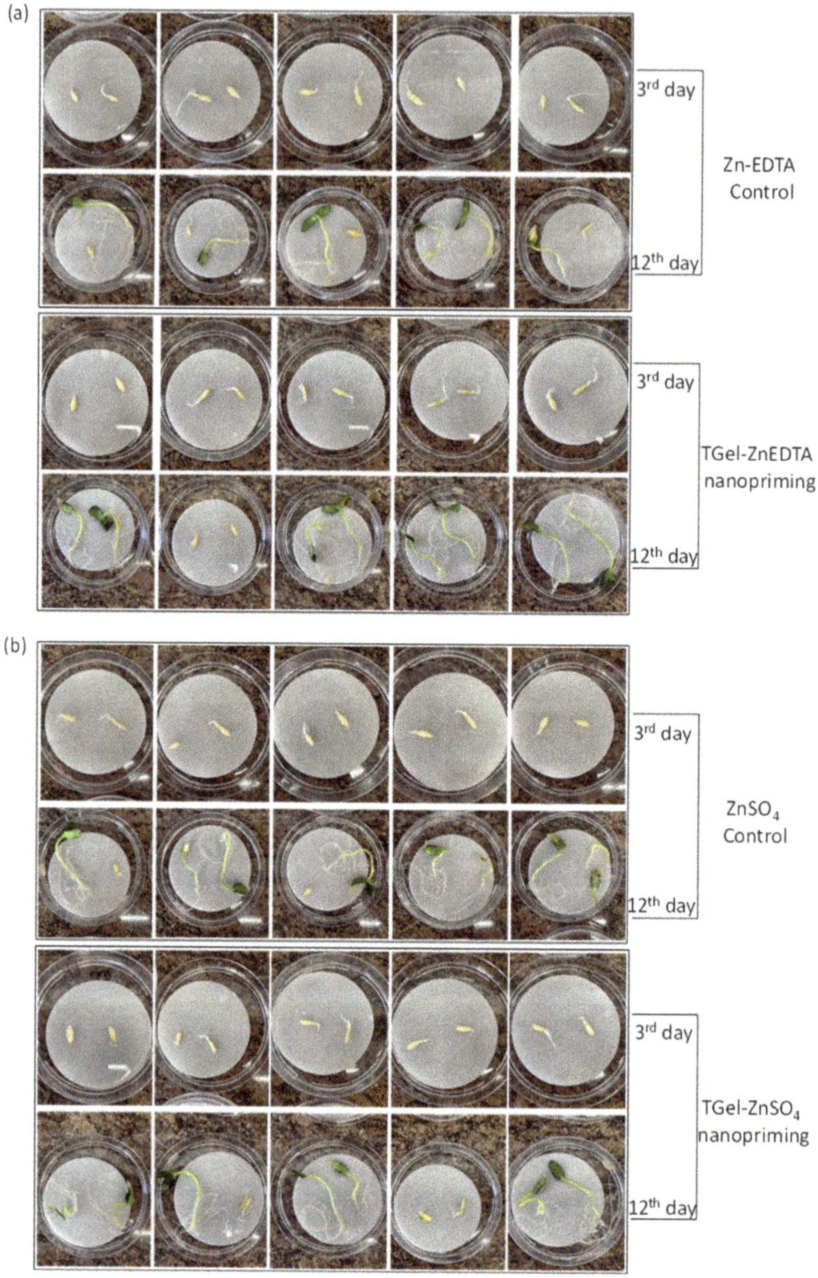


**Figure 5.** Comparison of cucumber seedling growth at 3rd and 12th days after priming with (**a**) ZnEDTA solution (controls) and TGels containing ZnEDTA source; (**b**) $ZnSO_4$ solutions (controls); and after nanopriming with TGels containing $ZnSO_4$ source. The images show five replicates of the assays. The germination assays showed two seeds per petri dish with replicates $n = 5$.

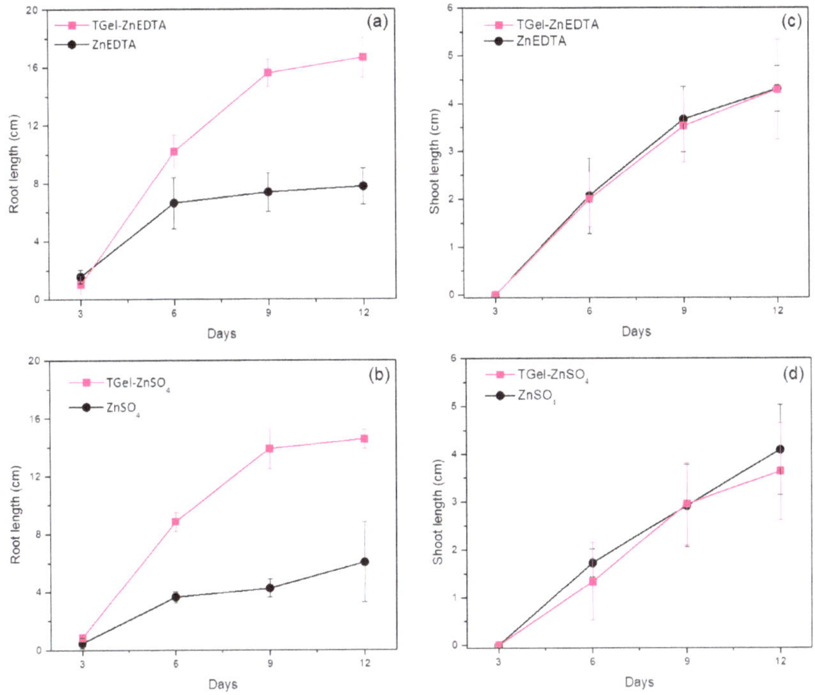


**Figure 6.** Evolution of (**a**,**b**) root length and (**c**,**d**) shoot length, as a function of time, up to 12 days after seed priming using the Zn-loaded TGels and the Zn control solutions. The data are shown as mean ± standard error.

### 2.3. Zn Distribution and Photosynthetic Efficacy

To evaluate the effects of the Zn-loaded TGel formulations on the absorption of Zn and its distribution in the cucumber seeds, specimens submitted to priming for 24 h were analyzed using micro X-ray fluorescence spectroscopy (μ-XRF). Figure 7 shows the Zn intensities recorded in scans of the embryo and cotyledon (Figure 7a), revealing higher Zn intensities for the seeds exposed to TGel-$ZnSO_4$ compared with treatment with the control solution (Figure 7b). This was confirmed by tissue-based comparisons showing that the Zn intensities were significantly higher for the tissues (seed coat, embryo, and cotyledon) of seeds primed with TGel-$ZnSO_4$ compared with those exposed to the $ZnSO_4$ solution (Figure 7c). Regardless of the sulfate-based treatment, the Zn intensities were in the order of coating > embryo > cotyledon, as shown by the 2D maps (Figures 7d and S1). Similar trends have been observed elsewhere for cucumber seeds treated with amine–epoxide and polyetheramine–epoxide gels with embedded micronutrients such as $Fe^{3+}$ [23,24], as well as for soybean seeds exposed to $ZnSO_4$ and bulk or nanosized ZnO solutions [25,26].

Conversely, no clear pattern was observed for the ZnEDTA-based treatments, where the inclusion of Zn in the gel led to virtually no differences in the Zn intensities for the seed coat and cotyledon tissues, while different intensities were found for the embryo. In addition, the Zn intensities for the cotyledon and embryo tissues were significantly higher than for the seed coat (details of this comparison are provided in Figure S2). Interestingly, the Zn intensities were around 10-fold lower than found for the sulfate-based treatments, suggesting either that there was lower adhesion of EDTA-chelated Zn to the cucumber seeds or that the EDTA-chelated Zn was readily absorbed towards the inner seed tissues (cotyledon and embryo), but at a slow rate, as observed elsewhere for soybean roots [27].

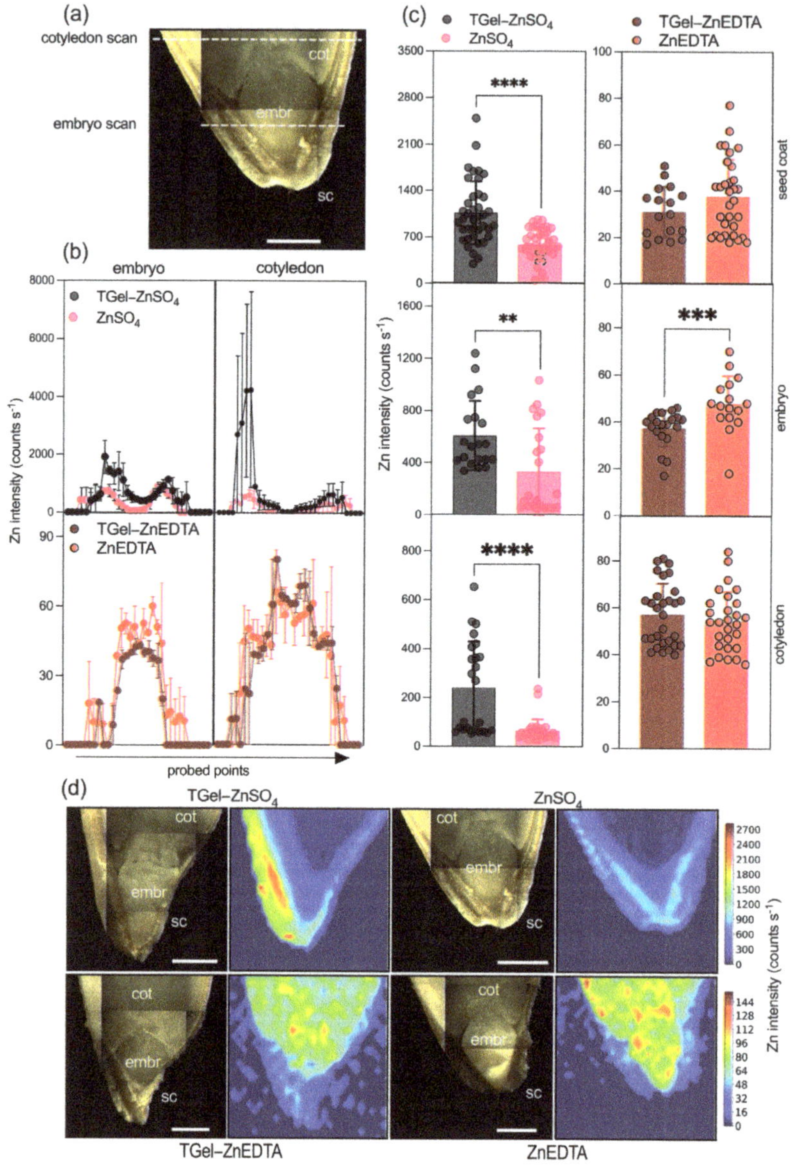


**Figure 7.** μ-XRF scanning of Zn distributions in cross-sections of cucumber seeds primed for 24 h with the Zn-loaded TGels or the $ZnSO_4$ and ZnEDTA solutions at a Zn concentration of 100 mg $L^{-1}$. (**a**) Photograph showing the scanned regions (indicated by the white dashed lines). (**b**) Zn intensities at the probed points in the embryo and cotyledon, with the data shown as mean ± maximum/minimum range. (**c**) Zn intensities obtained for the seed tissues (seed coat, embryo, and cotyledon) for the different treatments, where the bars show the mean ± standard deviation values for measurements using two independent biological replicates. Statistical analysis employed the Mann–Whitney *t*-test ($p < 0.05$). (**d**) Two-dimensional maps for the seed coat (sc), embryo (embr), and cotyledon (cot), with the highest Zn intensities obtained for the tissues exposed to the TGel-$ZnSO_4$ formulation. Scale bars: 1 mm. The *p*-values lower or equal to 0.01, 0.001, and 0.0001 are represented with two, three, or four asterisks, respectively.

Our findings (by μ-XRF analysis) indicate that loaded TGel with different Zn sources may significantly influence the permeability and diffusion dynamics of metal across the perisperm–endosperm envelope of the seed, thereby enhancing the germination process. The effects observed after seed priming with Zn-loaded TGel formulations can be related to the cells' osmotic equilibrium and membrane structure maintenance in which lignin present in seed coat cells is hydrophobic and acts as a natural barrier. The polyether backbone of TGel, composed of polyoxypropylene (-CH($CH_3$)-O-) units, enables potential interactions between the gel's functional groups and the lignin-rich phase, influencing the transport of the solute (polymeric gel) to the seed coat surface and facilitating the internalization of loaded polymeric chains. The absorbed polymeric gel remains accessible due to possible interactions such as hydrogen bonding, cation-π, and hydrophobic interactions between TGel and the seed. The underlying mechanism and interactions involving amine–epoxide-based polymeric gels and seeds were recently demonstrated for the first time, as reported in reference [24].

The higher Zn uptake by seeds treated with sulfate-loaded TGel aligns with theoretical insights into $Zn^{2+}$ cation coordination by functional groups within the polymeric gel network, as demonstrated by NOVC method calculations (Figure 3). The optimized $Polymer_{Frag}\cdots Zn^{2+}$ geometries reveal orbital interactions in close proximity to nitrogen from amine groups and oxygen from polyether chains, establishing $Zn^{2+}$ binding sites within the TGel framework. Given that the TGel structure comprises amide groups bonded to polyoxypropylene chains (three amides and three polyoxypropylene chains per structural unit), a high $Zn^{2+}$ encapsulation capacity is suggested. The $Polymer_{Frag}\cdots Zn^{2+}$ interactions, which exhibit partial covalent character, likely facilitate $Zn^{2+}$ entrapment within the cross-linked polymer matrix. The correlation between theoretical analysis and μ-XRF results supports the potential of TGel as an efficient nutrient delivery system, enhancing the uptake of hydrophilic ionic zinc by seeds.

In the seed priming assays, a high Zn concentration (100 mg $L^{-1}$) was used to facilitate the evaluation of the effectiveness of the Zn-loaded TGel formulations as systems to improve Zn uptake by seeds and enhance the subsequent germination process. The seeds primed with the control Zn solutions showed fairly rapid initial root growth (day 6), followed by stabilization (days 9 and 12). The use of the Zn-loaded polymeric gels resulted in substantially faster root development without any negative effects from the seed treatment on the processes of germination and plant growth (Figures 5 and 6). In addition, seed priming with the Zn-loaded TGels resulted in higher zinc concentrations in the shoots of the cucumber plants, with increases from 72 mg $kg^{-1}$ ($ZnSO_4$ control) to 92 mg $kg^{-1}$ (TGel-$ZnSO_4$) and from 56 mg $kg^{-1}$ (ZnEDTA control) to 78 mg $kg^{-1}$ (TGel-ZnEDTA) (Table 2). The observed increase in Zn uptake by seeds, followed by the enhanced metal concentration in cucumber plant shoots (as discussed above), demonstrates that the synthesized TGel effectively coordinates ionic zinc or Zn-EDTA complexes. The TGel characteristics enhance the polymer particles' diffusion ability to penetrate the seed coat barrier, facilitating controlled zinc release and uptake. An evaluation of the safety of using the TGel formulations in agricultural applications, without any negative effects on plant development, was conducted by determining photosynthesis parameters for the cucumber plants on day 18. The treatments using the Zn-loaded TGels and the Zn control solutions (all at the same Zn concentration of 100 mg $L^{-1}$) led to similar PSII activity, chlorophyll *a*, and anthocyanin indices (Figure 8, Table 2). The results showed that the use of the TGels to deliver Zn did not affect the physiological parameters of the plants. These findings revealed, for the first time, that the amine–polyether-based polymeric gels could (i) act as carrier systems to improve Zn absorption by seeds (as shown in the μ-XRF studies) and (ii) have a positive effect on root growth following the translocation of Zn during seedling development. The effects of the Zn-loaded TGels in potentiating plant development were significant since Zn plays important roles in improving cell integrity and stability, as well as in increasing root growth and acting as a defense factor against fungal infections [28,29]. Since positively charged nanoparticles tend to show greater adherence to the surfaces of soil particles and root tissues, which are both negatively charged [30], the use of the positively charged amine–epoxide particles as functional carriers opens possibilities for soil applications to combat fungal infections in cultivations such as corn and soybean, among others.

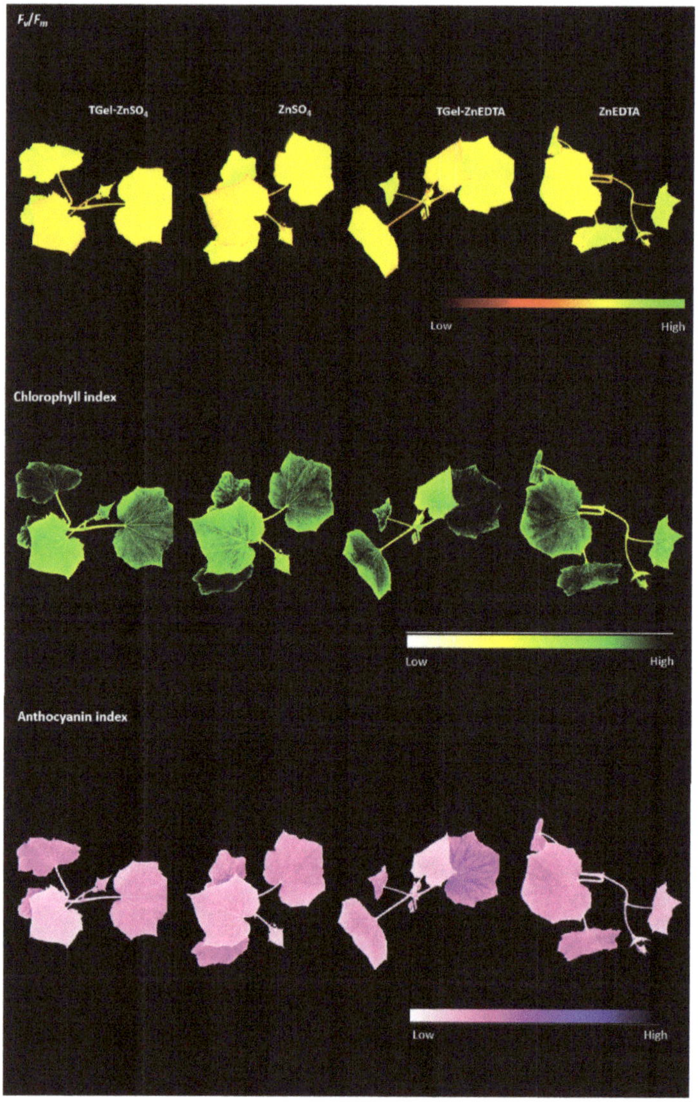


**Figure 8.** Multispectral images of the *C. sativus* plants after 18 days of growth for the seed treatments using the Zn-loaded TGels and the control Zn solutions. The plants were maintained in a hydroponic medium (Hoagland solution, without Zn). The values for the physiological parameter indices (PSII ($F_v/F_m$), chlorophyll *a*, and anthocyanin) are provided in the Supplementary Materials (Table S1).

**Table 2.** Zn concentration on cucumber shoot tissues from plants whose seeds were primed with TGel-based or pure $ZnSO_4$ and ZnEDTA solutions at 100 mg $L^{-1}$ determined by EDXRF. Data from EDXRF represent the mean ± standard deviation of two independent biological replicates and were not subjected to any statistical comparison. Photosynthetic parameters $F_v/F_m$, chlorophyll *a* and anthocyanin indices for 18-day growth cucumber plants. No statistical differences were observed for five independent biological replicates. Identical letters, regardless of case (uppercase or lowercase), indicate that there is no statistical significant difference between the data for the corresponding treatment.

| Zn Source | EDXRF Analysis Zn Amount Present in Shoot (mg $Kg^{-1}$) | | Photosynthetic Parameters PSII $F_v/F_m$ | | Chlorophyll *a* Index | | Anthocyanin Index | |
|---|---|---|---|---|---|---|---|---|
| | Loaded-TGel | Control | Loaded-TGel | Control | Loaded-TGel | Control | Loaded-TGel | Control |
| $ZnSO_4$ | 91.74 | 72.01 | 0.642Aa | 0.636Aa | 1.972Aa | 1.818Aa | 2.325Aa | 2.071Aa |
| ZnEDTA | 78.05 | 56.11 | 0.644Aa | 0.650Aa | 1.956Aa | 1.994Aa | 2.328Aa | 2.422Aa |

## 3. Conclusions

The use of polymer-based materials can have major effects on plants, as shown here for the seed priming formulations. The application of polymeric gels to improve seedling development represents a novel aspect of nanotechnology in agriculture. Seed priming with amine–epoxide particles based on TGel loaded with different zinc sources was shown to have positive effects on cucumber root growth and the absorption/translocation of Zn through the plant. Analysis of the $Polymer_{Frag}\cdots[Zn(EDTA)]^{2-}$ complex showed that the $Zn^{2+}$ cation interacted with the EDTA molecule by means of O⋯Zn noncovalent bonds, while the polymer fragment formed O–H⋯O· hydrogen bonds with the EDTA molecule. In contrast, in the $Polymer_{Frag}\cdots Zn^{2+}$ complex, the $Zn^{2+}$ cation directly interacted with the oxygen atoms of the polymer fragment by σ-bonds. Comprehensive μ-XRF analysis demonstrated that the levels of Zn were significantly higher in the tissues (seed coat, cotyledon, and embryo) of seeds primed with TGel-$ZnSO_4$ compared with those exposed to the $ZnSO_4$ control solution. Regardless of the Zn-loaded TGel treatment, the amounts of the micronutrient present in the cucumber shoots were higher than those obtained with the corresponding control solutions, with no negative effects on the physiological parameters of the plants. The experimental and theoretical findings of this study provide new perspectives on the use of polymeric nanomaterials as a green and safe seed priming technology. The insights from this work point towards the potential benefits that may be obtained from the application of polymeric nanogels in agriculture.

## 4. Materials and Methods

### 4.1. Chemicals

Diepoxy poly(ethylene glycol) (DPEG, $C_3H_5O_2$-$(C_2H_4O)_n$-$C_3H_5O$, average Mw = 500 g $mol^{-1}$, CAS number: 26403-72-5), disodium zinc ethylenediaminetetraacetate hydrate (Zn-EDTA), zinc sulfate heptahydrate ($ZnSO_4.(H_2O)_7$) and triamine-terminated polypropylene glycol (PPG)-based (known as Jeffamine T-403, $M_w$ = 440 g $mol^{-1}$, CAS number: 39423-51-3) were purchased from Sigma-Aldrich (São Paulo, Brazil). Jeffamine T-403 is a polyetheramine characterized by a backbone composed of repeating oxypropylene units. The amine groups are positioned on secondary carbon atoms at the end of aliphatic polyether chains. All reagents were used as received. Ultrapure water with a resistivity of 18.2 MΩ.cm was used in the synthesis of the polymer gels.

### 4.2. Preparation of Bisepoxide-Polyoxypropylenetriamine Gels: Unloaded and Zn-Loaded Polymeric Systems Chemicals

Amine–epoxide gels were synthesized according to a "click" reaction employing Jeffamine T-403 and DPEG, as described elsewhere [23,24,31], with adaptations. The initial step was the reaction of Jeffamine T-403 with DPEG in an aqueous medium (water) at a monomer concentration of 15 wt.% (mass of monomers in relation to the volume of water), ensuring complete solubilization of the reaction mixture. The molar ratio of amine to epoxide was maintained at 1:1 (Jeffamine T-403:DPEG). The 15 wt.% monomer solution was kept at 65 °C for 15 min to ensure initiation of the reaction. The solution was then diluted to 1.0 wt.% and allowed to react for a further 30 min. The final amine–epoxide solution was purified by dialysis in ultrapure water using regenerated cellulose membranes (denoted as TGel). In the next step, $ZnSO_4.(H_2O)_7$ or Zn-EDTA solutions (100 mg $L^{-1}$ of Zn) were individually mixed with 50 mg of TGel, lyophilized to ensure embedding of the Zn into the polymeric structure, and rehydrated with ultrapure water. The resulting Zn-loaded gels were denoted TGel-$ZnSO_4$ and TGel-ZnEDTA.

### 4.3. Characterization of TGels

Determination of the hydrodynamic diameter (Dh), polydispersity index (PdI), and zeta potential (ζ) of the unloaded and Zn-loaded TGels employed a ZSU3100 Zetasizer Lab Blue analyzer (Malvern Instruments, Malvern, Worcestershire, UK) equipped with an OBIS solid-state laser source emitting at a wavelength of 633 nm. The measurements were performed at room temperature (~25 °C), in triplicate, with the results expressed as mean ± standard deviation (SD) ($n \geq 3$). The morphologies of the gels were evaluated by transmission electron microscopy (TEM) using a JEM 100CXII instrument (JEOL, Peabody, MA, USA) operating at 100 kV. For acquisition of the images, a small droplet of the aqueous formulation was deposited onto a carbon-coated copper grid. The sample was then allowed to dry at room temperature in a desiccator to prevent contamination and ensure proper film formation. The polymeric gel used for the characterization was based on the solution diluted to 1.0 wt.%, denoted as TGel (see details in Section 4.2).

### 4.4. Computational Methods

All the molecular geometries were optimized without applying any geometric constraints. The vibrational frequency calculations were performed using the BP86 functional [32] and Grimme's D3(BJ) dispersion corrections, with Becke–Johnson damping [21]. The Def2-TZVP basis set was used for all the calculations [33]. To improve computational efficiency, the RIJCOSX approximation was employed, with Coulomb integrals handled using the RI-J [34] method and the Def2/J auxiliary basis set [35]. To ensure that each optimized geometry corresponded to a true energy minimum, vibrational frequency analysis was performed to confirm the absence of imaginary frequencies, ensuring the accuracy of the computational model. All the calculations were carried out using the ORCA software package v. 5.0.4 [36]. The choice of the BP86-D3(BJ)/Def2-TZVP level of theory for geometry optimization was consistent with recommendations reported in the literature for systems involving noncovalent interactions [37]. The chemical bonding mechanism was investigated using the EDA-NOCV methodology [22]. These calculations were carried out with Amsterdam Density Functional (ADF 2021.1) software [38], employing the BP86-D3(BJ) theory level and the TZ2P basis set [39]. Scalar relativistic corrections were applied in a self-consistent manner using the zero-order regular approximation (ZORA) [40]. The ZORA-BP86-D3(BJ)/TZ2P computational model has been shown to be effective in clarifying the bonding mechanisms in systems involving noncovalent interactions [41]. Table S1 shows the optimized Cartesian coordinates for the compounds analyzed in this study.

The NOCV method allows orbital interactions between interacting fragments, such as, for example, $\text{Polymer}_{\text{Frag}}$ and $[Zn(EDTA)]^{2-}$, to be decomposed into pairwise contributions of the most relevant molecular orbitals. The deformation density, $\Delta\rho(r)$, the density differences of $\text{Polymer}_{\text{Frag}}$ and $[Zn(EDTA)]^{2-}$ before and after the chemical bond establishment, can be constructed from pairs of complementary eigenfunctions $\psi_k$ and $\psi_{-k}$ with eigenvalues $\nu_k$ and $\nu_{-k}$, respectively [22]:

$$\Delta\rho(r) = \sum_k \nu_k \left[-\psi_{-k}^2(r) + \psi_k^2(r)\right] = \sum_k \Delta\rho_k(r) \tag{2}$$

This equation allows one to define the total charge deformation $\Delta\rho(r)$, created from the bond development, in terms of pairwise charge influences $\Delta\rho_k(r)$, which originated from specific pairs of NOCV orbitals. The total orbital interaction energy $\Delta E_{oi}$ can be obtained from pairwise orbital interaction energies $\Delta E_{oi}^k$ that are related to $\Delta\rho_k(r)$:

$$\Delta E_{oi} = \sum_k \Delta E_{oi}^k = \sum_k \nu_k \left[-F_{-k,-k}^{TS} + F_{k,k}^{TS}\right] \tag{3}$$

The components $F_{-k,-k}^{TS}$ and $F_{k,k}^{TS}$ are diagonal transition state (TS) Kohn–Sham matrix elements. The TS expression accounts for the charge density, which is intermediary between the final complex density, $\text{Polymer}_{\text{Frag}}$–$[Zn(EDTA)]^{2-}$, and the superimposed moiety densities of $\text{Polymer}_{\text{Frag}}$ and $[Zn(EDTA)]^{2-}$. The $\Delta E_{oi}^k$ term of a specific bond can be visualized from the deformation density shape, $\Delta\rho_k(r)$.

### 4.5. Seed Treatment

Seeds of cucumber (*Cucumis sativus*) were subjected to surface sterilization by sequential washes in 2% sodium hypochlorite (100 mL), followed by rinsing in deionized water (100 mL). The seeds were primed by application of the TGels loaded with the Zn sources (Zn-EDTA or $ZnSO_4$), with the corresponding Zn solutions used as controls, to evaluate the effects on subsequent growth of the cucumber plants (considering the shoot and root lengths) following seed germination. The seeds were immersed in flasks (3 seeds/mL of suspension, with a total of 21 seeds) containing the TGel-$ZnSO_4$ or TGel-ZnEDTA gel particles, with the Zn solutions as controls, allowing the flasks to stand in darkness at room temperature for 24 h, before germination on Petri dishes. The seeds primed with the Zn-loaded TGels or Zn solutions were arranged on filter papers in Petri dishes that were sealed to prevent water loss. For all seed assays, TGel formulations (both loaded and unloaded) were prepared based on the diluted system at a concentration of 1.0 wt.%, denoted as TGel. The zinc (Zn) concentration was maintained at 100 mg $L^{-1}$ for both control and Zn-loaded samples.

### 4.6. Micro X-Ray Fluorescence (μ-XRF) Analysis

The spatial distributions of Zn in the treated cucumber seed tissues were evaluated by micro X-ray fluorescence spectroscopy (μ-XRF), according to the procedure described elsewhere [24,42]. Briefly, the seeds treated for 24 h with TGel-$ZnSO_4$, TGel-ZnEDTA, or the positive controls ($ZnSO_4$ and ZnEDTA solutions) at a Zn concentration of 100 mg $L^{-1}$ were cut transversely, and cross-sections of the medial regions were fixed on sections of 6 μm thickness polypropylene film (FPPP25-R3, VHG, Manchester, NH, USA), mounted on XRF sample vials (no. 1530, Chemplex, Palm City, FL, USA). The samples were loaded into an XRF spectrometer (Orbis PC, EDAX, Mahwah, NJ, USA), and the seed cross-sections were investigated by 32-point line scanning and 800-pixel two-dimensional mapping, encompassing the seed coat, embryo, and cotyledon tissues. The samples were exposed to a 30 μm polycapillary-focused X-ray beam set at 45 kV and 500 μA, with a 250 μm thickness Al primary filter. The spectra were acquired using a 30 $mm^2$ silicon drift detector (SDD) with a dead time smaller than 10%. The dwell times were 15 s $point^{-1}$ and 1 s $pixel^{-1}$ for the line scans and maps, respectively. The analyses were carried out using two independent biological replicates, considering only the elemental intensities above the instrumental limit of detection (LOD) as valid, calculated as described in Equation (1) below. The intensities recorded for the seed coat, embryo, and cotyledon tissues were collected from the line scan data and were compared, as a function of the treatments, using the Mann–Whitney *t*-test at a 95% confidence level ($p < 0.05$). The data were processed using Prism v. 9.2.0 software (GraphPad, v. 10.1.1, Boston, MA, USA) and the Python-based (v. 3.9.6) Matplotlib library [43].

$$LOD = 3 \cdot \sqrt{\frac{BG}{t}}$$

where BG (cps) represents the average background counting rate, determined either from 10 randomly selected background points in elemental maps or from individual background values recorded along line scans. The acquisition time (t) is given in seconds (s).

### 4.7. Quantification of Zn in Cucumber Tissues, Using Energy-Dispersive X-Ray Fluorescence Spectroscopy (EDXRF) and Plant Photosynthesis Assessment

The concentrations of Zn in the cucumber shoot tissues were determined by EDXRF, as described by Montanha et al. [44], with minor modifications. Briefly, 100 mg portions of the dried and finely ground tissue samples were added to 6.3 mm diameter X-ray sample cups (SamplePrep no. 3577, SPEX, Livonia, MI, USA) sealed at the base with 6 μm thickness polypropylene film (FPPP25-R3, VHG, Manchester, NH, USA), gently pressed with a glass stick to remove void spaces, and covered with a spatula tip quantity of boric acid (Synth, Diadema, São Paulo, Brazil). The EDXRF analyses employed a spectrometer (model EDX-720, Shimadzu, Tokyo, Japan) with a 3 mm X-ray beam generated by a Rh anode-based X-ray tube at 50 kV and auto-tunable current up to 20% of the detector dead-time. The analyses were carried out under vacuum, and the acquisition time was 150 s. The X-ray spectra were recorded using a Si(Li) detector. The Zn K$\alpha$ intensities were normalized by the Compton scattering intensities, and Zn quantification was based on an external calibration curve utilizing a set of plant-certified reference materials (Figure S3). All the analyses were performed with at least two independent biological replicates.

High-resolution images (2448 $\times$ 2448 pixels) of the plants were acquired to determine the maximum quantum efficiency of photosystem II (PSII) ($F_v/F_m$, where $F_m$ is maximum fluorescence and $F_v$ is variable fluorescence). Prior to the measurement, the plants were dark-adapted for 30 min and then illuminated with a 0.8 s pulse of saturating light (6320 μmol $m^{-2}$ $s^{-1}$). Next, the plants were submitted to actinic light for 5 min for the determination of non-photochemical quenching (NPQ = $(F_m - F_m')/F_m$). Chlorophyll *a* fluorescence was measured using the excitation/emission combination of 620/730 nm. Multispectral images were also acquired to calculate the chlorophyll and anthocyanin indices [45,46]. Details of the procedures for obtaining these parameters can be found elsewhere [47]. All the parameters were obtained using a SeedReporter™ instrument (Enkhuizen, NL)and proprietary software (v. 5.5.1) (PhenoVation B.V., Wageningen, The Netherlands).

For these experiments (Zn quantification and photosynthesis studies), seeds were primed with the Zn-loaded TGels and the Zn control solutions (as described in Section 4.5). Following a 5-day germination period, uniform seedlings were identified and transferred to presterilized plastic pots (0.5 L) containing simulated Hoagland solution (without Zn, since this micronutrient was used in the seed priming treatments). After the 18-day growth period, the photosynthetic efficacy and the amounts of Zn in the plant shoots were determined as described above.

### 4.8. Statistical Analysis

GraphPad Prism v. 10.1.1 (GraphPad Software Inc., Boston, MA, USA) and OriginLab (v. 2022b, Northampton, MA, USA) were used for plotting graphs and performing statistical analyses. Evaluation of differences employed analysis of variance (ANOVA) combined with the Tukey test ($p$ = 0.05). The results are shown as mean values in the tables and figures.

**Supplementary Materials:** The following supporting information can be downloaded at https://www.mdpi.com/article/10.3390/gels11030167/s1, Figure S1: Two-dimensional XRF maps of the Zn spatial distribution in cross-sectioned cucumber seeds primed with the TGel-based or pure $ZnSO_4$ and ZnEDTA solutions at 100 mg $L^{-1}$ Zn for 24-h; Figure S2: XRF Zn intensities recorded at the seed coat, embryo, and cotyledon of cross-sectioned cucumber seeds primed with the TGel-based or pure $ZnSO_4$ and ZnEDTA solutions at 100 mg $L^{-1}$ Zn for 24-h; Figure S3: Certified reference material-based external calibration curve used for quantitative Zn determination through energy-dispersive X-ray fluorescence spectroscopy; Table S1: Optimized Cartesian coordinates for the compounds analyzed in this study using the BP86–D3(BJ)/Def2–TZVP computational model.

**Author Contributions:** Conceptualization, F.B.A., A.S.M.G., B.A.F. and E.F.M.; Methodology, F.B.A., A.S.M.G., B.A.F., V.S.A.S., R.P.O., R.L.T.P., H.E.A., G.S.M., H.J.F.A.d.S., E.d.A., N.C. and C.B.M.; Validation, R.P.O., R.L.T.P., E.d.A., H.W.P.d.C. and C.B.M.; Formal analysis, R.P.O., R.L.T.P., G.S.M., E.d.A., H.W.P.d.C. and C.B.M.; Investigation, H.W.P.d.C., C.B.M. and E.F.M.; Resources, H.W.P.d.C., C.B.M. and E.F.M.; Data curation R.P.O. and R.L.T.P.; Writing—original draft, R.P.O., R.L.T.P., G.S.M., E.d.A., C.B.M. and E.F.M.; Writing—review and editing, R.P.O., R.L.T.P., G.S.M., E.d.A., H.W.P.d.C., C.B.M. and E.F.M. All authors have read and agreed to the published version of the manuscript.

**Funding:** This research was funded by FAPESP (grants 2022/13408-7, 2021/06552-1, 2022/06507-9, 2020/07721-9, 2021/14619-9, 2018/03793-5, 2023/09543-9, 2024/11041-4 and 2024/01903-9), CAPES (Finance Code 001), and CNPq (grants 307696/2021-9 and 306185/2020-2).

**Institutional Review Board Statement:** Not applicable.

**Informed Consent Statement:** Not applicable.

**Data Availability Statement:** The original contributions presented in this study are included in the article/Supplementary Materials. Further inquiries can be directed to the corresponding author.

**Acknowledgments:** The authors acknowledge financial support from FAPESP, CAPES, and CNPq, with the respective grant numbers provided in the Funding section.

**Conflicts of Interest:** The authors declare no conflicts of interest.

## References

1. Khan, M.; Khan, A.U.; Hasan, M.A.; Yadav, K.K.; Pinto, M.M.C.; Malik, N.; Yadav, V.K.; Khan, A.H.; Islam, S.; Sharma, G.K. Agro-Nanotechnology as an Emerging Field: A Novel Sustainable Approach for Improving Plant Growth by Reducing Biotic Stress. *Appl. Sci.* **2021**, *11*, 2282. [CrossRef]
2. Zhang, X.; Cao, H.; Wang, H.; Zhao, J.; Gao, K.; Qiao, J.; Li, J.; Ge, S. The Effects of Graphene-Family Nanomaterials on Plant Growth: A Review. *Nanomaterials* **2022**, *12*, 936. [CrossRef]
3. Wang, Y.; Deng, C.; Rawat, S.; Cota-Ruiz, K.; Medina-Velo, I.; Gardea-Torresdey, J.L. Evaluation of the effects of nanomaterials on rice (*Oryza sativa* L.) responses: Underlining the benefits of nanotechnology for agricultural applications. *ACS Agric. Sci. Technol.* **2021**, *1*, 44–54. [CrossRef]
4. Kah, M.; Tufenkji, N.; White, J.C. Nano-enabled strategies to enhance crop nutrition and protection. *Nat. Nanotechnol.* **2019**, *15*, 533–547. [CrossRef] [PubMed]
5. Xu, Z.P. Material nanotechnology is sustaining modern agriculture. *ACS Agric. Sci. Technol.* **2022**, *2*, 232–239. [CrossRef]
6. Di Lorenzo, G.; Becagli, M.; Cardelli, R.; Pecchia, S. Preparation of microcapsules of wood distillate coated with chitosan and micronized biochar: A new material with potential for sustainable crop protection and precision agriculture. *Sustainability* **2025**, *17*, 706. [CrossRef]

7. Toma, D.-I.; Manaila-Maximean, D.; Fierascu, I.; Baroi, A.M.; Matei, R.I.; Fistos, T.; Chican, I.E.; Fierascu, R.C. Applications of natural polymers in the grapevine industry: Plant protection and value-added utilization of waste. *Polymers* **2025**, *17*, 18. [CrossRef]
8. Ali, K.; Asad, Z.; Agbna, G.H.D.; Saud, A.; Khan, A.; Zaidi, S.J. Progress and innovations in hydrogels for sustainable agriculture. *Agronomy* **2024**, *14*, 2815. [CrossRef]
9. Khan, F.; Atif, M.; Haseen, M.; Kamal, S.; Khan, M.S.; Shahid, S.; Nami, S.A.A. Synthesis, classification and properties of hydrogels: Their applications in drug delivery and agriculture. *J. Mater. Chem. B* **2022**, *10*, 170–203. [CrossRef]
10. Tariq, Z.; Iqbal, D.N.; Rizwan, M.; Ahmad, M.; Faheem, M.; Ahmed, M. Significance of biopolymer-based hydrogels and their applications in agriculture: A review in perspective of synthesis and their degree of swelling for water holding. *RSC Adv.* **2023**, *13*, 24731–24754. [CrossRef] [PubMed]
11. Tomadoni, B.; Salcedo, M.F.; Mansilla, A.Y.; Casalongué, C.A.; Alvarez, V.A. Macroporous alginate-based hydrogels to control soil substrate moisture: Effect on lettuce plants under drought stress. *Eur. Polym. J.* **2020**, *137*, 109953. [CrossRef]
12. Satriani, A.; Catalano, M.; Scalcione, E. The role of superabsorbent hydrogel in bean crop cultivation under deficit irrigation conditions: A case-study in Southern Italy. *Agric. Water Manag.* **2018**, *195*, 114–119. [CrossRef]
13. Ma, L.; Chai, C.; Wu, W.; Qi, P.; Liu, X.; Hao, J. Hydrogels as the plant culture substrates: A review. *Carbohydr. Polym.* **2023**, *305*, 120544. [CrossRef]
14. Palma, D.; Lagos, O.; Souto, C.; Pérez, A.; Quezada, L.; Hirzel, J.; Vera, M.; Ulloa, J.; Urbano, B. Evaluation of a natural superabsorbent polymer on water retention capacity in coarse-textured soils. *Water* **2024**, *16*, 3186. [CrossRef]
15. Sohail, M.; Pirzada, T.; Opperman, C.H.; Khan, S.A. Recent advances in seed coating technologies: Transitioning toward sustainable agriculture. *Green Chem.* **2022**, *24*, 6052–6085. [CrossRef]
16. Kabanov, A.V.; Vinogradov, S.V. Nanogels as pharmaceutical carriers: Finite networks of infinite capabilities. *Angew. Chem. Int. Ed.* **2009**, *48*, 5418–5429. [CrossRef]
17. Oh, J.K.; Drumright, R.; Siegwart, D.J.; Matyjaszewski, K. The development of microgels/nanogels for drug delivery applications. *Prog. Polym. Sci.* **2008**, *33*, 448–477. [CrossRef]
18. Takeshita, V.; Campos, E.V.R.; Rodrigues, J.S.; Fraceto, L.F. Opinion: Hybrid nanoparticle systems—Two-way delivery approach for agriculture. *Plant Nano Biol.* **2023**, *6*, 100053. [CrossRef]
19. Stropoli, S.J.; Elrod, M.J. Assessing the potential for the reactions of epoxides with amines on secondary organic aerosol particles. *J. Phys. Chem. A* **2015**, *119*, 10181–10189. [CrossRef] [PubMed]
20. Bickelhaupt, F.M.; Baerends, E.J. Kohn-Sham density functional theory: Predicting and understanding chemistry. In *Reviews in Computational Chemistry*; Wiley-VCH, Inc.: Hoboken, NJ, USA, 2000; Volume 15, pp. 1–86. [CrossRef]
21. Grimme, S.; Antony, J.; Ehrlich, S.; Krieg, H. A consistent and accurate ab initio parametrization of density functional dispersion correction (DFT-D) for the 94 elements H-Pu. *J. Chem. Phys.* **2010**, *132*, 154104. [CrossRef]
22. Mitoraj, M.P.; Michalak, A.; Ziegler, T. A Combined Charge and Energy Decomposition Scheme for Bond Analysis. *J. Chem. Theory Comput.* **2009**, *5*, 962–975. [CrossRef]
23. Tang, S.; Shi, Z.; He, W. Facile aqueous-phase synthesis of multi-responsive nanogels based on polyetheramines and bisepoxide. *J. Mater. Chem. B* **2013**, *1*, 1628–1634. [CrossRef] [PubMed]
24. Alves, F.B.; Andrada, H.E.; Fico, B.A.; Reinaldi, J.S.; Tavares, D.C.; Squarisi, I.S.; Montanha, G.S.; Nuevo, L.G.; de Carvalho, H.W.P.; Pérez, C.A.; et al. Facilitating seed iron uptake through amine-epoxide microgels: A novel approach to enhance cucumber (*Cucumis sativus*) germination. *J. Agric. Food Chem.* **2024**, *72*, 14570–14580. [CrossRef]
25. Savassa, S.M.; Duran, N.M.; Rodrigues, E.S.; De Almeida, E.; Van Gestel, C.A.; Bompadre, T.F.; de Carvalho, H.W.P. Effects of ZnO nanoparticles on *Phaseolus vulgaris* germination and seedling development determined by X-ray spectroscopy. *ACS Appl. Nano Mater.* **2018**, *1*, 6414–6426. [CrossRef]
26. Rohr, L.A.; França-Silva, F.; Corrêa, C.G.; de Carvalho, H.W.P.; Gomes-Junior, F.G. Soybean seeds treated with zinc evaluated by X-ray micro-fluorescence spectroscopy. *Crop Sci.* **2023**, *80*, e20210131. [CrossRef]
27. Doolette, C.L.; Read, T.L.; Li, C.; Scheckel, K.G.; Donner, E.; Kopittke, P.M.; Schjoerring, J.K.; Lombi, E. Foliar application of zinc sulphate and zinc EDTA to wheat leaves: Differences in mobility, distribution, and speciation. *J. Exp. Bot.* **2018**, *69*, 4469–4481. [CrossRef]
28. Khoshgoftarmanesh, A.H.; Kabiri, S.; Shariatmadari, H.; Sharifnabi, B.; Schulin, R. Zinc nutrition effect on the tolerance of wheat genotypes to Fusarium root-rot disease in a solution culture experiment. *Soil Sci. Plant Nutr.* **2010**, *56*, 234–243. [CrossRef]
29. Bastakoti, S. Role of zinc in management of plant diseases: A review. *Cogent Food Agric.* **2023**, *9*, 2194483. [CrossRef]
30. Su, Y.; Ashworth, V.; Kim, C.; Adeleye, A.S.; Rolshausen, P.; Roper, C.; White, J.; Jassby, D. Delivery, uptake, fate, and transport of engineered nanoparticles in plants: A critical review and data analysis. *Environ. Sci. Nano* **2019**, *6*, 2311–2331. [CrossRef]
31. Tang, S.; Huang, L.; Shi, Z.; He, W. Water-based synthesis of cationic hydrogel particles: Effect of the reaction parameters and in vitro cytotoxicity study. *J. Mater. Chem. B* **2015**, *3*, 2842–2852. [CrossRef]

32. Becke, A.D. Density-functional exchange-energy approximation with correct asymptotic behavior. *Phys. Rev. A* **1988**, *38*, 3098. [CrossRef] [PubMed]
33. Weigend, F.; Ahlrichs, R. Balanced basis sets of split valence, triple zeta valence and quadruple zeta valence quality for H to Rn: Design and assessment of accuracy. *Phys. Chem. Chem. Phys.* **2005**, *7*, 3297–3305. [CrossRef] [PubMed]
34. Neese, F. An improvement of the resolution of the identity approximation for the formation of the Coulomb matrix. *J. Chem. Phys.* **2003**, *24*, 1740–1747. [CrossRef]
35. Weigend, F. Accurate Coulomb-fitting basis sets for H to Rn. *Phys. Chem. Chem. Phys.* **2006**, *8*, 1057–1065. [CrossRef] [PubMed]
36. Neese, F. The ORCA program system. *Wiley Interdiscip. Rev. Comput. Mol. Sci.* **2012**, *2*, 73–78. [CrossRef]
37. Orenha, R.P.; Borges, A.; Andrade, A.L.O.; Ferreira, S.E.; Furtado, S.S.P.; Glitz, V.A.; Caramori, G.F.; Parreira, R.L.T. Cation recognition controlled by protonation or chemical reduction: A computational study. *Phys. Chem. Chem. Phys.* **2023**, *25*, 15518–15530. [CrossRef]
38. Baerends, E.J.; Ziegler, T.; Atkins, A.J.; Autschbach, J.; Baseggio, O.; Bashford, D.; Bérces, A.; Bickelhaupt, F.M.; Bo, C.; Boerrigter, P.M.; et al. *ADF 2021.1, SCM*; Theoretical Chemistry, Vrije Universiteit: Amsterdam, The Netherlands, 2021.
39. Van Lenthe, E.; Baerends, E.J. Optimized Slater-type basis sets for the elements 1–118. *J. Chem. Phys.* **2003**, *24*, 1142–1156. [CrossRef]
40. Zhao, Q.; Parr, R.G. Constrained-search method to determine electronic wave functions from electronic densities. *J. Chem. Phys.* **1993**, *98*, 543–548. [CrossRef]
41. Pramanik, U.; Kongasseri, A.A.; Shekhar, S.; Mathew, A.; Yadav, R.; Mukherjee, S. Structural compactness in hen egg white lysozyme induced by bisphenol S: A Spectroscopic and molecular dynamics simulation approach. *J. Phys. Chem. B* **2021**, *22*, 1745–1753. [CrossRef]
42. Cotrim, M.F.; Silva, J.B.D.; Lourenço, F.M.D.S.; Teixeira, A.V.; Gava, R.; Alves, C.Z.; Candido, A.C.D.S.; Campos, C.N.S.; Pereira, M.D.; Torres, S.B.; et al. Studying the link between physiological performance of Crotalaria ochroleuca and the distribution of Ca, P, K and S in seeds with X-ray fluorescence. *PLoS ONE* **2019**, *14*, e0222987. [CrossRef]
43. Hunter, J.D. Matplotlib: A 2D graphics environment. *Comput. Sci. Eng.* **2007**, *9*, 90–95. [CrossRef]
44. Montanha, G.S.; Perez, L.C.; Brandão, J.R.; de Camargo, R.F.; Tavares, T.R.; de Almeida, E.; de Carvalho, H.W.P. Profile of mineral nutrients and proteins in soybean seeds (*Glycine max* (L.) Merrill): Insights from 95 varieties cultivated in Brazil. *J. Food Compos. Anal.* **2024**, *134*, 106536. [CrossRef]
45. Gitelson, A.A.; Gritz, Y.; Merzlyak, M.N. Relationships between leaf chlorophyll content and spectral reflectance and algorithms for non-destructive chlorophyll assessment in higher plant leaves. *J. Plant Physiol.* **2003**, *160*, 271–282. [CrossRef]
46. Gitelson, A.A.; Merzlyak, M.N.; Chivkunova, O.B. Optical properties and nondestructive estimation of anthocyanin content in plant leaves. *Physiol. Plant.* **2001**, *74*, 38–45. [CrossRef]
47. Oliveira, N.M.; de Medeiros, A.D.; Nogueira, M.L.; Arthur, V.; Mastrangelo, T.A.; da Silva, C.B. Hormetic effects of low-dose gamma rays in soybean seeds and seedlings: A detection technique using optical sensors. *Comput. Electron. Agric.* **2021**, *187*, 106251. [CrossRef]

gels

MDPI

Article

# Fabrication of Functional Polymers with Gradual Release of a Bioactive Precursor for Agricultural Applications

Oscar G. Marambio [1,*], Rudy Martin-Trasancos [2], Julio Sánchez [3], Felipe A. Ramos [1] and Guadalupe del C. Pizarro [1,*]

[1] Departamento de Química, Facultad de Ciencias Naturales, Matemáticas y Medio Ambiente, Universidad Tecnológica Metropolitana (UTEM), J. P. Alessandri 1242, Santiago 7800002, Chile; framos@utem.cl
[2] Departamento de Química de los Materiales, Facultad de Química y Biología, Universidad de Santiago de Chile (USACH), Santiago 9170002, Chile; rudy.martin@usach.cl
[3] Departamento de Química Orgánica, Facultad de Química y de Farmacia, Pontificia Universidad Católica de Chile, Santiago 7820436, Chile; julio.sanchez@uc.cl
* Correspondence: omarambi@utem.cl (O.G.M.); gpizarro@utem.cl (G.d.C.P.)

**Abstract:** Biodegradable and biocompatible polymeric materials and stimulus-responsive hydrogels are widely used in the pharmaceutical, agricultural, biomedical, and consumer sectors. The effectiveness of these formulations depends significantly on the appropriate selection of polymer support. Through chemical or enzymatic hydrolysis, these materials can gradually release bioactive agents, enabling controlled drug release. The objective of this work is to synthesize, characterize, and apply two controlled-release polymeric systems, focusing on the release of a phyto-pharmaceutical agent (herbicide) at varying pH levels. The copolymers were synthesized via free radical polymerization in solution, utilizing tetrahydrofuran (THF) as the organic solvent and benzoyl peroxide (BPO) as the initiator, without the use of a cross-linking agent. Initially, the herbicide was grafted onto the polymeric chains, and its release was subsequently tested across different pH environments in a heterogeneous phase using an ultrafiltration (UF) system. The development of these two controlled-release polymer systems aimed to measure the herbicide's release across different pH levels. The goal is to adapt these materials for agricultural use, enhancing soil quality and promoting efficient water usage in farming practices. The results indicate that the release of the herbicide from the conjugate systems exceeded 90% of the bioactive compound after 8 days at pH 10 for both systems. Furthermore, the two polymeric systems demonstrated first-order kinetics for herbicide release in aqueous solutions at different pH levels. The kinetic constant was found to be higher at pH 7 and 10 compared to pH 3. These synthetic hydrogels are recognized as functional polymers suitable for the sustained release of herbicides in agricultural applications.

**Keywords:** controlled herbicide release system; ultrafiltration system; polymeric synthetic hydrogel

## 1. Introduction

Modern agriculture is under growing pressure to improve crop productivity while reducing its environmental impact. Although modern agriculture offers numerous solutions, the results can vary significantly because each farm is unique, with different landscapes, soil types, available technologies, and potential yields [1]. The past few decades have seen significant growth driven by plant protection products and other technological innovations. However, the excessive use of these inputs has a negative impact on the environment [2].

Traditional approaches to herbicide use can be beneficial when applied carefully. However, inappropriate or excessive application can lead to various issues, such as chemical residues, phytotoxicity in crops, and negative impacts on subsequent or susceptible plants and non-target organisms. Additionally, the repeated use of similar types of herbicides over the years can lead to accumulation in crops, soil, and groundwater, potentially posing health risks [3].

Weeds are a major factor in reducing the yields of many crops, and herbicides are the most commonly used pesticides [4]. The herbicide 2,4-D is a phenoxy herbicide that functions as a synthetic plant hormone. It effectively and selectively targets broadleaf weeds in cereal crops while leaving legumes and corn unharmed. Additionally, 2,4-D is compatible with most pesticides commonly used in agriculture [5].

Recently, increasing interest has been in designing and fabricating "smart" hydrogels that respond to external stimuli such as pH, pressure, light, temperature, ionic strength, and enzymes [6]. This interest stems from hydrogels' excellent biocompatibility, straightforward preparation, and a broad range of applications [7]. For example, they have been used as functional materials in drug delivery [8,9] and tissue engineering [10]. Polymers that support agricultural chemicals have been developed to address the significant environmental issues associated with conventional agrochemicals [9,11–23].

The delivery of herbicides through controlled-release formulations provides both ecological and economic benefits [13]. The effectiveness of these formulations relies on selecting an appropriate polymer support. Degradable polymer materials and hydrogels have various applications in pharmaceutical, agricultural, biomedical, and consumer-oriented fields, making them significant for controlled-release systems [24–28].

Conversely, numerous studies have examined how organic polymers affect soil physical properties [29]. Soil conservation has improved through the addition of polymeric materials to irrigation systems [30–32]. Hydrophilic polymers, including poly(vinyl alcohol) (PVA), carboxymethylcellulose, poly(acrylamide) (PAM), and hydrolyzed starch-g-poly(acrylonitrile) copolymers (HSPAN). These polymeric materials have been suggested as soil conditioners [33,34]. The successful use of polyacrylamide (PAM) in irrigation water has generated interest in exploring other polymers with similar properties, especially since acrylamide, the monomer used to create PAM, is known to be a neurotoxin. Although using PAM that contains less than 0.05% monomer helps alleviate some concerns, it does not fully address the potential risk of monomer formation as a degradation product, particularly because the amine group can be removed from the polymer backbone during degradation [35]. Interest in agricultural polymers' final fate and their degradation products' potential ecotoxicity is increasing [36–38]. These materials have interconnected polymer networks that respond to specific stimuli [39–42]. Furthermore, it has been observed that hydrogels' hydration and adsorption capacity is influenced by factors such as the medium's pH, ionic strength, and osmotic pressure.

Our research group has shown that acrylate-based hydrogels excel in this type of application due to their high reactivity, obtained through free radical polymerization [43–45]. Drug release from hydrogel networks is influenced by various mechanisms, such as matrix swelling, drug dissolution and diffusion, and hydrogel erosion. The two main chemical controlled-release systems are erodible drug delivery systems [46,47] and pendant chain systems [48,49]. The drug is connected to the polymer backbone via degradable linkages in pendant chain systems. As these linkages break down, the drug is released gradually. The drug can be attached either directly to the polymer or through a "spacer" group. It is uniformly dispersed throughout the polymer, and the drug is released slowly as the polymer disintegrates [49]. The degradation rate of polymer–drug linkages affects drug release. Typically, these linkages undergo hydrolysis, allowing the degradation and release rates to be described using first-order kinetic relationships [48]. In certain applications, the drug–polymer linkages may be designed to be degradable by enzymes, resulting in more complex release kinetics [27,49].

In this report, we present two chemical controlled-release systems based on synthetic poly(2-hydroxyethylmethacrylate-*alt*-maleic anhydride)[P(HEMA-*alt*-MAn)] and poly(2-hydroxypropylmethacrylate-*alt*-maleic anhydride)[P(HPMA-*alt*-MAn)] as physical hydrogels, in which the functional groups in the polymer support and hydrogels' unique network structure allow for high levels of hydrophilicity. This work presents several key contributions. First, we designed two controlled-release polymers for a herbicide as a pharmaceutical agent to assess its release in various pH environments. The goal is to adapt these materials for agricultural use, improving soil quality and ensuring efficient water usage, particularly during drought conditions.

Second, the polymers were synthesized using free radical polymerization in solution. An organic solvent was employed, with benzoyl peroxide (BPO) acting as the initiator. The functional groups of the hydrophilic monomers were activated in a basic environment to enhance physical bonds through hydrogen bonding between the chains, leading to the formation of a physical hydrogel. Subsequently, we characterized its structure and incorporated the herbicide into the main chain through an esterification reaction. Finally, we utilized an ultrafiltration system to investigate the controlled release of the herbicide from the prepared system. We studied the release characteristics of these systems in neutral, alkaline, and acidic media.

These degradable polymeric hydrogels hold potential for a wide range of agricultural applications, making them particularly significant for controlled-release systems. The copolymer hydrogels and copolymer–herbicide conjugates were analyzed using FTIR, $^{1}$H NMR spectroscopy, and TG analyses. Additionally, their excellent soft physical properties linked to the degradation process make them ideal agricultural materials.

## 2. Results and Discussion

### 2.1. Characterization of the Alternating Copolymers

It is widely recognized that interesting alternating structures can be formed through the copolymerization of maleic anhydride (MAn) with vinyl monomers. The effectiveness of this process depends on the reactivity of the double bonds in the vinyl monomers and the specific reaction conditions used. When MAn is copolymerized with vinyl monomers that have electron-donating characteristics using conventional radical techniques, it can create alternating copolymers. The copolymers were synthesized through free radical polymerization, following the conditions outlined in Table 1. These copolymers resulted in alternating structures, P(HEMA-*alt*-MAn) and P(HPMA-*alt*-MAn), as depicted in Figure 1. The molecular weight (Mw) of Poly(HEMA-*alt*-MAn) was 14,200, while that of Poly (HPMA-*alt*-MAn) was 16,580. With a polydispersity (Đ = Mw/Mn) of 1.45 and 1.53, respectively, the molecular weight (Mw) ratios were determined using GPC.

**Table 1.** Experimental conditions for the copolymerization reaction in tethahydrofuran (THF) (5.0 mL).

| Systems | HEMA | HPMA | MAn | BPO | Time | Yield |
|---|---|---|---|---|---|---|
| P(HEMA-*alt*-MAn) | 3.40 mL 28.0 mmol | | 2.78 g (28.00 mmol) | 69.1 mg (0.5059 mol%) | 6.0 h | 51.7% |
| P(HPMA-*alt*-MAn) | | 3.95 mL 28.00 mmol | 2.78 g (28.00 mmol) | 67.90 mg (0.4971 mol%) | 6.0 h | 34.2% |

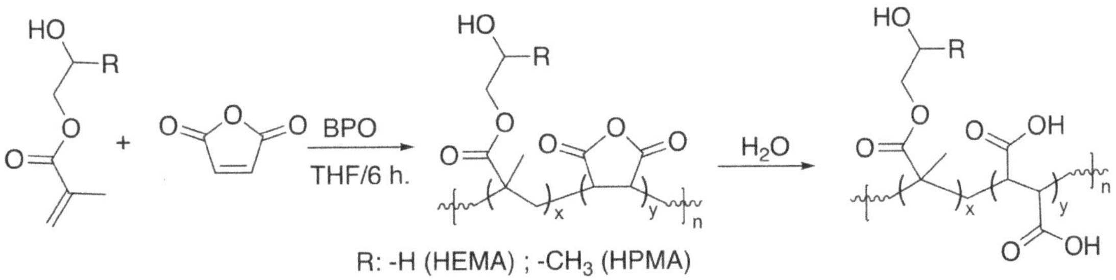


**Figure 1.** Polymerization reaction of the P(HEMA-*alt*-MAn) and P(HPMA-*alt*-MAn) and copolymers under hydrolysis conditions.

The hydrolysis of the anhydride group leads to carboxylic acid, which is responsible for the hydrogel properties and swelling behavior.

The characterization of the P(HEMA-*alt*-MAn) by FTIR (KBr, $cm^{-1}$) exhibited the following signals: 3449.7 ν(OH); 2930.9 ν(CH, $CH_2$); 1734.0 ν(C=O, for maleic anhydride, MAn); 1634.7.ν(C=O, for HEMA); and 1482.3 and 1390.7 ν($CH_2$-; $CH_3$-); see Figure 2a. FTIR (KBr, $cm^{-1}$) of P(HPMA-*alt*-MAn) exhibited the following signals: 3415.9. ν(OH); 2931.8 ν(CH, $CH_2$); 1718.6 ν(C=O, ester for MAn); 1632.7 ν(C=O, ester for HPMA); and 1480.3 and 1387.8 ν($CH_2$-; $CH_3$-); see Figure 3a.

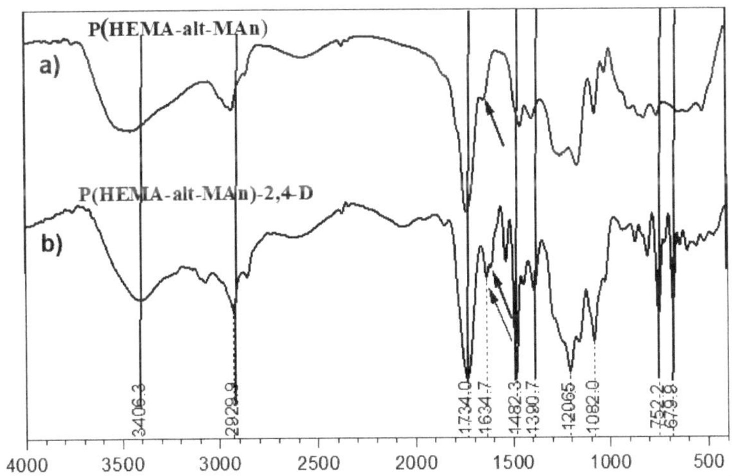


**Figure 2.** (**a**) FTIR spectrum of P(HEMA-*alt*-MAn) and (**b**) its copolymer conjugate 2,4-D.

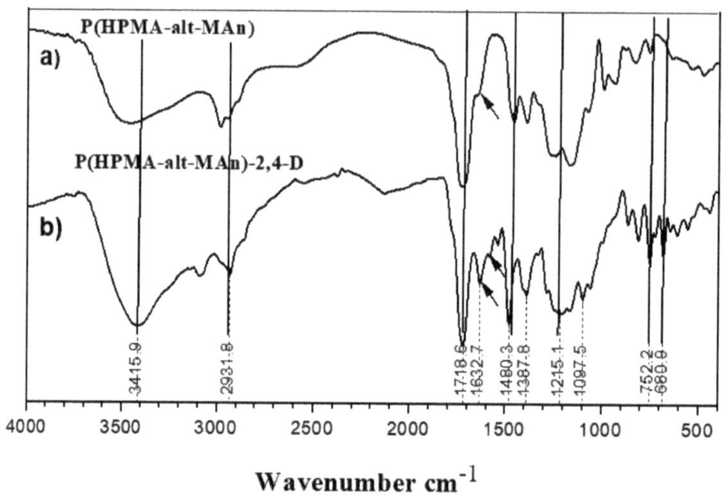


**Figure 3.** (**a**) FTIR spectrum of P(HPMA-*alt*-MAn) and (**b**) its copolymer conjugate 2,4-D.

### 2.2. Characterization of the Copolymers Grafted with the Herbicide

The grafting procedure using the herbicide 2,4-dichlorophenoxyacetic chloride was carried out according to the reaction shown in Figure 4 and the experimental conditions in Table 2.

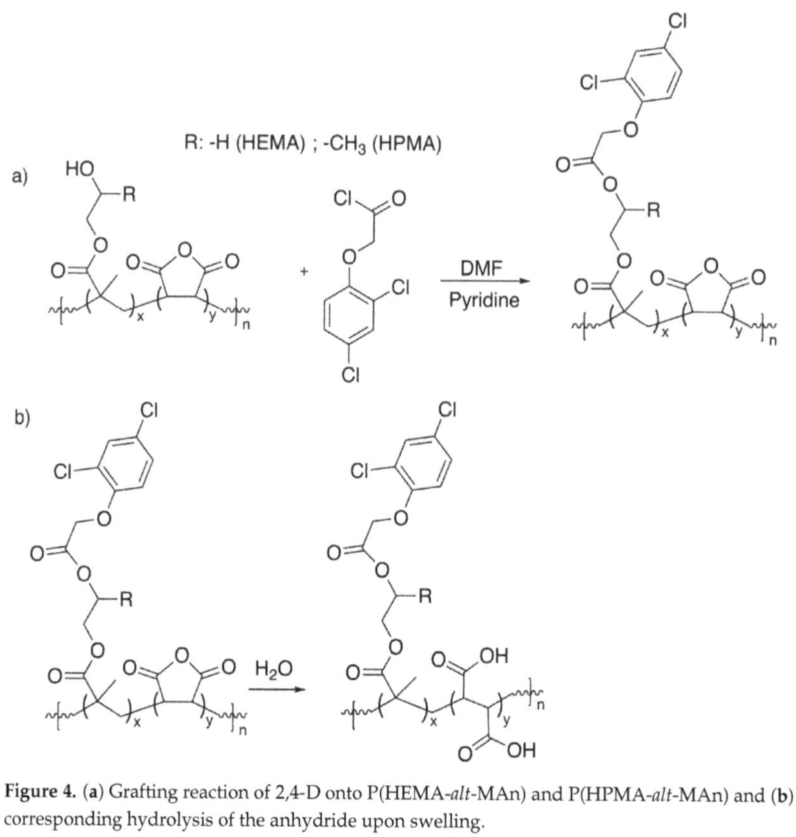


**Figure 4.** (**a**) Grafting reaction of 2,4-D onto P(HEMA-*alt*-MAn) and P(HPMA-*alt*-MAn) and (**b**) the corresponding hydrolysis of the anhydride upon swelling.

**Table 2.** Experimental conditions for the esterification reaction of poly(HEMA-*alt*-MAn) and poly(HPMA-*alt*-MAn).

| System | Copolymer | 2,4-D Chloride | DMF | Pyridine | Yield |
|---|---|---|---|---|---|
| P(HEMA-*alt*-MAn)-2,4-D | 356 mg<br>0.021 mmol | 609 mg<br>2.4 mmol | 6.0 mL | 1.0 mL | 355 mg (49.7%) |
| P(HPMA-*alt*-MAn)-2,4-D | 300 mg<br>0.021 mmol | 547 mg<br>2.23 mmol | 11.0 mL | 1.0 mL | 315 mg (55.0%) |

FTIR (KBr, $cm^{-1}$) of P(HEMA-*alt*-MAn)-2,4-D exhibited the following signals: 3406.3 ν(OH); 2929.9 ν(CH, $CH_2$); 1734.0, 1634.7, and 1616.3 assigned at ν(C=O, ester from MAn, HEMA-2,4-D, respectively), and an increase in signal intensity was observed, along with a slight shift towards a shorter wavenumber; and 1482.3 and 1390.7 ν($CH_2$-; $CH_3$-); and 752.2 and 679.9 ν(C=C, aromatic ring) from herbicide; see Figure 3b. FTIR (KBr, $cm^{-1}$) of P(HPMA-*alt*-MAn)-2,4-D: 3415.9 ν(OH); 2931.8 ν(CH, $CH_2$); 1781.1, 1632.8, and and 1622.6 assigned at ν(C=O, ester from MAn, HPMA-2,4-D, respectively), and an increase in signal intensity was observed, along with a slight shift towards a shorter wavenumber; 1480.3 and 1387.8 ν($CH_2$-; $CH_3$-); 752.2 and 680.9 ν(C=C, aromatic ring) from herbicide; see Figure 4b [9,50,51].

$^1$H-NMR (δ in ppm) showed the following signals: 0.3–2.2, [-$CH_3$, -$CH_2$ from backbone of HEMA], 2.6–5.4 [-CH-CO- of MAn], [-$CH_2$-O- from lateral chain of HEMA], and 6.4 [-OH lateral chain]. On the other hand, the characterization of the P(HPMA-*alt*-MAn) by $^1$H-NMR (δ in ppm) showed the following signals: 0.6–1.6, [-$CH_3$, -$CH_2$ from backbone, and -$CH_3$, of lateral chain of HPMA] and 3.4–4.2 [−$CH_2$-O, and -CH-O- from HPMA and -CH-CO- from MA]; see Figure 5. The $^1$H-NMR spectrum (δ in ppm) of P(HPMA-*alt*-MAn)-2,4-D exhibited the following signals: 0.6–1.2, [-$CH_3$, -$CH_2$ from backbone of HEMA], 3.6 [-CH-CO- of MAn], 4.16 [-$CH_2$-O- from lateral chain of HPMA], 4.83–4.93 [-$CH_2$O-Ar from 2,4-D unit], and 6.8–7.7 [-CH aromatic ring from the herbicide, 2,4-D]; see Figure 5 [9,50,51].

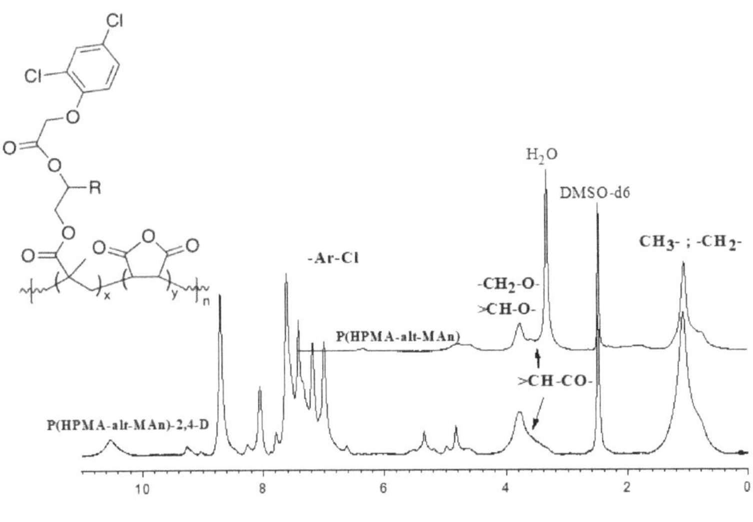


**Figure 5.** $^1$H-NMR spectrum of P(HPMA-*alt*-MAn) and its copolymer conjugate 2,4-D.

The $^1$H-NMR analyses also showed increasing concentration or intensity of the C=O band of the esterification, since the signal at 4.83–4.93 is assigned to -$CH_2O$-Ar from the 2,4-D unit and 7.0–7.6 is assigned to the -CH aromatic ring from the herbicide; see Figure 5. Conversely, the FTIR analyses of the copolymers showed typical bands corresponding to carbonyl groups (see Figures 2 and 3). Once the esterification reaction was achieved, the FTIR analyses showed an increase in the intensity of the C=O in the esterification reaction, since the band corresponding to carbonyl groups appears at around 1634.7 $cm^{-1}$, and bending bands corresponding to the aromatic group appear at 752.2 and 679.9 $cm^{-1}$.

### 2.3. Studies of the Thermal Behavior of the Hydrogels

The thermal behavior of P(HEMA-*alt*-MAn) and P(HPMA-*alt*-MAn), along with their conjugates with 2,4-D, was examined using thermogravimetric analysis (TGA). As shown in Figure 6, both the copolymers and their conjugates exhibited consistent degradation patterns, occurring in one-stage and two-stage processes, respectively. The copolymers remained stable up to approximately 300 °C, with a weight loss (WL) of less than 20%. In contrast, the herbicide conjugates exhibited lower thermal stability than their corresponding copolymers, indicating that the covalent bond formed by the ester functional group between the polymer and the herbicide is relatively weak. When comparing the thermal stability of both copolymers, P(HEMA-*alt*-MAn) demonstrated a higher thermal decomposition temperature (TDT) of 369.2 °C, with an average weight loss of 6.2%. In contrast, the TDT for P(HPMA-*alt*-MAn) copolymers was 311.3 °C, accompanied by a weight loss of 19.13%. Furthermore, when examining the thermal stability of the copolymer–herbicide conjugates, the P(HPMA-*alt*-MAn)–herbicide combination exhibited a lower TDT of 62.1 °C (average weight loss of 10.03%) compared to the P(HEMA-*alt*-MAn)–herbicide, which had a TDT, of 130.6 °C (average weight loss of 14.5%). This indicates that the P(HEMA-*alt*-MAn)–herbicide conjugate has higher thermal stability.

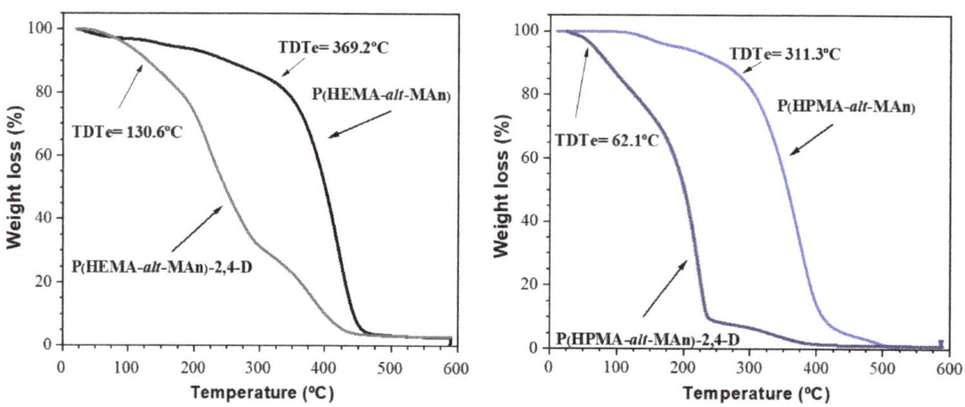


**Figure 6.** Thermograms of P(HEMA-*alt*-MAn) and its conjugate P(HEMA-*alt*-MAn)-2,4-D (**left**), and P(HPMA-*alt*-MAn) and its conjugate P(HPMA-*alt*-MAn)-2,4-D (**right**). Heating rate, 10 °C $min^{-1}$, in inert environment (gaseous $N_2$).

### 2.4. Effect of the pH on Swelling Studies

The swelling behavior of P(HEMA-*alt*-MAn) and P(HPMA-*alt*-MAn) copolymers were studied based on their behavior at different pH levels by immersing the gels in buffered solutions at different pH values (3, 7, and 10), all at 25 °C. The results, displayed in Figure 7, show the swelling times corresponding to each condition.

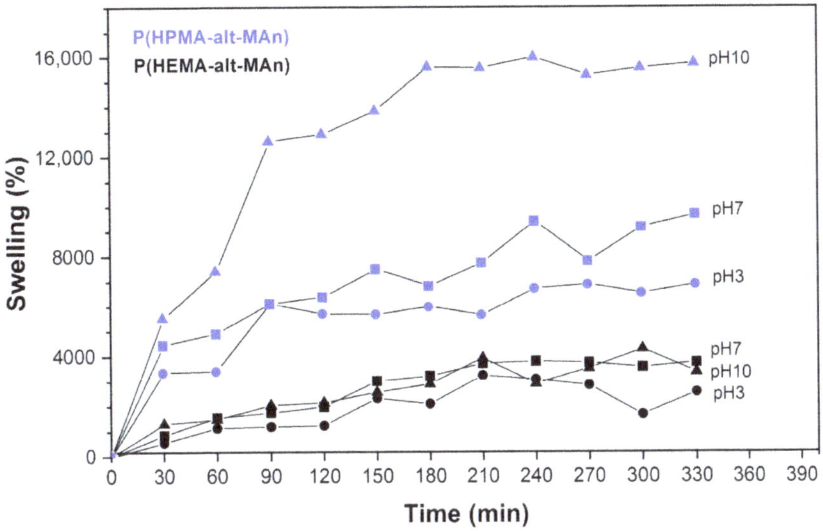


**Figure 7.** Swelling isotherm of P(HEMA-*alt*-MAn) and P(HPMA-*alt*-MAn) as a function of time in buffered solutions at specific pH levels: 3 (•), 7 (■), and 10 (▲) at 25 °C.

The swelling capacity of P(HPMA-*alt*-MAn) was higher than that of P(HEMA-*alt*-MAn). This can be attributed to the fact that in the HPMA residue, there is a greater balance of hydrophilic and hydrophobic residues, which favors the swelling properties, where the hydroxyl group interacts with the acidic carboxylic group of MAn (acid medium) or carboxylate (basic medium), causing a larger cell size and greater flexibility, which increases the hydration capacity. On the other hand, similar behavior was found for P(HEMA-*alt*-MAn) at pH 3, 7, and 10, while for the P(HPMA-*alt*-MAn) hydrogel, the swelling capacity increased when increasing the pH from 3 to 10. The swelling properties of these polymers depend significantly on the balance between hydrophilic and hydrophobic components. In the case of P(HPMA-*alt*-MAn), the observed pH dependency can be explained by an increase in the hydrolysis of the MAn residues, which generates charged carboxylate groups at higher (basic) pH levels. Conversely, the swelling properties of P(HEMA-*alt*-MAn) do not depend on pH due to a lower hydrophobic–hydrophilic balance of the HEMA residue. Additionally, after the hydrolysis of the MAn residues, the carboxylic groups may stabilize through hydrogen bonding with the hydroxyl groups of HEMA. It is also suggested that this stabilizing effect may be reduced by steric hindrance in the case of HPMA.

### *2.5. Release of 2,4-D Through Heterogeneous Hydrolysis*

An ultrafiltration system was employed to analyze the release 2,4-D from copolymer–herbicide conjugates in buffer solutions with pH levels of 3, 7, and 10. To effectively analyze the resulting data, several assumptions need to be established:

(a) The kinetics of hydrolysis of the conjugates is slower than the time taken to collect the filtration fractions, which occurs daily over a period of 3 to 6 min. Therefore, the amount of herbicide released during this short duration is considered negligible.

(b) The herbicide is released from the conjugates into the solution filtrate freely from the ultrafiltration cell. In this case, the concentration of herbicide in each filtration fraction can be related to the concentration of the herbicide free in solution every day ($c^{free\text{-}day}$). In the washing method of the LPR technique, as illustrated in Figure 6, when a low-molecular-weight species is filtered out of the ultrafiltration system and no interaction occurs between this species and the components of the ultrafiltration cell, including the ultrafiltration membrane, the instantaneous concentration of this species in the filtrate is determined by the following Equations (1)–(9), previously reported [9]. These represent the mass balance in releasing the 2,4-D herbicide in ultrafiltration.

$$c^{filtrate} = c^{filtrate-init} \exp(-F) \quad (1)$$

When $F$ is defined as the filtration factor $F = V^{filtrate}/V^{cell}$, it refers to the measure of how well a substance or fluid can pass through a filter or porous material.

The concentration of 2,4-D in the filtration fractions ($c^{filtrate}$) is a mean value considering the instantaneous concentrations in the filtration process ($c^{filtrate}$), which decrease during filtration. Making a mass balance, we have that

$$\int_{F=0}^{F} c^{filtrate} dF = < c^{filtrate} > \Delta F \quad (2)$$

substituting (1) in (2) we obtain

$$\int_{F=0}^{F} c^{filtrate-init} \exp(-F) dF = < c^{filtrate} > \Delta F \quad (3)$$

and integrating both equations

$$c^{filtrate-init} = \frac{< c^{filtrate} > \Delta F}{1 - e^{-F}} \quad (4)$$

As the filtration fraction obtained every day consists of 20 mL, so that final $F = 1$ and $\Delta F = 1$, it is found that

$$c^{free-day} = c^{filtrate-init} = \frac{< c^{filtrate} >}{1 - e^{-1}} \quad (5)$$

Thus, $^{free\text{-}day}$ corresponds to the concentration of the first differential volume obtained in the collection process ($c^{filtrate\text{-}init}$), and is calculated from the value of $<c^{filtrate}>$ corrected by Equation (5).

The concentration of herbicide bound to the polymer every day ($c^{bound\text{-}day}$) is given by

$$c^{bound-day} = c^{cell-day} - c^{free-day} = c^{cell-day} - \frac{< c^{filtrate} >}{1 - e^{-1}} \quad (6)$$

where $c^{cell\text{-}day}$ is the concentration of herbicide in the cell every day and is calculated following

$$c^{cell-day} = c^{cell-init} - \sum_{day} (< c^{filtrate} > \Delta F) \quad (7)$$

where $c^{cell\text{-}init}$ is the initial concentration of herbicide in the cell. If the kinetics of the hydrolysis of the herbicide from the conjugate systems is of order one, then there should be found an exponential decay of $c^{bound\text{-}day}$ with time, since

$$-\frac{dc^{bound-day}}{dt} = kc^{bound-day} \quad (8)$$

and then

$$c^{bound-day} = c^{bound-day-init} \exp(-kt) \quad (9)$$

where $k$ is the release constant. Release profiles of P(HEMA-*alt*-MAn)-2,4-D and P(HPMA-*alt*-MAn)-2,4-D are exhibited in Figure 8. The release rates of both conjugates were notably higher at pH 10, exceeding 85% over a duration of 7 days.

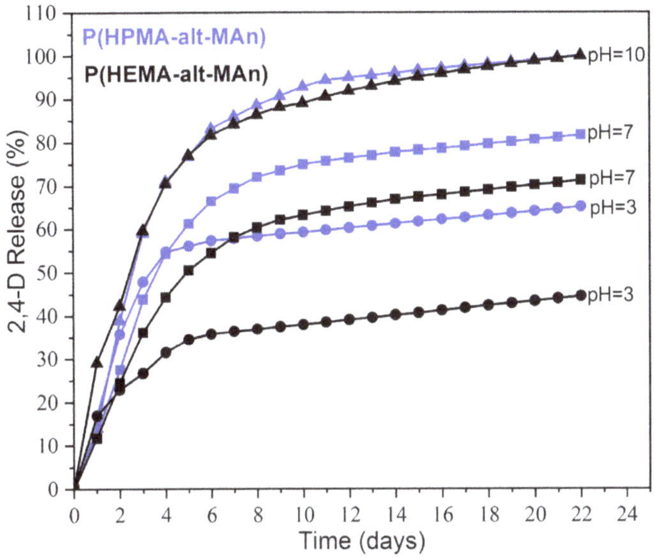


**Figure 8.** Release of 2,4-D (%) from copolymer–herbicide conjugates in an aqueous solution at pH 3, 7, and 10.

The release of drugs linked by covalent bonds depends on the degradation rate of the polymer–drug connection. Typically, these connections degrade through hydrolysis, allowing the rates of degradation and release to follow simple first-order kinetic relationships [48]. In specific applications, the drug–polymer linkages can be designed to degrade enzymatically, leading to more complex release kinetics [49,52,53].

First-order kinetics occur when a constant proportion of the phytopharmaceutical is eliminated per unit of time. The elimination rate is proportional to the amount of phytopharmaceutical in the copolymeric systems, but is pH-dependent; at basic pH, it is increasing. The higher the concentration, the greater the amount of drug eliminated per unit of time at higher pH. For each half-life that passes, the concentration of the drug is halved. The elimination rate decreases proportionally as the drug concentration decreases while maintaining a constant elimination rate, k. The reaction rates are influenced by the olefinic comonomer used, whether it is HEMA or HPMA, to which 2,4-D is grafted. In this way, the positive release of the drug is affected by the increased hydrophilicity and conformational flexibility of the polymer, with the HPMA comonomer. This enhancement occurs due to the extension of the lateral chain of HPMA and the steric hindrance caused by the methyl group, which prevents the formation of inter- or intra-dipole–dipole interactions (hydrogen bonds) between the highly electronegative oxygen (O) atoms of the O–H groups, which is different for the HEMA comonomer; see Figure 9. Additionally, for the HPMA monomer, an increase in the content of hydrophilic co-units, as seen in copolymers containing MAc, contributes to this effect. This suggests that the release of herbicides significantly increases with higher pH levels, primarily because hydrolysis becomes more effective at elevated pH. The notable increase in release at pH 10 indicates that the basic hydrolysis of acetate groups may impact the breakdown of certain herbicides. Lower release values were observed at pH levels of 7 and 3, which aligns with the mechanism of hydrolytic cleavage of the herbicide–polymer bond. The differences in release rates between the two copolymer conjugates at pH 3 and 7 can be explained by their varying swelling capacities, which differ between the HEMA and HPMA comonomers. The release rate typically increases with higher swelling from aqueous solutions, associated with greater polymer hydrophilicity for the P(HPMA-*alt*-MAn) system. The MAn copolymer reacts easily with water to form MAn-MAc terpolymers at basic pH (the maleic anhydride ring is open). With a sufficient amount of water, MAn hydrolyzes completely to MAc copolymers at pH 10. The reaction rates also depend on the olefinic comonomer. Additionally, an increase in the content of hydrophilic co-units, as seen in copolymers containing MAc, contributes to this effect.

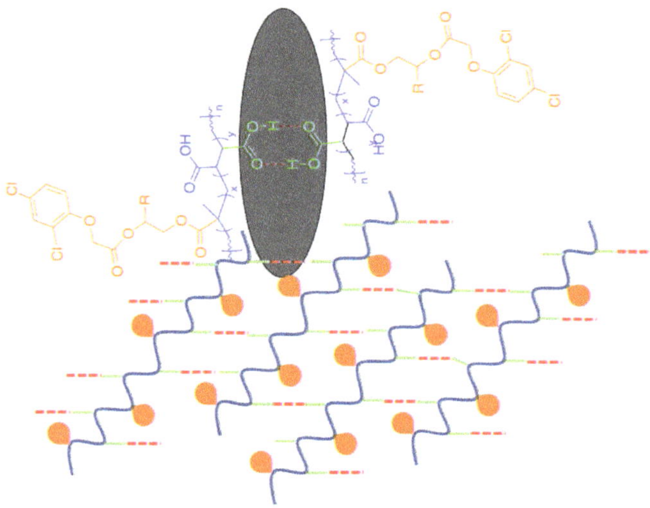

**Figure 9.** Schematic representation of inter- or intra-hydrogen bonds in fabricating stimulus-responsive hydrogels.

Conversely, the positive release is influenced by the increased hydrophilicity and conformational flexibility of the polymer.

The kinetic constants for the various systems at different pH values were obtained from the slopes of the plots shown in Figure 10, as indicated by Equation (9). These values are presented in Table 1. The exponential decay observed in the release profiles confirms that the hydrolysis of the copolymer conjugates follows first-order kinetics. It can be seen in Figure 8 the evolution of ln $c^{\,bound\text{-}day}$ with time during the first 6 days.

The values of R2 for n = 1 for the release of herbicide for P(HPMA-*alt*-MAn) y P(HEMA-*alt*-MAn) at pH 3.0, 7.0, and 10 are shown in the Table 3. These results show that the value of n = 1 is maintained with an increase in pH.

**Table 3.** Kinetic constants ($k_1$; in day$^{-1}$) during the first 6 days.

| System | $k_1$ (pH 3) | R2 | $k_1$ (pH 7) | R2 | $k_1$ (pH 10) | R2 |
|---|---|---|---|---|---|---|
| P(HEMA-*alt*-MAn)-2,4-D | 0.0353 | 0.9802 | 0.1276 | 0.8694 | 0.267 | 0.9114 |
| P(HPMA-*alt*-MAn)-2,4-D | 0.0999 | 0.9673 | 0.1929 | 0.8447 | 0.3349 | 0.9176 |

$k_1$ first-order release constant; $n$: order of kinetic release.

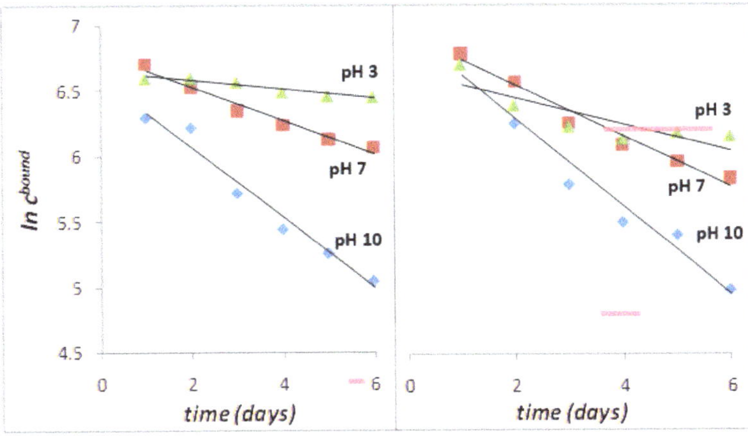


**Figure 10.** Determination of kinetic release constants at different pH. Plot of ln $c^{bound}$ vs. time (days).

## 3. Conclusions

A series of two pH-sensitive copolymers, P(HEMA-*alt*-MAn) and P(HPMA-*alt*-MAn), have been designed as controlled-release systems using the herbicide 2,4-D. These copolymers demonstrate pH sensitivity, exhibiting a higher swelling ratio in aqueous solutions at elevated pH levels. The release values measured at pH 10 indicate that the cleavage of the herbicide may be influenced by the basic hydrolysis of the acetate groups. In contrast, lower release values were observed at pH 7 and pH 3, consistent with the mechanism of hydrolytic cleavage of the herbicide–polymer bond. The release rate typically increases with a greater swelling from the water solution, which corresponds to enhanced hydrophilicity of the polymer. This phenomenon can be attributed to the increased hydrolysis of the MAn residues and the dissociation of the acidic groups. The release is positively influenced by the increased hydrophilicity of the lateral chain in HPMA (due to the presence of the methyl group). This enhancement occurs because the steric hindrance created by the methyl group prevents the formation of inter- or intra-molecular hydrogen bonds. Additionally, there is an increase in the content of the hydrophilic co-unit (MAc), which results from the esterification of the MAn-containing copolymer. In contrast, the swelling capacity of P(HPMA-*alt*-MAn) is greater than that of P(HEMA-*alt*-MAn). This difference may be due to the higher steric hindrance caused by the methyl group in HPMA. The copolymer functionalized with 2,4-D displayed labile ester bonds that can be cleaved by heating, as demonstrated by TGA analyses. The hydrolysis of the herbicide from the conjugates follows first-order kinetics, with higher kinetic constants observed at increasing pH levels. This indicates that the hydrolysis reaction is base-catalyzed. These findings suggest that pH-sensitive hydrogels could be used as intelligent drug delivery carriers in agriculture.

## 4. Materials and Methods

### *4.1. Materials*

2-hydroxyethylmethacrylate (HEMA) and 2-hydroxypropylmethacrylate (HPMA) (Merck, Darmstadt, Germany) were purified by distillation. Benzoyl peroxide (BPO) (Sigma Aldrich, St. Louis, MO, USA) was used as an initiator of the copolymerization reaction between HEMA or HPMA with maleic anhydride (MAn) (Sigma Aldrich, St. Louis, MO, USA). All the other reagents, including the herbicide 2,4-dichlorophenoxy acetic acid (2,4-D) (Sigma Aldrich, St. Louis, MO, USA), were used as received. Thionyl chloride ($SOCl_2$) (Sigma Aldrich, St. Louis, MO, USA) was used as received.

### 4.2. Equipment

The structural characterization of the polymers was carried out by proton nuclear magnetic resonance ($^1$H-NMR) on a Bruker 200 MHz spectrometer, Karlsruhe, Germany. in DMSO-$d_6$, at room temperature. Fourier transform infrared spectroscopy (FTIR) spectra were acquired by a Bruker Vector 22 (Bruker Optics GmbH, Karlsruhe, Germany), in the range of 400 to 4000 $cm^{-1}$. Gel permeation chromatography (GPC) was conducted to determine the number-averaged molecular weight (Mn) and weight-averaged molecular weight (Mw) under the following conditions: a WATERS 600E instrument (Kioto, Japan) equipped with UV and RI detectors, using tetrahydrofuran (THF) as the solvent at a flow rate of 1.0 mL/min. The samples were analyzed at a concentration of 4 mg/mL and at a temperature of 25 °C. Calibration was performed using poly(methyl methacrylate) (PMMA).

The UV–VIS spectra were obtained using a spectrophotometer at 25 °C between 250–700 nm using a Perkin Elmer Lambda 35, Waltham, MA, USA. The "Lyph-lock" freeze-dry system (Lab condo 6L) was used. The pH was determined with a pH-meter Hanna 211, Woonnsocket, USA. The copolymers' thermal analysis was performed by recording TGA using a Star System 1 thermogravimetric analyzer (TGA), Barcelona, Spain, under gaseous nitrogen (at 150 and 50 $cm^3$ $min^{-1}$, respectively). The ultrafiltration system comprised a filtration cell equipped with a magnetic stirrer (Amicon 8010, 10 mL capacity), a membrane (cellulose) with a molecular weight cut-off of 10,000 Daltons (Ultracel PLCC, 5 mm diameter), a reservoir system, a flow selector, and a pressure source. Figure 11 shows the experimental set-up of the membrane filtration system used for the release studies in aqueous solution at different pH.

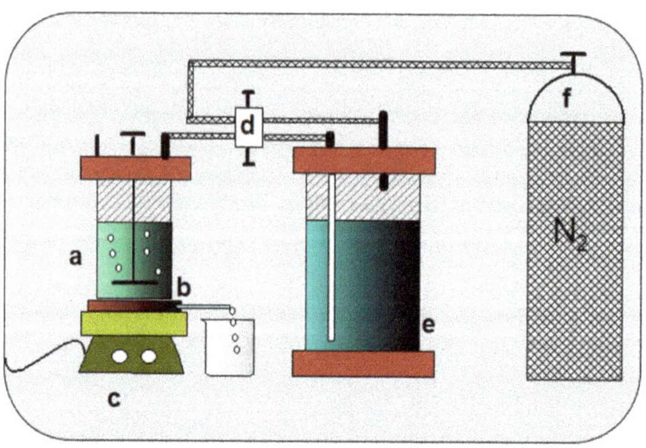


**Figure 11.** Experimental set-up of ultrafiltration system: (a) filtration cell, (b) filtrate, (c) magnetic stirrer, (d) selector, (e) reservoir, and (f) pressure source.

### 4.3. Experimental Procedures

#### 4.3.1. Preparation of the Copolymer Hydrogels

P(HEMA-*alt*-MAn) and P(HPMA-*alt*-MAn) were synthesized by free radical copolymerization (1:1 feed monomer ratio) in solution using 0.5 mol% of BPO in tetrahydrofuran (THF). The polymerization reaction was kept at 70 °C for 6 h (see Table 1).

The polymers were precipitated with n-heptane and dried under reduced pressure at 50 °C, with yields of 51.7% and 34.2%, respectively. The copolymers are soluble in dimethyl sulfoxide and exhibit swelling properties in water. Figure 1 shows the structure of the copolymers.

4.3.2. Graft Reaction of the Copolymers with the Herbicide 2,4-D

The grafting procedure was carried out by obtaining 2,4-dichlorophenoxyacetyl chloride, previously carried out with $SOCl_2$ to obtain the controlled-release system, purified by precipitation using hexane as solvent. According to the reaction shown in Figure 4, P(HEMA-*alt*-MAn), P (HPMA-*alt*-MAn) and 2,4-dichlorophenoxy acetyl chloride were dissolved in 5 mL of DMF as shown in Table 2. The solution was placed in a three-necked flask, which was equipped with a nitrogen inlet and outlet, a dropping funnel, a magnetic stirrer, and a thermometer. Pyridine (1 mL) was added drop-wise while stirring at approximately 0 °C. The reaction mixture was then heated to 25 °C for 5 h. After this period, the solution was poured into a large volume of cold 0.5 M HCl to precipitate the product. The product was subsequently filtered and washed several times with cold distilled water. Purification was achieved by precipitation, using THF as a solvent and cold distilled water as the precipitating agent. Finally, the product was dried under reduced pressure at 30 °C until a constant weight was obtained.

4.3.3. Swelling Studies Procedure

The studies were carried out at 25 °C in buffered solutions. Before swelling and release studies, the polymer hydrogels were dispersed and purified by a membrane ultrafiltration system using a membrane with an exclusion limit of a molecular weight of 10,000 g $mol^{-1}$. The polymeric fraction over 10,000 g $mol^{-1}$ was freeze-dried and selected for further experiments. Dried samples of the copolymers were placed in water at a defined pH (3, 7, or 10) in buffered solutions using either 0.01 M citrate buffer (pH $<$ 7) or phosphate buffer (pH $\geq$ 7). Every half hour, the hydrogel samples were quickly removed from the solution and weighed. The swelling values ($S_w$) were calculated using the following Equation (10) [9].

$$S_W(\%) = \frac{(W_S - W_d)}{W_d} \times 100 \quad (10)$$

$W_s$ is the weight of the swollen hydrogel at an equilibrium state, and $W_d$ is the weight of the dried hydrogel (Xerogel).

4.3.4. Heterogeneous Hydrolysis Procedure

This study was conducted over a month at ambient temperature. A filtration fraction was collected daily for 30 days (Z ranging from 1 to 30), using the ultrafiltration system in buffer solutions with pH levels of 3, 7, and 10. The concentration of the bioactive agent present in each filtration fraction was quantitatively assessed through UV spectroscopy, utilizing a wavelength specifically suited for detecting 2,4-D acid, measured at 282 nm. This analytical technique allowed for the precise determination of the agent's concentration, across the different filtration fractions.

**Author Contributions:** Conceptualization, O.G.M. and G.d.C.P.; Methodology, O.G.M., F.A.R. and G.d.C.P.; Validation, J.S.; Formal analysis, F.A.R. and G.d.C.P.; Investigation, O.G.M., F.A.R., R.M.-T. and G.d.C.P.; Resources, O.G.M. and G.d.C.P.; Data curation, R.M.-T. and J.S.; Writing—original draft, O.G.M., G.d.C.P. and G.d.C.P.; Writing—review & editing, O.G.M., R.M.-T., J.S. and G.d.C.P. All authors have read and agreed to the published version of the manuscript.

**Funding:** Research Continuity Project Fund, year of 2023, code PY LCLI23-02, UTEM; FONDECYT, grant number 1240357 and ANID InES Género INGE210029.

**Institutional Review Board Statement:** Not applicable.

**Informed Consent Statement:** Not applicable.

**Data Availability Statement:** The original contributions presented in this study are included in the article. Further inquiries can be directed to the corresponding author.

**Acknowledgments:** The authors gratefully acknowledge the support provided by the project funded by the Research Continuity Project Fund, year 2021, code PY LCLI23-02, Universidad Tecnológica Metropolitana, ANID InES Género INGE210029, FONDECYT grant number 1240357 and 1231498.

**Conflicts of Interest:** The authors declare no conflict of interest.

## References

1. Ball, M.; Quintart, A.-D. *The Future of Sustainable Food Production in Europe: A Concept Paper*; Syngenta: Brussels, Belgium, 2023.
2. Amodeo, G.; Giacometti, R.; Spagnoletti, F.; Santagapita, P.R.; Perullini, M. Eco-Friendly Routes for Obtaining Nanoparticles and Their Application in Agro-Industry. In *Nano-Enabled Agrochemicals in Agriculture*; Elsevier: Amsterdam, The Netherlands, 2022; pp. 49–62. [CrossRef]
3. Nath, C.P.; Singh, R.G.; Choudhary, V.K.; Datta, D.; Nandan, R.; Singh, S.S. Challenges and Alternatives of Herbicide-Based Weed Management. *Agronomy* **2024**, *14*, 126. [CrossRef]
4. Cooper, J.; Dobson, H. The Benefits of Pesticides to Mankind and the Environment. *Crop Prot.* **2007**, *26*, 1337–1348. [CrossRef]
5. Magnoli, K.; Carranza, C.S.; Aluffi, M.E.; Magnoli, C.E.; Barberis, C.L. Herbicides Based on 2,4-D: Its Behavior in Agricultural Environments and Microbial Biodegradation Aspects. A Review. *Environ. Sci. Pollut. Res.* **2020**, *27*, 38501–38512. [CrossRef]
6. Nie, G.; Hong, K.; Zhang, E.; Liu, N.; Wang, M.; Wang, L.; Zang, Y. Fabrication of a Porous Chitosan/Poly-(γ-Glutamic Acid) Hydrogel with a High Absorption Capacity by Electrostatic Contacts. *Int. J. Biol. Macromol.* **2020**, *159*, 986–994. [CrossRef]
7. Doghish, A.S.; Hashem, A.H.; Shehabeldine, A.M.; Sallam, A.-A.M.; El-Sayyad, G.S.; Salem, S.S. Nanocomposite Based on Gold Nanoparticles and Carboxymethyl Cellulose: Synthesis, Characterization, Antimicrobial, and Anticancer Activities. *J. Drug Deliv. Sci. Technol.* **2022**, *77*, 103874. [CrossRef]
8. Asai, D.; Kanamoto, T.; Takenaga, M.; Nakashima, H. In Situ Depot Formation of Anti-HIV Fusion-Inhibitor Peptide in Recombinant Protein Polymer Hydrogel. *Acta Biomater.* **2017**, *64*, 116–125. [CrossRef]
9. Marambio, O.G.; Muñoz, A.; Martin-Trasancos, R.; Sánchez, J.; Pizarro, G.D.C. The Design of a Controlled-Release Polymer of a Phytopharmaceutical Agent: A Study on the Release in Different PH Environments Using the Ultrafiltration Technique. *Polymers* **2024**, *16*, 3492. [CrossRef]
10. Nisar, S.; Pandit, A.H.; Nadeem, M.; Pandit, A.H.; Rizvi, M.M.A.; Rattan, S. γ-Radiation Induced L-Glutamic Acid Grafted Highly Porous, PH-Responsive Chitosan Hydrogel Beads: A Smart and Biocompatible Vehicle for Controlled Anti-Cancer Drug Delivery. *Int. J. Biol. Macromol.* **2021**, *182*, 37–50. [CrossRef]
11. Cardarelli, N. *Controlled Release Pesticides Formulations*; Zweig, G., Ed.; CRC Press: Boca Raton, FL, USA, 2018. [CrossRef]
12. Kydonieus, A.F. *Controlled Release Technologies: Methods, Theory, and Applications*; CRC Press: Boca Raton, FL, USA, 1980; Volume I.
13. Kenawy, E.R.; Sherrington, D.C.; Akelah, A. Controlled Release of Agrochemical Molecules Chemically Bound to Polymers. *Eur. Polym. J.* **1992**, *28*, 841–862. [CrossRef]
14. Schnatter, W.F.K. Functionalized Polymers and Their Applications, by A. Akelah and A. Moet, Chapman and Hall, London, 1990, 345 pp. Price: $79.95. *J. Polym. Sci. A Polym. Chem.* **1992**, *30*, 2473. [CrossRef]
15. Kenawy, E.-R. Polymers for Agricultural Applications: Controlled-Release Polymeric Formulations with Pendant 2,6-Dichlorobenzaldehyde. *Polym. Plast. Technol. Eng.* **2001**, *40*, 437–450. [CrossRef]
16. Rehab, A.; Akelah, A.; El-Gamal, M.M. Controlled-release Systems Based on the Intercalation of Polymeric Metribuzin onto Montmorillonite. *J. Polym. Sci. A Polym. Chem.* **2002**, *40*, 2513–2525. [CrossRef]
17. Kandil, S.; Kenawy, E.; El-Maghraby, A. Recycling of Pharmaceutical Waste Gelatin for Controlled-release Applications. I. A 2,4-dicholorphenoxy Acetic Acid Based System. *J. Appl. Polym. Sci.* **2004**, *91*, 2313–2319. [CrossRef]
18. Kandil, S.; Kenawy, E.-R.; El-Maghraby, A.; ElAshry, E.H. Recycling of Pharmaceutical Waste Gelatin for Controlled Release Applications II: A Tri-Fluralin Based System. *Polym. Plast. Technol. Eng.* **2004**, *43*, 1695–1709. [CrossRef]
19. Signori, F.; Chiellini, F.; Solaro, R. New Self-Assembling Biocompatible–Biodegradable Amphiphilic Block Copolymers. *Polymer* **2005**, *46*, 9642–9652. [CrossRef]
20. Chiellini, E.E.; Chiellini, F.; Solaro, R. Bioerodible Polymeric Nanoparticles for Targeted Delivery of Proteic Drugs. *J. Nanosci. Nanotechnol.* **2006**, *6*, 3040–3047. [CrossRef]
21. Kenawy, E.-R.; El-Newehy, M.; Abdel-Hay, F.; Ottenbrite, R.M. A New Degradable Hydroxamate Linkage for PH-Controlled Drug Delivery. *Biomacromolecules* **2007**, *8*, 196–201. [CrossRef]
22. Kenawy, E.; Abdel-Hay, F.; El-Newehy, M.; Ottenbrite, R.M. Effect of PH on the Drug Release Rate from a New Polymer–Drug Conjugate System. *Polym. Int.* **2008**, *57*, 85–91. [CrossRef]
23. He, C.; Kim, S.W.; Lee, D.S. In Situ Gelling Stimuli-Sensitive Block Copolymer Hydrogels for Drug Delivery. *J. Control. Release* **2008**, *127*, 189–207. [CrossRef]

24. Akelah, A.; Kenawy, E.R.; Sherrington, D.C. Hydrolytic Release of Herbicides from Modified Polyamides of Tartrate Derivatives. *Eur. Polym. J.* **1995**, *31*, 903–909. [CrossRef]
25. Phillips, M.A.; Gran, M.L.; Peppas, N.A. Targeted Nanodelivery of Drugs and Diagnostics. *Nano Today* **2010**, *5*, 143–159. [CrossRef] [PubMed]
26. Lao, L.L.; Venkatraman, S.S.; Peppas, N.A. A Novel Model and Experimental Analysis of Hydrophilic and Hydrophobic Agent Release from Biodegradable Polymers. *J. Biomed. Mater. Res. A* **2009**, *90A*, 1054–1065. [CrossRef] [PubMed]
27. Betancourt, T.; Pardo, J.; Soo, K.; Peppas, N.A. Characterization of PH-responsive Hydrogels of Poly(Itaconic Acid-*g*-ethylene Glycol) Prepared by UV-initiated Free Radical Polymerization as Biomaterials for Oral Delivery of Bioactive Agents. *J. Biomed. Mater. Res. A* **2010**, *93A*, 175–188. [CrossRef] [PubMed]
28. Lauten, E.H.; Peppas, N.A. Intelligent Drug Release Using Molecular Imprinting Methods Recognitive Systems for Angiotensin II. *J. Drug Deliv. Sci. Technol.* **2009**, *19*, 391–399. [CrossRef]
29. Lentz' And, R.D.; Sojka', R.E. Field Results Using Polyacrylamide to Manage Furrow Erosion and Infiltration. *Soil Sci.* **1994**, *158*, 274–282. [CrossRef]
30. Fritz, J.; Sandhofer, M.; Stacher, C.; Braun, R. Strategies for Detecting Ecotoxicological Effects of Biodegradable Polymers in Agricultural Applications. *Macromol. Symp.* **2003**, *197*, 397–410. [CrossRef]
31. Šimanský, V. Soil Structure Stability and Distribution of Carbon in Water-Stable Aggregates in Different Tilled and Fertilized Haplic Luvisol. *Acta Univ. Agric. Silvic. Mendel. Brun.* **2013**, *60*, 173–178. [CrossRef]
32. Kenawy, E.; Cinelli, P.; Corti, A.; Miertus, S.; Chiellini, E. Biodegradable Composite Films Based on Waste Gelatin. *Macromol. Symp.* **1999**, *144*, 351–364. [CrossRef]
33. Cinelli, P.; Chiellini, E.; Corti, A. Developments and Future Trends for Environmentally Degradable Plastics. In *Renewable Resources and Renewable Energy*; CRC Press: Boca Raton, FL, USA, 2006; pp. 63–119. [CrossRef]
34. Cinelli, P.; Chiellini, E.; Imam, S.H. Hybrid Composite Based on Poly(Vinyl Alcohol) and Fillers from Renewable Resources. *J. Appl. Polym. Sci.* **2008**, *109*, 1684–1691. [CrossRef]
35. Finley, J.H. Spectrophotometric Determination of Alcohol in Paper Coatings. *Anal. Chem.* **1961**, *33*, 1925–1927. Available online: https://pubs.acs.org/sharingguidelines (accessed on 16 January 2025). [CrossRef]
36. Chiellini, E.; Corti, A.; D'Antone, S.; Solaro, R. Biodegradation of Poly (Vinyl Alcohol) Based Materials. *Prog. Polym. Sci.* **2003**, *28*, 963–1014. [CrossRef]
37. Takei, T.; Yoshihara, R.; Danjo, S.; Fukuhara, Y.; Evans, C.; Tomimatsu, R.; Ohzuno, Y.; Yoshida, M. Hydrophobically-Modified Gelatin Hydrogel as a Carrier for Charged Hydrophilic Drugs and Hydrophobic Drugs. *Int. J. Biol. Macromol.* **2020**, *149*, 140–147. [CrossRef] [PubMed]
38. Dreiss, C.A. Hydrogel Design Strategies for Drug Delivery. *Curr. Opin. Colloid Interface Sci.* **2020**, *48*, 1–17. [CrossRef]
39. Oyarce, E.; Pizarro, G.D.C.; Oyarzún, D.P.; Martin-Trasanco, R.; Sánchez, J. Adsorption of Methylene Blue in Aqueous Solution Using Hydrogels Based on 2-Hydroxyethyl Methacrylate Copolymerized with Itaconic Acid or Acrylic Acid. *Mater. Today Commun.* **2020**, *25*, 101324. [CrossRef]
40. Yang, Z.; Chen, L.; McClements, D.J.; Qiu, C.; Li, C.; Zhang, Z.; Miao, M.; Tian, Y.; Zhu, K.; Jin, Z. Stimulus-Responsive Hydrogels in Food Science: A Review. *Food Hydrocoll.* **2022**, *124*, 107218. [CrossRef]
41. Rizwan, M.; Yahya, R.; Hassan, A.; Yar, M.; Azzahari, A.; Selvanathan, V.; Sonsudin, F.; Abouloula, C. PH Sensitive Hydrogels in Drug Delivery: Brief History, Properties, Swelling, and Release Mechanism, Material Selection and Applications. *Polymers* **2017**, *9*, 137. [CrossRef]
42. Alasvandian, S.; Shahgholi, M.; Karimipour, A. Investigating the Effects of Chitosan Atomic Ratio and Drug Type on Mechanical Properties of Silica Aerogel/Chitosan Nanocomposites Using Molecular Dynamics Approach. *J. Mol. Liq.* **2024**, *401*, 124639. [CrossRef]
43. del Carmen Pizarro, G.; Marambio, O.G.; Jeria-Orell, M.; Huerta, M.R.; Rivas, B.L.; Habicher, W.D. Metal Ion Retention Using the Ultrafiltration Technique: Preparation, Characterization of the Water-soluble Poly(1-vinyl-2-pyrrolidone-*Co*-itaconic Acid) and Its Metal Complexes in Aqueous Solutions. *J. Appl. Polym. Sci.* **2008**, *108*, 3982–3989. [CrossRef]
44. del Carmen Pizarro, G.; Marambio, O.G.; Jeria-Orell, M.; Huerta, M.R.; Sánchez, J.; Rivas, B.L.; Habicher, W.D. Preparation, Characterization, and Thermal Properties of Hydrophilic Copolymers: *P*-chlorophenylmaleimides with Hydroxylethyl Methacrylate and B-methyl Itaconate. *Polym. Int.* **2007**, *56*, 1166–1172. [CrossRef]
45. Jeria-Orell, M.; Pizarro, G.D.C.; Marambio, O.G.; Huerta, M.; Geckeler, K.E. Synthesis of *N*-hydroxymethyl Acrylamide with B-methyl Hydrogen Itaconate and Itaconic Acid Hydrogels: Effects of the PH, Composition, and Ionic Strength on the Swelling Behavior. *J. Appl. Polym. Sci.* **2006**, *100*, 1735–1741. [CrossRef]
46. Ozturk, V.; Okay, O. Temperature Sensitive Poly(N-t-Butylacrylamide-Co-Acrylamide) Hydrogels: Synthesis and Swelling Behavior. *Polymer* **2002**, *43*, 5017–5026. [CrossRef]
47. Paul, D.R.; Harris, F.W. (Eds.) *Controlled Release Polymeric Formulations*; American Chemical Society: Washington, DC, USA, 1976; Volume 33. [CrossRef]

48. Lin, C.-C.; Metters, A.T. Hydrogels in Controlled Release Formulations: Network Design and Mathematical Modeling. *Adv. Drug Deliv. Rev.* **2006**, *58*, 1379–1408. [CrossRef] [PubMed]
49. Sintov, A.; Levy, R.J. Polymeric Drug Delivery of Enzymatically Degradable Pendant Agents: Peptidyl-Linked Procainamide Model System Studies. *Int. J. Pharm.* **1997**, *146*, 55–62. [CrossRef]
50. del Carmen Pizarro, G.; Marambio, O.G.; Jeria-Orell, M.; Flores, M.E.; Rivas, B.L. Amphiphilic Diblock Copolymers Poly(2-hydroxyethylmethacrylate)-*b*-(*N*-phenylmaleimide) and Poly(2-hydroxyethylmethacrylate)-*b*-(Styrene) Using the Macroinitiator Poly(HEMA)-Cl by ATRP: Preparation, Characterization, and Thermal Properties. *J. Appl. Polym. Sci.* **2010**, *118*, 3649–3657. [CrossRef]
51. del Carmen Pizarro, G.; Marambio, O.G.; Jeria-Orell, M.; Valdés, D.T.; Geckeler, K.E. Self-Assembled Nanostructures: Preparation, Characterization, Thermal, Optical and Morphological Characteristics of Amphiphilic Diblock Copolymers Based on Poly(2-Hydroxyethyl Methacrylate-*Block-N*-Phenylmaleimide). *Polym. Int.* **2013**, *62*, 1528–1538. [CrossRef]
52. Bauli, C.R.; Lima, G.F.; de Souza, A.G.; Ferreira, R.R.; Rosa, D.S. Eco-Friendly Carboxymethyl Cellulose Hydrogels Filled with Nanocellulose or Nanoclays for Agriculture Applications as Soil Conditioning and Nutrient Carrier and Their Impact on Cucumber Growing. *Colloids Surf. A Physicochem. Eng. Asp.* **2021**, *623*, 126771. [CrossRef]
53. Lakshani, N.; Wijerathne, H.S.; Sandaruwan, C.; Kottegoda, N.; Karunarathne, V. Release Kinetic Models and Release Mechanisms of Controlled-Release and Slow-Release Fertilizers. *ACS Agric. Sci. Technol.* **2023**, *3*, 939–956. [CrossRef]

*gels*

*Article*

# Development of Variable Charge Cationic Hydrogel Particles with Potential Application in the Removal of Amoxicillin and Sulfamethoxazole from Water

**Francisca L. Aranda [1,2], Manuel F. Meléndrez [3], Mónica A. Pérez [2], Bernabé L. Rivas [2], Eduardo D. Pereira [4] and Daniel A. Palacio [2,*]**

1 Departamento de Ingeniería de Materiales, Facultad de Ingeniería, Universidad de Concepción, Concepción 4070371, Chile
2 Departamento de Polímeros, Facultad de Ciencias Químicas, Universidad de Concepción, Concepción 3349001, Chile
3 Facultad de Ciencias para el Cuidado de la Salud, Universidad San Sebastián, Campus Las Tres Pascualas, Concepción 4060000, Chile
4 Departamento de Química Analítica e Inorgánica, Facultad de Ciencias Químicas, Universidad de Concepción, Casilla 160-C, Concepción 3349001, Chile
* Correspondence: dapalacio@udec.cl

**Abstract:** Cationic hydrogel particles (CHPs) crosslinked with glutaraldehyde were synthesized and characterized to evaluate their removal capacity for two globally consumed antibiotics: amoxicillin and sulfamethoxazole. The obtained material was characterized by FTIR, SEM, and TGA, confirming effective crosslinking. The optimal working pH was determined to be 6.0 for amoxicillin and 4.0 for sulfamethoxazole. Under these conditions, the CHPs achieved over 90.0% removal of amoxicillin after 360 min at room temperature, while sulfamethoxazole removal reached approximately 60.0% after 300 min. Thermodynamic analysis indicated that adsorption occurs through a physisorption process and is endothermic. The $\Delta H°$ values of 28.38 kJ $mol^{-1}$, 12.39 kJ $mol^{-1}$, and $\Delta S°$ 97.19 J $mol^{-1}$ $K^{-1}$, and 33.94 J $mol^{-1}$ $K^{-1}$ for AMX and SMX, respectively. These results highlight the potential of CHPs as promising materials for the removal of such contaminants from aqueous media.

**Keywords:** cationic hydrogels particles; wastewater treatment; adsorption mechanism; antibiotics; pharmaceutical compounds

## 1. Introduction

Water is an essential and indispensable resource for any ecosystem [1,2]. However, it is currently severely impacted by domestic wastewater, agricultural and industrial residues, and pharmaceutical compounds [3,4]. Among these, emerging organic contaminants such as antibiotics have become a major concern. It is estimated that between 20% and 90% of the total antibiotics consumed are not metabolized and are excreted either unchanged or as derivatives and metabolites, which directly enter aquatic systems [5,6].

The most widely used antibiotic families worldwide include β-lactams and sulfonamides, particularly amoxicillin and sulfamethoxazole, respectively. Amoxicillin, a penicillin derivative, has 6-aminopenicillanic acid as its core structure, consisting of a thiazolidine ring fused with a β-lactam ring and a side chain. This structure contains three main functional groups (Figure 1b): COOH ($pKa_1$ = 2.25), $NH_2$ ($pKa_2$ = 7.22), and OH ($pKa_3$ = 9.48) [7–9]. Sulfamethoxazole, used to treat various diseases and infections [10], contains a basic amine group (-$NH_2$) and an acidic sulfonamide group (-$SO_2NH$-), resulting in two distinct dissociation sites within the molecule (Figure 1a), with $pKa_1$ and $pKa_2$ values of 1.97 and 5.86, respectively [11,12].

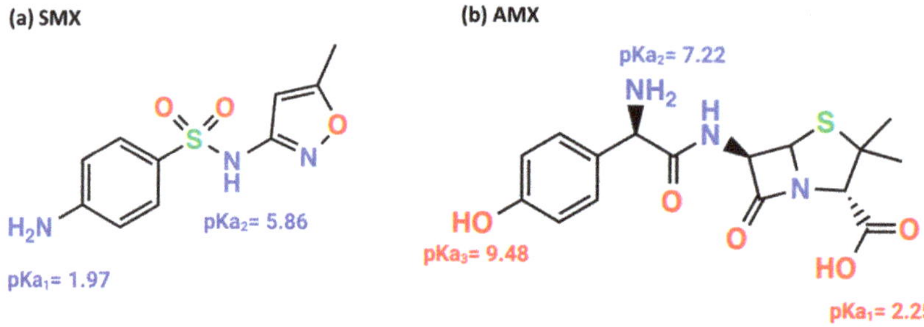


**Figure 1.** Structures of the antibiotics sulfamethoxazole (**a**), amoxicillin (**b**), and their respective pKa values.

Currently, a wide variety of materials are available for the removal of such contaminants [7,13], including carbon-based materials and metal–organic frameworks (MOFs) with removal efficiencies ranging from 75.3% to 82.52% [14,15]. Other compounds, such as $In_2S_3$, achieve 66% removal [16], while materials like UiO-66@$WO_3$/GO exhibit an 84% removal rate [17]. Porous organic polymers, such as triazine-based materials, demonstrate removal efficiencies between 73% and 88% [18].

Among polymeric materials with promising applications, chitosan [(1,4)-2-amino-2-deoxy-D-glucan] stands out. Chitosan is a polysaccharide derived from chitin, one of the most abundant natural substances after cellulose, and serves as the primary structural component of the exoskeletons of shrimp, lobsters, and crabs [19]. It exhibits excellent biological properties and finds extensive applications in medical and food sectors. Chemically, chitosan is a cationic polysaccharide composed of glucosamine units linked by glycosidic bonds. Its biocompatibility, biodegradability, and antibacterial activity further enhance its versatility [20–22]. Chemical modification of chitosan improves its absorption properties, solubility, porosity, and permeability. Its high nitrogen content, present as amine groups, enables interactions with various compounds through chelation mechanisms, offering bifunctional capabilities for the removal of both cationic and anionic contaminants [23,24].

Chitosan can undergo various chemical modifications, including etherification, carboxylation, crosslinking, alkylation, and the formation of Schiff bases, enhancing its antibacterial properties, hydrophilicity, and water solubility [25]. Due to its versatility, chitosan is widely used in wastewater treatment as an additive or in composites with other materials such as montmorillonite, polyurethanes, zeolites, cellulose, magnetite, cotton, calcium alginate, and alumina, among others.

A variety of chitosan-based materials have been reported for contaminant removal. For instance, chitosan nanocomposites with $MnFe_2O_4$ nanoparticles have shown removal capacities of 20.85 mg $g^{-1}$, while those with $Fe_3O_4$ nanoparticles have achieved 78.11 mg $g^{-1}$ [26]. Photocatalytic removal of sulfamethoxazole has also been reported using chitosan/alginate nanocomposites doped with $Fe_3O_4$/ZnO, with antibiotic degradation rates of 93.31% [27]. The employment of cationic hydrogel particles (CHP) based on chitosan crosslinked with glutaraldehyde for the removal of pharmaceutical antibiotics, specifically amoxicillin and sulfamethoxazole from aqueous solutions is a promising approach. This is due to the greater advantages cationic materials exhibit over other materials. This is a consequence of the tendency of antibiotics to ionize, which increases their interaction with the active sites of the particles. Concurrently, they can interact with polar groups, thereby enhancing the efficacy of contaminant removal through van der Waals interactions. Consequently, CHPs demonstrate considerable potential for the elimination of emerging organic pollutants in water, outperforming other materials.

## 2. Results and Discussion

### 2.1. Synthesis of Cationic Hydrogel Particles (CHPs)

The synthesis of CHPs was conducted in accordance with the methodology delineated in Table 2. The NaOH solution was delivered via a peristaltic pump, and once the CHPs were formed, the crosslinking agent, glutaraldehyde (Glu), was introduced under agitation [28]. The crosslinking capacity of Glu is attributed to the nucleophilic reaction between its aldehyde groups and the free amino groups of chitosan, enhancing the mechanical, thermal, and water resistance properties of the resulting hydrogel [29–31]. The Glu concentration was set at 5%, as determined by FTIR spectra (Figure 2), which showed the availability of active amino groups and a decrease in the intensity of the band associated with these groups as the Glu concentration increased; furthermore, an elevated Glu concentration enhances the band signal of carbonyl and amine groups associated with an excess of Glu. The FTIR spectra of the CHPs displayed the following characteristic bands: 3437 $cm^{-1}$ (N–H and O–H stretching vibrations), 2925 $cm^{-1}$ (symmetric stretching of $CH_3$), 1660–1670 $cm^{-1}$ (C=O stretching vibration), and 1150 $cm^{-1}$ (glycosidic bond) [32].

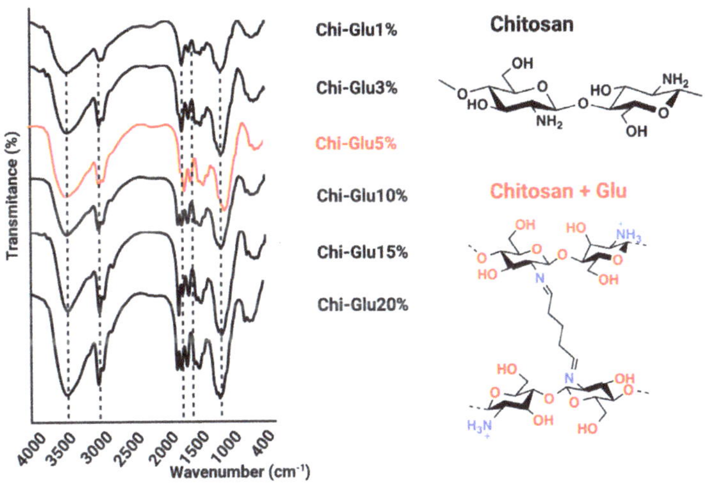


**Figure 2.** FTIR spectra of CHPs.

Between 1560 and 1600 $cm^{-1}$, an increase in intensity is observed, associated with the C=C double bond, attributed to the aldol condensation of the Glu molecule [33–35]. The application of glutaraldehyde-mediated crosslinking resulted in an observable increase in the signals between 1700 and 1740 $cm^{-1}$, which can be attributed to the C=O signal of glutaraldehyde. This phenomenon indicates that as the concentration of glutaraldehyde increases, A greater proportion of these groups remain unreacted, as evidenced by the presence of signals between 1630 and 1690 $cm^{-1}$. These are associated with the imine bonds (C=N) formed by the reaction of the amino groups of chitosan and glutaraldehyde [36]. The morphological characteristics of the CHP are illustrated in Figure 3a. These studies reveal a rough surface with a spherical geometry, which is predominantly populated by particles measuring between 874 and 970 μm for the crosslinked CHPs, while control CHPs (Chi-Control) are predominantly by particles measuring between 600 and 700 μm. The difference can be attributed to the glutaraldehyde employed. The surface roughness of the CHPs provides an increased number of adsorption sites, enhancing the removal of the target compounds [32]. Furthermore, the roughness is closely related to the degree of crosslinking.

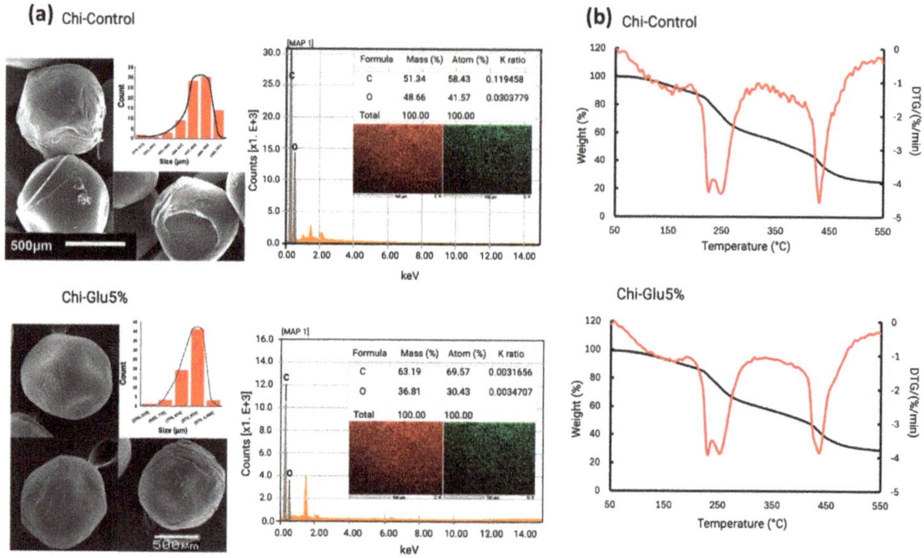


**Figure 3.** (**a**) SEM-EDS microstructure of CHIPs (Chi-Control and Chi-Glu5%) and size distribution; (**b**) thermogravimetric analysis of CHIPs (Chi-Control and Chi-Glu5%).

Figure 3b shows the thermogravimetric analysis (TGA) of the CHPs (Chi-Control and Chi-Glu5%), revealing that the crosslinked CHPs (Chi-Glu5%) exhibit greater thermal stability than the non-crosslinked ones, with two distinct stages of mass loss. The first stage shows a slight mass loss associated with the evaporation of adsorbed and chemically bound water in the structure of the CHPs. It is also noted that non-crosslinked particles experience greater water loss, while the crosslinked CHPs retain less water due to the hydrophobic nature acquired after the crosslinking reaction. The derivative thermogravimetry (DTG) results indicate that the thermal decomposition temperature of the Glu-crosslinked CHPs occurs at a higher temperature, attributed to the formation of a chemically crosslinked network, which improves the material's thermal stability [37,38]. In the second stage, decomposition is observed in both the control and crosslinked CHPs. Although greater decomposition is seen in the control material, and the residual mass is lower compared to the Glu-crosslinked material, no significant difference is observed. The residual percentages of the crosslinked and non-crosslinked beads are found to be similar, which can be attributed to the low concentration of glutaraldehyde employed (5%). The quantity in question is insufficient to affect a notable alteration in the material's structure.

### *2.2. Determination of Water Absorption Capacity*

The absorption and diffusion of water in polymeric materials are influenced by the degree of molecular crosslinking [39]. As shown in Figure 4a, lower concentrations of glutaraldehyde result in higher water absorption. The maximum absorption equilibrium was reached after 8 h, with no significant changes observed beyond this point. This behavior is attributed to the crosslinking process between amino groups and Glu, which reduces the material's interaction with water as the concentration of the crosslinking agent increases [40,41]. This phenomenon likely occurs due to a reduction in the internal cavities of the material, limiting the diffusion of water into the interior and resulting in decreased water absorption.

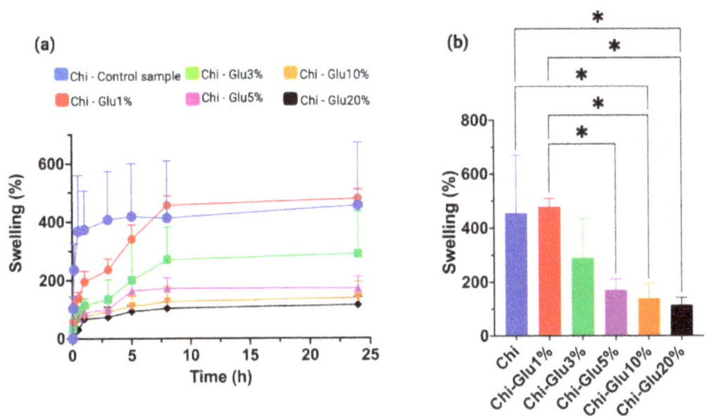


**Figure 4.** (**a**) CHPs crosslinked with glutaraldehyde at different concentrations subjected to a water absorption process over time and (**b**) comparison of water absorption at equilibrium time 8 h. * Significant at $p < 0.01$.

## 2.3. Evaluation of pH Effect on Removal and Point of Zero Charge (PZC)

By definition, the point of zero charge (PZC) is the pH at which the surface charge of the adsorbent is neutralized [42]. For CHPs, the PZC was determined to be 5.7 (Figure 5a), meaning that at this pH, the surface charge of the CHPs is zero, and no electrostatic repulsions occur due to the absence of charged particles. At pH values below the PZC, protonation of surface hydroxyl groups and available amino groups occurs, resulting in a positively charged surface. Conversely, at pH values above the PZC, deprotonation of hydroxyl groups takes place, leading to a negatively charged surface [43].

Figure 5c,d shows the effect of pH on the adsorption capacity of crosslinked CHPs for the antibiotics AMX and SMX, respectively, with a contact time of 6 h at room temperature. The highest adsorption of AMX occurs at pH 6.0, while for SMX, it is at pH 4.0. Considering that the PZC is 5.7, at pH values below this point, the surface of the CHPs becomes positively charged. Additionally, it is important to note that AMX exhibits zwitterionic behavior, with $pKa_1 = 2.25$, $pKa_2 = 7.22$, and $pKa_3 = 9.48$ [7]. Near neutral pH, the adsorption of AMX is favored by electrostatic interactions (Figure 5b) and diffusion processes associated with the material's water absorption capacity [44]. In the case of SMX, the presence of an aromatic amine and sulfonamide groups makes its behavior pH dependent. At specific pH values, SMX exists in a neutral state, and its removal relies on both electrostatic interactions and hydrogen bonding [45,46]. Thus, the highest removal of SMX was achieved at pH 4.0, where the molecule is in its neutral form, suggesting that hydrogen bonding (Figure 5b) and diffusion processes played key roles in the adsorption.

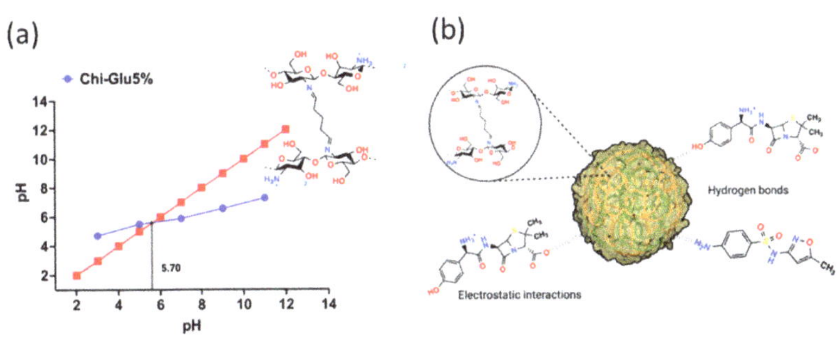


**Figure 5.** *Cont.*

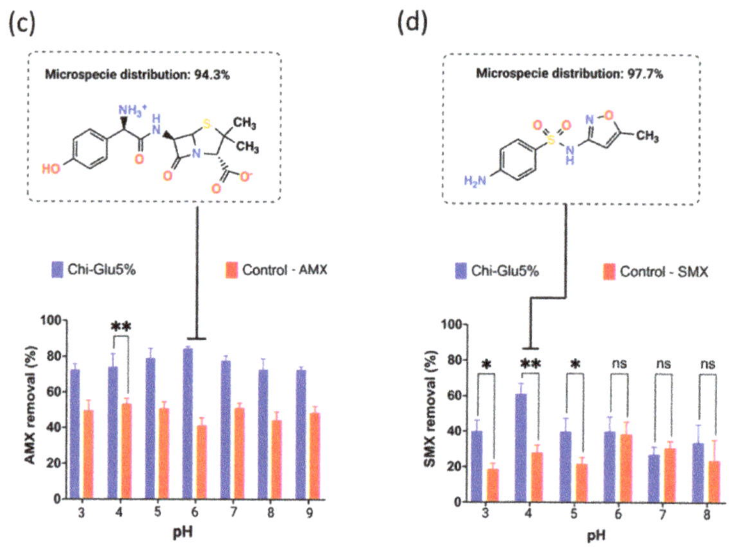


**Figure 5.** (**a**) Identification of the point of zero charge (PZC) of chitosan CHPs; (**b**) adsorption mediated by electrostatic interactions and hydrogen bonding between the crosslinked chitosan CHPs and the study antibiotics; removal of the antibiotics (**c**) amoxicillin and (**d**) sulfamethoxazole at different pH values. ** Significant at $p < 0.05$, * Significant at $p < 0.01$, ns = no significant difference.

*2.4. Evaluation of Ionic Strength in Antibiotic Removal*

In general, the presence of dissolved ions in aqueous solutions is reported to act as interfering agents, potentially reducing removal capacity due to various interactions among different species. Thus, the effect of monovalent and divalent ions on the removal capacity of antibiotics was evaluated [47]. For AMX, a significant difference in removal efficiency was observed. As the NaCl concentration increased, the rate of retention decreased (Figure 6b). Given the dual nature of AMX as a charged molecule at pH 6.0, its behavior in solution is impacted by the presence of $Na^+$ and $Cl^-$ ions. This results in a competitive interaction between the AMX molecule and the active sites of the ECCs, effectively preventing AMX from binding to these sites and, consequently, from exerting any influence on the removal of ions in solution. The rates exceed 80% in the absence of NaCl, falling to 45% when a concentration of 0.025 M NaCl is applied. Further increases in concentration result in a continued decrease in removal, falling to approximately 17% at a concentration of 1.00 M.

In contrast, at pH 4.0, SMX is relatively neutral (Figure 6a), although positively charged species may also be present. Despite being mostly neutral, the interaction between SMX and the CHPs is hindered by chloride ions, which interact with the few positively charged sites on the antibiotic and the protonated amino groups of the CHPs. For both antibiotics, the presence of $Na^+$ and $Cl^-$ ions play a significant role in the removal process, directly affecting electrostatic interactions and hydrogen bonding between the CHPs and the antibiotics AMX and SMX, ultimately reducing removal capacity.

As in the previous case, an increase in salt concentration leads to a decrease in removal capacity. However, the effect of divalent ions is much less pronounced for AMX (Figure 6d), with removal remaining similar despite the increase in $MgSO_4$ concentration. A similar trend is observed for SMX at 0.0025 mol $L^{-1}$ and in the absence of salt (Figure 6c). This effect can be attributed to the interaction of $Mg^{2+}$ ions with the -OH groups present in chitosan. Additionally, $Mg^{2+}$ ions are smaller than antibiotic molecules, allowing them to diffuse into the structure more easily. As with monovalent ions, increasing the salt concentration reduces removal capacity due to the higher availability of competing ions.

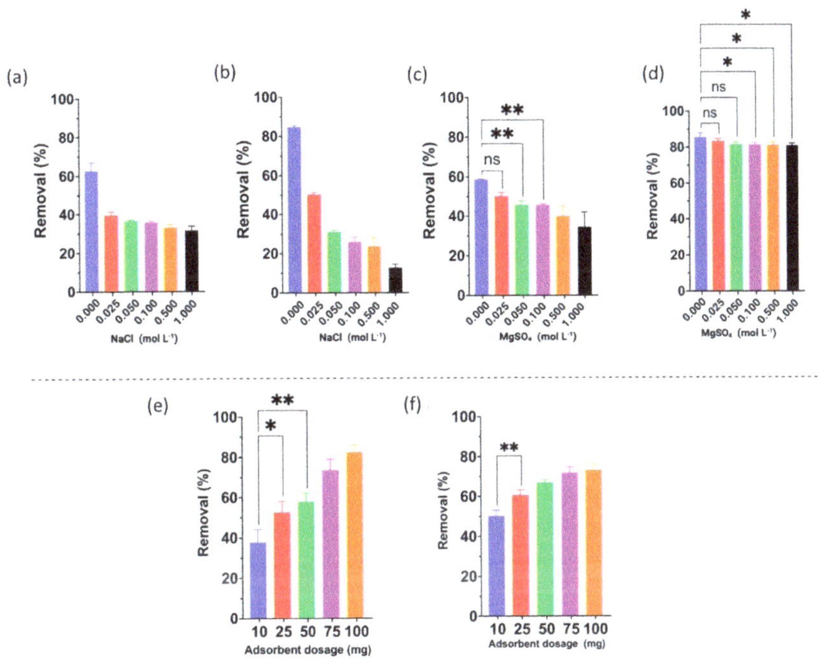


**Figure 6.** Effect of the ionic strength of monovalent and divalent ions on the removal of the antibiotics (**a**,**c**) sulfamethoxazole and (**b**,**d**) amoxicillin; effect of the adsorbent dose on the removal of (**e**) amoxicillin and (**f**) sulfamethoxazole. Those values that are very significant are not expressed graphically, only those of low significance and those that are not significant are detailed. ** Significant at $p < 0.05$, * Significant at $p < 0.01$, ns = no significant difference.

For both antibiotics, the effect of adsorbent dose followed the same trend: removal improved with increasing adsorbent dosage (Figure 6e,f). This behavior is attributed to the higher number of active sites available in the crosslinked CHPs, which enhances the attraction of more antibiotic molecules to the greater quantity of CHPs present in the system. Additionally, it was observed that while higher doses increase removal efficiency, the effect becomes less pronounced at higher adsorbent concentrations due to the saturation of the active adsorption sites [48,49].

### *2.5. Evaluation of Retention Kinetics at Different Temperatures*

Adsorption generally improves with increasing temperature; however, AMX exhibits a reverse effect at 40 °C (Figure 7a) compared to SMX (Figure 7b). SMX shows higher adsorption at elevated temperatures, as the chains of the crosslinked CHPs become more relaxed. Additionally, the smaller molecular size of SMX allows it to penetrate more easily into the structure of the crosslinked CHPs, facilitating stronger interactions [50].

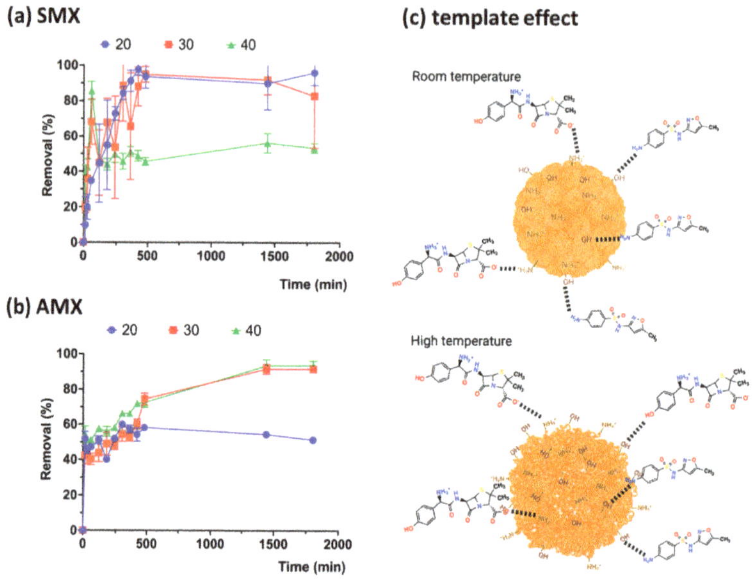


**Figure 7.** Absorption kinetics of the antibiotics amoxicillin (**a**) and sulfamethoxazole (**b**) by Chi-Glu CHPs as a function of temperature, (**c**) template effect.

*2.6. Effect of Initial Antibiotic Concentration at Different Temperatures*

The antibiotics were allowed to interact with the crosslinked CHPs (Chi-Glu5%) for the maximum adsorption time determined in the previous section. The effect varied between the two antibiotics: SMX showed higher adsorption at elevated temperatures, whereas AMX reached similar adsorption percentages around 80% across all temperatures. The removal capacities recorded were 11.65, 11.67, and 11.49 mg $g^{-1}$ of CHPs at 20, 30, and 40 °C, respectively. These values remained consistent regardless of temperature, indicating that the effect is primarily influenced by concentration. Similar values have been reported for cellulose-derived beads (10.8 mg $g^{-1}$) [51], chitosan/biochar beads (7.64 mg $g^{-1}$) [52], and silica nanostructures (24.15 mg $g^{-1}$), as well as other results shown in Table S1, which are directly related to removing AMX and SMX [53]. In each case, adsorption capacity increased with rising temperature (Figure 8a,b), albeit not significantly. This increase can be attributed to the enhanced molecular mobility in solution and the greater flexibility of the polymer chains in the crosslinked CHPs, resulting in an increased number of active adsorption sites [54,55].

Thermodynamic parameters, such as the standard Gibbs free energy change ($\Delta G^\circ$), standard enthalpy change ($\Delta H^\circ$), and standard entropy change ($\Delta S^\circ$), provide insights into the adsorption mechanism, distinguishing between physisorption and chemisorption [56]. This evaluation was performed using Freundlich isotherm models, yielding $\Delta H^\circ$ values of 28.38 kJ $mol^{-1}$, 12.39 kJ $mol^{-1}$, and $\Delta S^\circ$ 97.19 J $mol^{-1}$ $K^{-1}$, and 33.94 J $mol^{-1}$ $K^{-1}$ for AMX and SMX, respectively (Figures S1 and S2). The results are summarized in Table 1. For AMX, the negative $\Delta G^\circ$ values at all temperatures indicate that the adsorption process is spontaneous. Additionally, the decrease in $\Delta G^\circ$ with increasing temperature suggests that higher temperatures facilitate adsorption. This phenomenon occurs because the enhanced molecular mobility at elevated temperatures promotes interactions between the CHPs and the antibiotic molecules. The positive $\Delta H^\circ$ values indicate that the adsorption of AMX onto CHPs is an endothermic process. Furthermore, the adsorption mechanism is identified as physisorption, as chemisorption generally predominates only when $\Delta H^\circ$ values exceed 30 kJ $mol^{-1}$ [57].

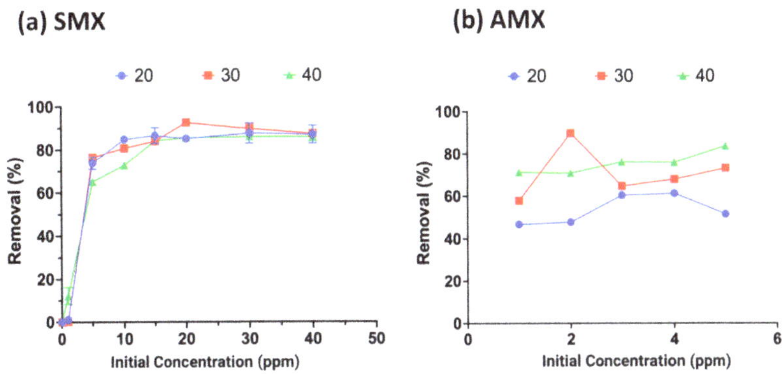


**Figure 8.** Effect of concentration variation on the adsorption of antibiotics amoxicillin (**a**) and sulfamethoxazole (**b**) by Chi-Glu at different temperatures.

**Table 1.** Gibbs energy values for antibiotics AMX and SMX at different temperatures.

| | | $\Delta G$ (KJ mol$^{-1}$) | |
|---|---|---|---|
| Temperature (K) | 293.15 | 303.15 | 313.15 |
| AMX | −0.113 | −1.085 | −2.057 |
| SMX | 2.447 | 2.107 | 1.768 |

For AMX, the positive values indicate the irreversibility and stability of the adsorption process [58], as the antibiotic binds strongly to the active sites of the CHPs. For sulfamethoxazole, both $\Delta H^\circ$ and $\Delta S^\circ$ values are positive, indicating that, similar to AMX, the adsorption process is endothermic, irreversible, and involves strong binding to the active sites of the CHPs, though to a lesser extent than with AMX. Although these values are positive, they decrease with increasing temperature, suggesting that, as with AMX, higher temperatures enhance adsorption. However, unlike AMX, the adsorption process for sulfamethoxazole is not spontaneous.

## 3. Conclusions

Cationic hydrogel particles (CHPs) are promising adsorbent materials for the treatment of water contaminated with emerging pollutants. The CHPs were characterized using spectroscopic and thermal techniques, confirming the presence of key functional groups. Adsorption of the studied antibiotics occurred at pH 6.0 for AMX and pH 4.0 for SMX, with a contact time of 6 h. While the adsorbent dosage, set at 30 mg, allowed a removal efficiency of 90% for AMX and 60% for SMX, primarily through adsorption processes, with physisorption being the dominant mechanism between the CHPs and the antibiotics. The interactions were affected by the increase in ion concentrations (both monovalent and divalent), with higher ion concentrations resulting in decreased antibiotic adsorption, so that the analysis of the removal was favored in the absence of ionic compounds. The most significant impact was observed for amoxicillin, with a reduction of over 60% in removal efficiency under ion interference. Current and future perspectives highlight the urgent need for improved technologies to treat and remediate water bodies contaminated with antibiotics. In this context, CHPs present a viable alternative for various processes involved in wastewater treatment and a preliminary step for the degradation of these antibiotics.

## 4. Materials and Methods

### *4.1. Reagents*

Chitosan (Chi) with low molecular weight, 75–85% deacetylated, was obtained from Sigma-Aldrich; glacial acetic acid for analysis (HAc); sodium hydroxide (NaOH) for analysis; glutaraldehyde (50% in water) for synthesis (Glu); absolute ethanol EMSURE (EtOH); amoxicillin (AMX) and sulfamethoxazole (SMX) of analytical-grade standard (Titripur,); 0.1N HCl standard and 0.1N NaOH standard (Titripur). All reagents were purchased from Sigma-Aldrich Chile, Santiago, Chile.

### 4.2. Synthesis and Optimization of Cationic Hydrogel Particles (CHPs)

Chitosan is dissolved in 5% (*w*/*v*) acetic acid, and the viscous solution is stirred until complete dissolution. The solution is then added dropwise into NaOH solutions at 20% and 25%, using a peristaltic pump. CHPs form upon contact between the viscous mixture and the alkaline solution. The mixture is stirred for 30 min, and the CHPs are washed until reaching a neutral pH (Figure 9) [59]. The obtained CHPs are exposed to glutaraldehyde at different concentrations (Table 2) to determine which formulation yields the best results for subsequent analyses.

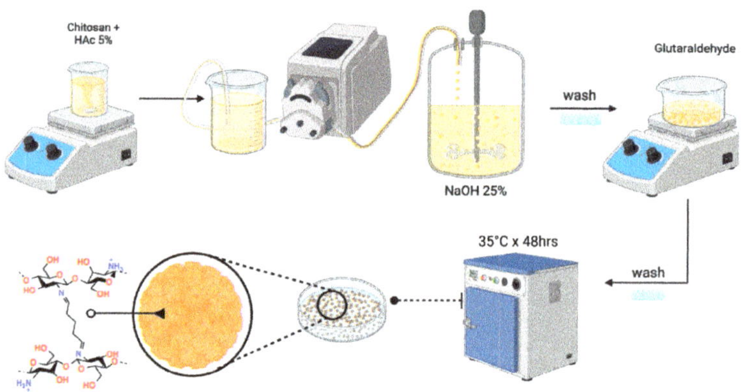


**Figure 9.** Process for obtaining cationic hydrogel particles.

**Table 2.** Variables for obtaining CHPs.

| NaOH (%) | Glutaraldehyde (%) |
|---|---|
| 25 | 1 |
| | 3 |
| | 5 |
| | 10 |
| | 20 |

### 4.3. Characterization

Fourier Transform Infrared Spectroscopy (FTIR): Spectra were obtained to examine the presence of characteristic functional groups of chitosan and its modifications. The samples were recorded in the frequency range of 400 to 4000 $cm^{-1}$ using a Nicolet spectrometer equipped with a DTGS-KBr detector.

Scanning Electron Microscopy (SEM): The surface characteristics of the CHPs were studied using a JEOL-SEM-PROBE CAMECA SU-30 microscope equipped with an EDS detector.

Thermogravimetric Analysis (TGA): Thermogravimetric spectra of the samples were recorded using a NETZSCH 209 F1 Iris thermogravimetric analyzer. Measurements were conducted from room temperature to 550 °C, with a heating rate of 10 °C $min^{-1}$ under a nitrogen atmosphere.

Water Absorption Capacity: A sample of known mass was used to assess water absorption capacity, evaluating the effect over time from 0 to 24 h under agitation.

Point of Zero Charge (PZC) Determination: Fifty milliliters of distilled water was adjusted to pH values between 3 and 11. To each solution, 0.5 g of the adsorbent material was added, and the mixtures were stirred for 48 h at room temperature. The final pH was measured, and the PZC was determined as the point where the final pH curve intersects the diagonal line representing the initial pH.

### 4.4. Removal Studies

Antibiotic solutions were prepared with concentrations of 20 mg $L^{-1}$ for AMX and 5 mg $L^{-1}$ for SMX at pH values of 3, 4, 5, 6, 7, and 8. Each solution was mixed with 30 mg of the adsorbent sample and agitated for 6 h. Afterward, the solutions were measured using a Thermo Fisher Evolution One Plus UV–vis spectrophotometer. Calibration curves (Figure 10a,b) were obtained for each antibiotic at different pH values to determine the removal percentages.

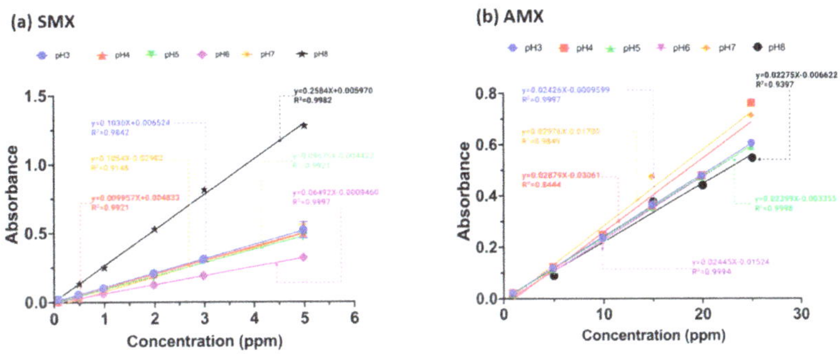


**Figure 10.** Calibration curves of the antibiotics (**a**) sulfamethoxazole and (**b**) amoxicillin at pH values 3.0–8.0.

With the pH value determined in the previous section, antibiotic solutions were prepared at different ionic strengths (0.025, 0.050, 0.1, 0.5, and 1 M) using $Na^{+}$, $Mg^{2+}$ cations, and $Cl^{-}$, $SO_4^{2-}$ anions. The solutions were in contact with the CHPs for 6 h in a horizontal shaker and subsequently analyzed using a UV–vis spectrophotometer. For the effect of CHP dosage, 0.005, 0.025, 0.05, 0.075, and 0.1 g of material were used under the optimal experimental conditions described previously.

The effect of temperature and time on the removal process was evaluated at temperatures of 25, 30, 40, and 50 °C and at time intervals of 0, 15, 30, 60, 120, 180, 240, 300, 360, 420, 480, 1440, and 1800 min, following the previously established experimental conditions. To assess the effect of antibiotic concentration at different temperatures, optimal experimental conditions (pH, ionic strength, adsorbent dosage, and contact time) were employed. Thermodynamic parameters ($\Delta H^{\circ}$, $\Delta G^{\circ}$, and $\Delta S^{\circ}$) were also evaluated based on these conditions.

**Supplementary Materials:** The following supporting information can be downloaded at: https://www.mdpi.com/article/10.3390/gels10120760/s1, Figure S1. Determination of thermodynamic parameters using the Freundlich model for the antibiotic amoxicillin, Figure S2. Determination of thermodynamic parameters using the Freundlich model for the antibiotic Sulfamethoxazole, Table S1. Comparison of the adsorption capacity of the various adsorbents from the literature and with this study [60–70].

**Author Contributions:** F.L.A. investigation, formal analysis, and writing—review and editing; M.F.M., M.A.P., B.L.R. and E.D.P. validation, writing—review and editing; D.A.P. conceptualization, validation, formal analysis, resources, supervision, project administration, funding acquisition, and writing—review and. All authors have read and agreed to the published version of the manuscript.

**Funding:** This research was funded by FONDEF IDEA 2022, grant number ID22I10092.

**Institutional Review Board Statement:** Not applicable.

**Informed Consent Statement:** Not applicable.

**Data Availability Statement:** Data are contained within the article.

**Acknowledgments:** The authors acknowledge the financial support from FONDEF Grant N° ID22I10092 and Fondecyt Postdoctoral Grant N° 3220108 of the Chilean Agency for Research and Development -Chile, for its Spanish acronym ANID. Francisca L. Aranda thanks the postgraduate Direction, University of Concepcion for the Doctoral Scholarship.

**Conflicts of Interest:** The authors declare no conflicts of interest. The funders had no role in the design of the study; in the collection, analysis, and interpretation of data; in the writing of the manuscript; or in the decision to publish the results.

## References

1. Fida, M.; Li, P.; Wang, Y.; Alam, S.M.K.; Nsabimana, A. Water contamination and human health risks in Pakistan: A review. *Expo. Health* **2023**, *15*, 619–639. [CrossRef]
2. Shahra, E.Q.; Wu, W. Water contaminants detection using sensor placement approach in smart water networks. *J. Ambient. Intell. Humaniz. Comput.* **2023**, *14*, 4971–4986. [CrossRef]
3. Akowanou, A.V.O.; Deguenon, H.E.J.; Balogoun, K.C.; Daouda, M.M.A.; Aina, M.P. The combined effect of three floating macrophytes in domestic wastewater treatment. *Sci. Afr.* **2023**, *20*, e01630. [CrossRef]
4. Saravanan, A.; Kumar, P.S.; Duc, P.A.; Rangasamy, G. Strategies for microbial bioremediation of environmental pollutants from industrial wastewater: A sustainable approach. *Chemosphere* **2023**, *313*, 137323. [CrossRef] [PubMed]
5. Ashiq, A.; Walpita, J.; Vithanage, M. Functionalizing non-smectic clay via methoxy-modification for enhanced removal and recovery of oxytetracycline from aqueous media. *Chemosphere* **2021**, *276*, 130079. [CrossRef]
6. Palacio, D.A.; Becerra, Y.; Urbano, B.F.; Rivas, B.L. Antibiotics removal using a chitosan-based polyelectrolyte in conjunction with ultrafiltration membranes. *Chemosphere* **2020**, *258*, 127416. [CrossRef]
7. Aranda, F.L.; Rivas, B.L. Removal of amoxicillin through different methods, emphasizing removal by biopolymers and its derivatives. An overview. *J. Chil. Chem. Soc.* **2022**, *67*, 5643–5655. [CrossRef]
8. Hirte, K.; Seiwert, B.; Schüürmann, G.; Reemtsma, T. New hydrolysis products of the beta-lactam antibiotic amoxicillin, their pH-dependent formation and search in municipal wastewater. *Water Res.* **2016**, *88*, 880–888. [CrossRef]
9. Foti, C.; Giuffrè, O. Interaction of Ampicillin and Amoxicillin with $Mn^{2+}$: A Speciation Study in Aqueous Solution. *Molecules* **2020**, *25*, 3110. [CrossRef]
10. Ying, G.-G.; He, L.-Y.; Ying, A.J.; Zhang, Q.-Q.; Liu, Y.-S.; Zhao, J.-L. China must reduce its antibiotic use. *Environ. Sci. Technol.* **2017**, *51*, 1072–1073. [CrossRef]
11. Teixeira, S.; Delerue-Matos, C.; Santos, L. Application of experimental design methodology to optimize antibiotics removal by walnut shell based activated carbon. *Sci. Total Environ.* **2019**, *646*, 168–176. [CrossRef]
12. El Hani, O.; García-Guzmán, J.J.; Palacios-Santander, J.M.; Digua, K.; Amine, A.; Cubillana-Aguilera, L. Development of a molecularly imprinted membrane for selective, high-sensitive, and on-site detection of antibiotics in waters and drugs: Application for sulfamethoxazole. *Chemosphere* **2024**, *350*, 141039. [CrossRef] [PubMed]
13. Palacio, D.A.; Aranda, F.L.; Rivas, B.L. Removal of antibiotic emerging pollutants: An overview. *J. Chil. Chem. Soc.* **2022**, *67*, 5547–5561. [CrossRef]
14. Wu, J.; Fang, X.; Zhu, Y.; Ma, N.; Dai, W. Well-designed $TiO_2$@ UiO-66-$NH_2$ nanocomposite with superior photocatalytic activity for tetracycline under restricted space. *Energy Fuels* **2020**, *34*, 12911–12917. [CrossRef]
15. Gao, Y.; Wu, J.; Wang, J.; Fan, Y.; Zhang, S.; Dai, W. A novel multifunctional p-type semiconductor@ MOFs nanoporous platform for simultaneous sensing and photodegradation of tetracycline. *ACS Appl. Mater. Interfaces* **2020**, *12*, 11036–11044. [CrossRef] [PubMed]
16. Fang, Y.; Zhu, S.-R.; Wu, M.-K.; Zhao, W.-N.; Han, L. MOF-derived $In_2S_3$ nanorods for photocatalytic removal of dye and antibiotics. *J. Solid State Chem.* **2018**, *266*, 205–209. [CrossRef]
17. Fakhri, H.; Bagheri, H. Highly efficient Zr-MOF@ $WO_3$/graphene oxide photocatalyst: Synthesis, characterization and photodegradation of tetracycline and malathion. *Mater. Sci. Semicond. Process.* **2020**, *107*, 104815. [CrossRef]
18. Matias, P.M.; Nunes, S.C.; Rodrigues, A.C.B.; Ltayef, M.; Sellaoui, L. Efficient removal of sulfonamide and tetracycline antibiotics using triazine-based porous organic polymers. *Sep. Purif. Technol.* **2024**, *355*, 129731. [CrossRef]
19. Lessa, E.F.; Gerhardt, R.; Arabidian, V.; Junior, T.R.S.A.C.; de Almeida Pinto, L.A. Synthesis of chitosan/carbon nanotubes composite films as potential removal of anionic and cationic dyes in aqueous solutions. *Res. Sq.* **2024**. [CrossRef]
20. Alyasi, H.; Mackey, H.; McKay, G. Adsorption of Methyl Orange from Water Using Chitosan Bead-like Materials. *Molecules* **2023**, *28*, 6561. [CrossRef]
21. Alsamman, M.T.; Sánchez, J. Chitosan-and alginate-based hydrogels for the adsorption of anionic and cationic dyes from water. *Polymers* **2022**, *14*, 1498. [CrossRef] [PubMed]

22. Ngah, W.W.; Endud, C.; Mayanar, R. Removal of copper(II) ions from aqueous solution onto chitosan and cross-linked chitosan beads. *React. Funct. Polym.* **2002**, *50*, 181–190. [CrossRef]
23. Rasool, A.; Rizwan, M.; Islam, A.; Abdullah, H.; Shafqat, S.S.; Azeem, M.K.; Rasheed, T.; Bilal, M. Chitosan-Based Smart Polymeric Hydrogels and Their Prospective Applications in Biomedicine. *Starch-Stärke* **2024**, *76*, 2100150. [CrossRef]
24. Zhou, Q.; Lan, W.; Xie, J. Phenolic acid-chitosan derivatives: An effective strategy to cope with food preservation problems. *Int. J. Biol. Macromol.* **2024**, *254*, 127917. [CrossRef] [PubMed]
25. Aranaz, I.; Alcántara, A.R.; Civera, M.C.; Arias, C.; Elorza, B.; Caballero, A.H.; Acosta, N. Chitosan: An Overview of Its Properties and Applications. *Polymers* **2021**, *13*, 3256. [CrossRef]
26. Ahamad, T.; Ruksana; Chaudhary, A.A.; Naushad, M.; Alshehri, S.M. Fabrication of MnFe2O4 nanoparticles embedded chitosan-diphenylureaformaldehyde resin for the removal of tetracycline from aqueous solution. *Int. J. Biol. Macromol.* **2019**, *134*, 180–188. [CrossRef]
27. Roy, N.; Kannabiran, K.; Mukherjee, A. Studies on photocatalytic removal of antibiotics, ciprofloxacin and sulfamethoxazole, by $Fe_3O_4$-ZnO-Chitosan/Alginate nanocomposite in aqueous systems. *Adv. Powder Technol.* **2022**, *33*, 103691. [CrossRef]
28. Guan, L.; Yu, W.; Asghar, M.R.; Zhang, W.; Su, H.; Li, C.; Xing, L.; Xu, Q. Effect of graphene aerogel as a catalyst layer additive on performance of direct methanol fuel cell. *Fuel* **2024**, *360*, 130503. [CrossRef]
29. Guisán, J. Aldehyde-agarose gels as activated supports for immobilization-stabilization of enzymes. *Enzym. Microb. Technol.* **1988**, *10*, 375–382. [CrossRef]
30. Farris, S.; Song, J.; Huang, Q. Alternative Reaction Mechanism for the Cross-Linking of Gelatin with Glutaraldehyde. *J. Agric. Food Chem.* **2010**, *58*, 998–1003. [CrossRef]
31. Migneault, I.; Dartiguenave, C.; Bertrand, M.J.; Waldron, K.C. Glutaraldehyde: Behavior in aqueous solution, reaction with proteins, and application to enzyme crosslinking. *BioTechniques* **2004**, *37*, 790–802. [CrossRef] [PubMed]
32. Billah, R.E.K.; Islam, M.A.; Nazal, M.K.; Bahsis, L.; Soufiane, A.; Abdellaoui, Y.; Achak, M. A novel glutaraldehyde cross-linked chitosan@acid-activated bentonite composite for effectivePb (II) and Cr (VI) adsorption: Experimental and theoretical studies. *Sep. Purif. Technol.* **2024**, *334*, 126094. [CrossRef]
33. Li, T.-T.; Liu, Y.-G.; Peng, Q.-Q.; Hu, X.-J.; Liao, T.; Wang, H.; Lu, M. Removal of lead (II) from aqueous solution with ethylenediamine-modified yeast biomass coated with magnetic chitosan microparticles: Kinetic and equilibrium modeling. *Chem. Eng. J.* **2013**, *214*, 189–197. [CrossRef]
34. Monteiro Jr, O.A.; Airoldi, C. Some studies of crosslinking chitosan–glutaraldehyde interaction in a homogeneous system. *Int. J. Biol. Macromol.* **1999**, *26*, 119–128. [CrossRef] [PubMed]
35. Worzakowska, M. TG/DSC/FTIR/QMS analysis of environmentally friendly poly (citronellyl methacrylate)-co-poly (benzyl methacrylate) copolymers. *J. Mater. Sci.* **2023**, *58*, 2005–2024. [CrossRef]
36. Wahba, M.I. Glutaraldehyde-copper gelled chitosan beads: Characterization and utilization as covalent immobilizers. *Biocatal. Agric. Biotechnol.* **2023**, *50*, 102668. [CrossRef]
37. Saruchi; Kumar, V. Separation of crude oil from water using chitosan based hydrogel. *Cellulose* **2019**, *26*, 6229–6239. [CrossRef]
38. Sahu, S.; Bishoyi, N.; Patel, R.K. Cerium phosphate polypyrrole flower like nanocomposite: A recyclable adsorbent for removal of Cr (VI) by adsorption combined with in-situ chemical reduction. *J. Ind. Eng. Chem.* **2021**, *99*, 55–67. [CrossRef]
39. Abdelkader, A.F.; White, J.R. Water absorption in epoxy resins: The effects of the crosslinking agent and curing temperature. *J. Appl. Polym. Sci.* **2005**, *98*, 2544–2549. [CrossRef]
40. Martinez, L.; Agnely, F.; Leclerc, B.; Siepmann, J.; Cotte, M.; Geiger, S.; Couarraze, G. Cross-linking of chitosan and chitosan/poly (ethylene oxide) beads: A theoretical treatment. *Eur. J. Pharm. Biopharm.* **2007**, *67*, 339–348. [CrossRef]
41. Ellis, T.; Karasz, F. Karasz, Interaction of epoxy resins with water: The depression of glass transition temperature. *Polymer* **1984**, *25*, 664–669. [CrossRef]
42. Kamari, K.; Taheri, A. Preparation and evaluation of magnetic core–shell mesoporous molecularly imprinted polymers for selective adsorption of amitriptyline in biological samples. *J. Taiwan Inst. Chem. Eng.* **2018**, *86*, 230–239. [CrossRef]
43. Bée, A.; Obeid, L.; Mbolantenaina, R.; Welschbillig, M.; Talbot, D. Magnetic chitosan/clay beads: A magsorbent for the removal of cationic dye from water. *J. Magn. Magn. Mater.* **2017**, *421*, 59–64. [CrossRef]
44. Adriano, W.; Veredas, V.; Santana, C.; Gonçalves, L. Adsorption of amoxicillin on chitosan beads: Kinetics, equilibrium and validation of finite bath models. *Biochem. Eng. J.* **2005**, *27*, 132–137. [CrossRef]
45. Gong, Y.; Liu, L.; Wang, F.; Pei, Y.; Liu, S.; Lyu, R.; Luo, X. Aminated chitosan/cellulose nanocomposite microspheres designed for efficient removal of low-concentration sulfamethoxazole from water. *J. Mol. Liq.* **2021**, *339*, 116407. [CrossRef]
46. Yu, K.; Ahmed, I.; Won, D.-I.; Lee, W.I.; Ahn, W.-S. Highly efficient adsorptive removal of sulfamethoxazole from aqueous solutions by porphyrinic MOF-525 and MOF-545. *Chemosphere* **2020**, *250*, 126133. [CrossRef]
47. da Silva, P.M.M.; Camparotto, N.G.; de Figueiredo Neves, T.; Lira, K.T.G.; Mastelaro, V.R.; Picone, C.S.F.; Prediger, P. Effective removal of basic dye onto sustainable chitosan beads: Batch and fixed-bed column adsorption, beads stability and mechanism. *Sustain. Chem. Pharm.* **2020**, *18*, 100348. [CrossRef]
48. Dutta, P.K.; Ravikumar, M.N.V.; Dutta, J. Chitin and chitosan for versatile applications. *J. Macromol. Sci. Part C Polym. Rev.* **2002**, *42*, 307–354. [CrossRef]

49. Ahsan, M.A.; Islam, M.T.; Hernandez, C.; Castro, E.; Katla, S.K.; Kim, H.; Lin, Y.; Curry, M.L.; Gardea-Torresdey, J.; Noveron, J.C. Biomass conversion of saw dust to a functionalized carbonaceous materials for the removal of Tetracycline, Sulfamethoxazole and Bisphenol A from water. *J. Environ. Chem. Eng.* **2018**, *6*, 4329–4338. [CrossRef]
50. Vakili, M.T.; Rafatullah, M.; Ibrahim, M.H.; Abdullah, A.Z.; Salamatinia, B.; Gholami, Z. Preparation of chitosan beads for the adsorption of reactive blue 4 from aqueous solutions. *Iran. J. Energy Environ.* **2016**, *7*, 124–128.
51. Wang, L.; Yang, C.; Lu, A.; Liu, S.; Pei, Y.; Luo, X. An easy and unique design strategy for insoluble humic acid/cellulose nanocomposite beads with highly enhanced adsorption performance of low concentration ciprofloxacin in water. *Bioresour. Technol.* **2020**, *302*, 122812. [CrossRef] [PubMed]
52. Afzal, M.Z.; Sun, X.-F.; Liu, J.; Song, C.; Wang, S.-G.; Javed, A. Enhancement of ciprofloxacin sorption on chitosan/biochar hydrogel beads. *Sci. Total Environ.* **2018**, *639*, 560–569. [CrossRef] [PubMed]
53. Nassar, M.Y.; Ahmed, I.S.; Raya, M.A. A facile and tunable approach for synthesis of pure silica nanostructures from rice husk for the removal of ciprofloxacin drug from polluted aqueous solutions. *J. Mol. Liq.* **2019**, *282*, 251–263. [CrossRef]
54. Balarak, D.; Khatibi, A.D.; Chandrika, K. Antibiotics Removal from Aqueous Solution and Pharmaceutical Wastewater by Adsorption Process: A Review. *Int. J. Pharm. Investig.* **2020**, *10*, 106–111. [CrossRef]
55. Li, X.; Chen, S.; Fan, X.; Quan, X.; Tan, F.; Zhang, Y.; Gao, J. Adsorption of ciprofloxacin, bisphenol and 2-chlorophenol on electrospun carbon nanofibers: In comparison with powder activated carbon. *J. Colloid Interface Sci.* **2015**, *447*, 120–127. [CrossRef]
56. Lima, E.C.; Hosseini-Bandegharaei, A.; Moreno-Piraján, J.C.; Anastopoulos, I. A critical review of the estimation of the thermodynamic parameters on adsorption equilibria. Wrong use of equilibrium constant in the Van't Hoof equation for calculation of thermodynamic parameters of adsorption. *J. Mol. Liq.* **2019**, *273*, 425–434. [CrossRef]
57. Wang, B.; Mo, Q.; Qin, B.; Song, L.; Li, J.; Sheng, G.; Shi, D.; Xu, X.; Hou, L. Adsorption behaviors of three antibiotics in single and co-existing aqueous solutions using mesoporous carbon. *Environ. Res.* **2022**, *215*, 114375. [CrossRef]
58. Sahmoune, M.N. Evaluation of thermodynamic parameters for adsorption of heavy metals by green adsorbents. *Environ. Chem. Lett.* **2019**, *17*, 697–704. [CrossRef]
59. Marques Neto, J.D.O.; Bellato, C.R.; Milagres, J.L.; Pessoa, K.D.; Alvarenga, E.S.D. Preparation and evaluation of chitosan beads immobilized with Iron (III) for the removal of As (III) and As (V) from water. *J. Braz. Chem. Soc.* **2013**, *24*, 121–132. [CrossRef]
60. Yeo, J.Y.J.; Khaerudini, D.S.; Soetaredjo, F.E.; Waworuntu, G.L.; Ismadji, S.; Putranto, A.; Sunarso, J. Experimental and modelling study of adsorption isotherms of amoxicillin, ampicillin and doripenem on bentonite-chitosan composite. *S. Afr. J. Chem. Eng.* **2023**, *43*, 38–45. [CrossRef]
61. Putra, E.K.; Pranowo, R.; Sunarso, J.; Indraswati, N.; Ismadji, S. Performance of activated carbon and bentonite for adsorption of amoxicillin from wastewater: Mechanisms, isotherms and kinetics. *Water Res.* **2009**, *43*, 2419–2430. [CrossRef] [PubMed]
62. Danalıoğlu, S.T.; Bayazit, Ş.S.; Kerkez Kuyumcu, Ö.; Salam, M.A. Efficient removal of antibiotics by a novel magnetic adsorbent: Magnetic activated carbon/chitosan (MACC) nanocomposite. *J. Mol. Liq.* **2017**, *240*, 589–596. [CrossRef]
63. Mirizadeh, S.; Solisio, C.; Converti, A.; Casazza, A.A. Efficient removal of tetracycline, ciprofloxacin, and amoxicillin by novel magnetic chitosan/microalgae biocomposites. *Sep. Purif. Technol.* **2024**, *329*, 125115. [CrossRef]
64. Pinheiro, C.P.; Tokura, B.K.; Germano, N.S.; de Moraes, M.A.; Bresolin, I.T.L. Adsorption of amoxicillin by chitosan and alginate biopolymers composite beads. *Environ. Sci. Pollut. Res.* **2024**, 1–20. [CrossRef] [PubMed]
65. Zhou, A.; Yang, K.; Wu, X.; Liu, G.; Zhang, T.C.; Wang, Q.; Luo, F. Functionally-Designed Chitosan-based hydrogel beads for adsorption of sulfamethoxazole with light regeneration. *Sep. Purif. Technol.* **2022**, *293*, 120973. [CrossRef]
66. Khumalo, S.M.; Bakare, B.F.; Rathilal, S. Single and multicomponent adsorption of amoxicillin, ciprofloxacin, and sulfamethoxazole on chitosan-carbon nanotubes hydrogel beads from aqueous solutions: Kinetics, isotherms, and thermodynamic parameters. *J. Hazard. Mater. Adv.* **2024**, *13*, 100404. [CrossRef]
67. Li, Y.; Wang, B.; Shang, H.; Cao, Y.; Yang, C.; Hu, W.; Feng, Y.; Yu, Y. Influence of adsorption sites of biochar on its adsorption performance for sulfamethoxazole. *Chemosphere* **2023**, *326*, 138408. [CrossRef]
68. El Messaoudi, N.; El Mouden, A.; Fernine, Y.; El Khomri, M.; Bouich, A.; Faska, N.; Ciğeroğlu, Z.; Américo-Pinheiro, J.H.P.; Jada, A.; Lacherai, A. Green synthesis of Ag2O nanoparticles using Punica granatum leaf extract for sulfamethoxazole antibiotic adsorption: Characterization, experimental study, modeling, and DFT calculation. *Environ. Sci. Pollut. Res.* **2023**, *30*, 81352–81369. [CrossRef]
69. Mashile, P.P. *Biopolymer-Based Nanocomposite as Recyclable Adsorbents for Removal of Pollutants in Wastewater Treatment*; University of Johannesburg: Johannesburg, South Africa, 2023.
70. Son Tran, V.; Hao Ngo, H.; Guo, W.; Ha Nguyen, T.; Mai Ly Luong, T.; Huan Nguyen, X.; Lan Anh Phan, T.; Trong Le, V.; Phuong Nguyen, M.; Khai Nguyen, M. New chitosan-biochar composite derived from agricultural waste for removing sulfamethoxazole antibiotics in water. *Bioresour. Technol.* **2023**, *385*, 129384. [CrossRef]

*gels*

Article

# Development of High-Efficiency Fertilizer by Hydrogels Obtained from Cassava Starch and Citric Acid for Slow Release of Ammonium and Potassium

**Andrés F. Chamorro [1,*], Manuel Palencia [2,*] and Álvaro A. Arrieta [3]**

1 Research Group of Electrochemistry and Environment (GIEMA), Faculty of Basic Sciences, Universidad Santiago de Cali, Cali 760035, Colombia
2 Research Group in Science with Technological Applications (GICAT), Department of Chemistry, Faculty of Natural and Exact Science, Universidad del Valle, Cali 760032, Colombia
3 Department of Biology and Chemistry, Faculty of Education and Sciences, Universidad de Sucre, Sincelejo 700003, Colombia; alvaro.arrieta@unisucre.edu.co
* Correspondence: andres.chamorro03@usc.edu.co (A.F.C.); manuel.palencia@correounivalle.edu.co (M.P.); Tel.: +57-3105012327 (A.F.C.); +57-3205271934 (M.P.)

**Abstract:** Fertilizers with enhanced efficiency or high-efficiency fertilizers increase the nutrient availability, minimize losses, and reduce costs, thereby increasing crop yields and food production while mitigating environmental impacts. This research evaluates the synthesis of biodegradable hydrogels from cassava starch and citric acid for agrochemical applications. Hydrogels were synthesized using water as the solvent and applied for the controlled release of macronutrients (N and K). Four concentrations of nutrient-containing salts were tested (0.5 to 10.0% w/w). Materials were analyzed using ATR-FTIR spectroscopy and swelling studies. The presence of nutrients reduced both the crosslinking efficacy and the water absorption capacity, with the latter dropping from 183.4 $\pm$ 0.6% to 117.9 $\pm$ 3.7% and 157.4 $\pm$ 25.0% for hydrogels loaded with $NH_4Cl$ and KCl, respectively. The cumulative release of K and N from the hydrogel was monitored for 144 h and examined using kinetics models, revealing that the releases follow Fickian's diffusion and anomalous diffusion, respectively. Additionally, the material was formed using cassava with peel previously milled to reduce the production costs, and its potential for nutrient-controlled delivery was evaluated, with the finding that this hydrogel decreases the release rate of nitrogen. The results suggest that these biomaterials may have promising applications in the agrochemical industry in the making of high-efficiency fertilizers.

**Keywords:** cassava starch; hydrogel; high-efficiency fertilizers; sustainable agriculture

## 1. Introduction

The continuous rise in global food demand requires an increase of approximately 60% in agricultural production to feed an estimated 9.7 billion people by 2050 [1]. Traditionally, fertilizers have significantly increased crop yields over the last 50 years. However, the excessive and inefficient use of fertilizers has revealed several drawbacks, including economic losses, increased food prices, and environmental contamination [2,3]. Currently, fertilizers exhibit poor nutrient use efficiency due to the loss of many nutrients thought various processes such as leaching, runoff, volatilization, and lixiviation [3,4]. These processes result in economic losses for agricultural producers and an increase in the cost of food. Additionally, the loss of nutrients contributes to the environmental contamination of water, soil, and air and can even impact human health. For instance, the nitrogen use efficiency from urea is approximately 40% [5], indicating that more than 50% of the applied nitrogen is lost and transported into the environment. This leads to the formation of nitrite in water, which reduces oxygen levels and hinders the development of aquatic species. Nevertheless, a new type of fertilizer has been proposed as an alternative to overcome the limitations of conventional fertilizers. These are the called "fertilizers with enhanced efficiency" or "high-efficiency fertilizers". These products are designed to increase nutrient availability, minimize losses, and reduce the costs of agricultural production. Consequently, their objective is to increase crop yields, enhance food production, and reduce the negative environmental impacts of agriculture. Among the most promising matrices for these fertilizers are polymeric hydrogels, particularly those of natural origin (biopolymeric hydrogels).

Polymeric hydrogels are three-dimensional structures with crosslinking, characterized by their high water absorption capacity (WAC), low solubility, high durability, and adequate stability. These materials have demonstrated their ability to reduce soil erosion and runoff by altering the hydrophysical properties of soil [6,7]. Furthermore, they can trap nutrients and promote their slow release in soils. Synthetic polymers such as acrylamide (AAm), acrylic acid (AAc), and copolymers of AAm have been utilized to form hydrogels and entrap macronutrients like nitrogen, phosphorus, and potassium (NPK). For example, Liu et al. (2007) reported the application of AAc hydrogel formed from N,N′-methylene bisacrylamide as a crosslinker and ammonium persulfate as an initiator in the presence of urea. This hydrogel exhibited a slow-release profile, delivering approximately 80% of the entrapped nitrogen over a period of approximately 28 days [8]. The same material has been used to trap phosphorous, using $KH_2PO_4$ as the source, demonstrating a controlled release of this nutrient [9]. However, despite the effectiveness of hydrogels formed using synthetic polymers in facilitating slow nutrient release, their high production costs and low biodegradability make them unsuitable for agriculture applications [10]. To address these drawbacks, researchers have explored combining synthetic polymers with naturals biopolymers (e.g., cellulose, starch, and chitosan) and clays (e.g., bentonite, kaolin, and montmorillonite) to enhance the properties of hydrogels and reduces production costs [11,12].

Biopolymers, which are biodegradable, non-toxic, and biocompatible materials derived from natural sources, show a robust alternative for forming hydrogels. They contribute to reducing biomass waste and protecting the environment [13]. For instance, a carboxymethyl starch-g-polyacrylamide hydrogel was developed and used for the slow release of phosphorous. This hydrogel delivered approximately 87% of the nutrient over 30 days, highlighting its potential as a material for controlled nutrient supply [14]. However, the literature has explored only a limited range of biopolymers, such as cellulose, chitosan, and starch, to enhance the biodegradability of hydrogels. Starch is one of the most abundant polysaccharides worldwide, found in high concentrations in vegetables and cereals. It is composed of a co-polymer chain consisting of amylose (AM) and amylopectin (AP), typically comprising approximately 20–30% AM and 70–80% AP. Additionally, starch is also a biodegradable, biocompatible, and non-toxic biopolymer [15,16]. Several crosslinking compounds have been reported for producing starch hydrogels, such as glutaraldehyde, epichlorohydrin, and sodium hypophosphite. However, their high cost and negative environmental impacts have limited their use in agriculture applications. Hence, to achieve an eco-friendly and cost-effective hydrogel material, citric acid (2-hydroxypropane-1,2,3 tricarboxylic acid, $C_6H_8O_7$) was investigated as a crosslinking agent for cassava starch biopolymer. The crosslinking reaction was monitored using ATR-FITR measurements, and the swelling behavior and water retention properties were determined and compared. Additionally, the materials were employed for the entrapment of potassium and nitrogen, and their slow-release profiles were assessed. This work provides valuable insights into the production of cassava starch–citric acid hydrogels, which have the potential to enhance the macronutrient release while mitigating the negative environmental impacts of the fertilizer.

## 2. Results and Discussion

### 2.1. Hydrogel Characterization

The crosslinking reaction between cassava starch (AGCY) and citric acid (CA) to form starch hydrogel (AGCY-CA(10)) was performed using a concentration of 15% and 10% *w/w* of AGCY and CA, respectively. Figure 1 shows the main bands in the ATR-FITR spectrum of AGCY, CA, and AGCY-CA(10). In cassava starch (Figure 1A), a band around

3348 cm$^{-1}$ can be observed and is attributed to the stretching of the hydroxyl groups. The band at 1640 cm$^{-1}$, indicating the presence of water in the biopolymer, suggests the high hydrophilicity of cassava starch [17]. Additionally, a band around 2914 cm$^{-1}$ appears, corresponding to the stretching of C–H groups in the biopolymer chain. Bellow 1700 cm$^{-1}$, bands related to AM and AP units are observed: the band at 1330 cm$^{-1}$ is attributed to the symmetric deformation and scissoring vibration of the $CH_2$ groups in the glucose units. Furthermore, bands a (1139 cm$^{-1}$) and b (1078 cm$^{-1}$) correspond to the C–O stretching, while bands c (1016 cm$^{-1}$), d (918 cm$^{-1}$), and e (846 cm$^{-1}$) are attributed to the C–O–C ring vibration in the monomer units [17,18].

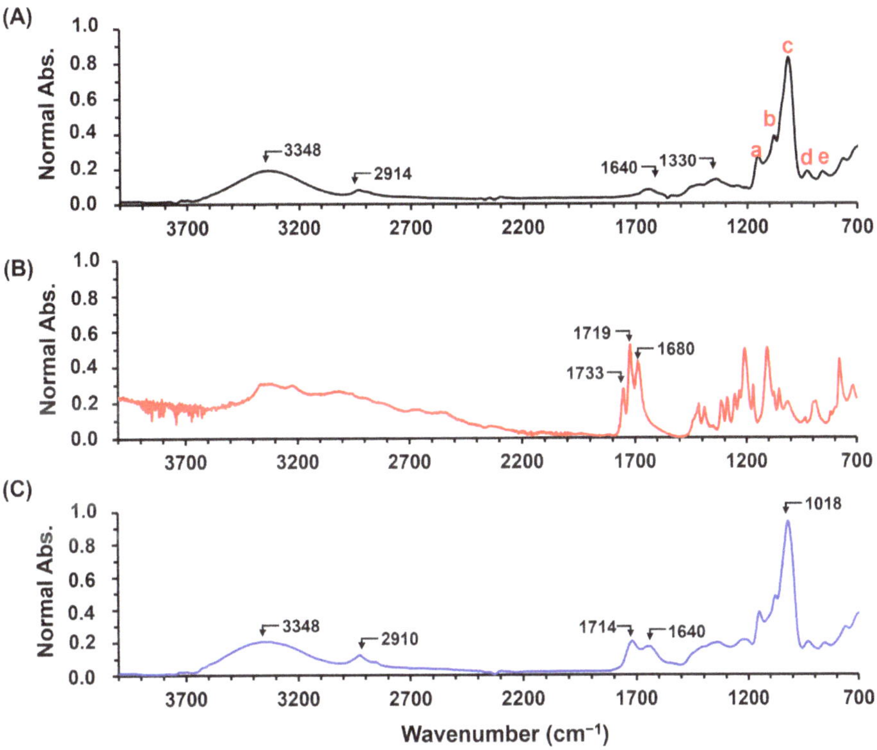


**Figure 1.** ATR-FITR spectra of CA and hydrogels: (**A**) AGCY, (**B**) CA, and (**C**) AGCY-CA(10). In spectrum A, bands related to monomeric units are identified with a, b, c, d and e (see details in Section 2.1).

On the other hand, the ATR-FITR spectrum of CA (Figure 1B) exhibits three bands between 1700 cm$^{-1}$ and 1600 cm$^{-1}$, attributed to the stretching of the carboxylic (C=O) groups of CA [19]. In the ATR-FITR spectrum of the AGCY-CA(10) hydrogel, similar bands were observed in the fingerprint region, attributed to the vibration of the AM and AP units. Additionally, a band appears at 1714 cm$^{-1}$ (Figure 1C), confirming the crosslinking reaction between the carboxyl groups of the CA and the hydroxyl groups of the cassava starch to form ester groups. Due to the steric hindrance, the hydroxyl at position C6 of the starch is more reactive than the hydroxyl at positions C2 and C3 (Figure 2) [20,21]. During the production of AGCY-CA(10) hydrogel, CA is converted to its anhydride, forming ester bonds with the hydroxyl groups of anhydroglucoses from neighboring chains of cassava starch [22]. This process results in greater stabilization of the starch structure in the material. These results are consistent with other studies on starch esterification reactions with CA [23,24].

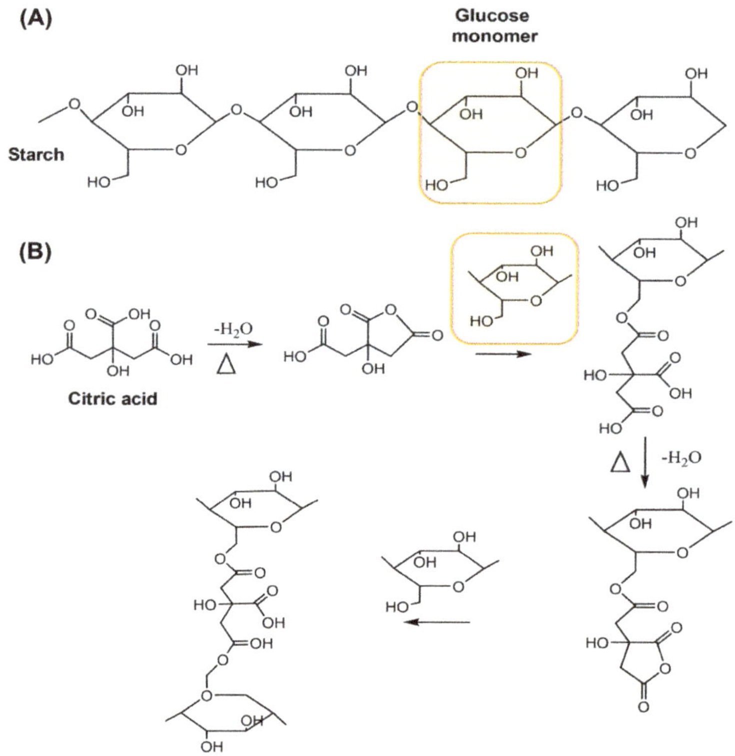


**Figure 2.** (**A**) Starch structure. (**B**) Mechanism of crosslinking reaction between CA and starch.

Furthermore, the intense bands around 1640 cm$^{-1}$ indicate the presence of water in the formed material. The higher intensity of this band compared to the cassava starch spectrum suggests a significant presence of water, likely entrapped within the hydrogel's polymeric matrix. This band could also be attributed to the carbonyl groups of the citric acid that are not crosslinked with starch [25]. Incomplete crosslinking can leave citrate groups available to interact electrostatically with ions such as $K^+$ and NH4$^+$, promoting favorable interactions to retain theses fertilizer ions in the polymeric network and enable slow-release behavior. Moreover, CA is recognized as a safe food additive and is considered safe for the environment and human health [26,27]. This is an important characteristic for materials applied in nutrient release in soils to enhance agricultural productivity, as it does not negatively impact the environment. Additionally, starch esterification with CA increases starch resistance to amylolysis (i.e., transformation of starch into sugar by the effect of acids or enzymes) [28]. Consequently, it could increase degradation resistance in soils, prolonging the time for fertilizer release, and inhibiting nutrient losses through processes such as leaching, runoff, volatilization, and lixiviation.

To facilitate the slow delivery of nutrients, hydrogel was applied using different concentrations (0.5 to 1.0% *w*/*w*) of KCl and $NH_4Cl$ as sources of potassium and nitrogen, respectively (Figure 3). The material AGCY-CA(10) without nutrients exhibited a reddish-brown color and a dry, rigid appearance, which remained after water absorption (swollen), indicating a high degree of crosslinking and a lower WAC. These findings are congruent with the results reported by Mei et al. (2015), who observed that citrate-crosslinked starch granules showed lower WAC in comparison with native starch. This reduction in WAC is attributed to changes in the AM and AP structures, which hinder granule swelling [29]. However, upon incorporating nutrient salt sources, the materials became more malleable as the nutrient concentration increased. Even at concentrations of 5.0% and 10.0% KCl, the materials did not form a rigid structure and dispersed in water after immersion. In contrast, the materials formed with 0.5% and 1.0% KCl maintained their structure after water absorption. On the other hand, when the hydrogel was formed with $NH_4Cl$, all the materials maintained their structure after water absorption, except for the hydrogel formed with 10.0% $NH_4Cl$, which dispersed in water. Additionally, as the $NH_4Cl$ concentration increased, the material became softer after water absorption, suggesting that the addition of nutrient salts inhibits starch crosslinking or modifies the starch's chemical structure. This is further supported by the WAC results, as the increase in nutrient source salts led to a decrease in WAC (Figure 4), attributed to a reduction in the crosslinking reaction, which reduces the material's capacity to absorb water. It is important to highlight that the AGCY-CA(10) hydrogel exhibited a lower WAC (125.1 ± 13.9%) compared to the materials formed with 0.5% KCl (173.9 ± 28.4%) and 0.5% $NH_4Cl$ (141.5 ± 25.5%).

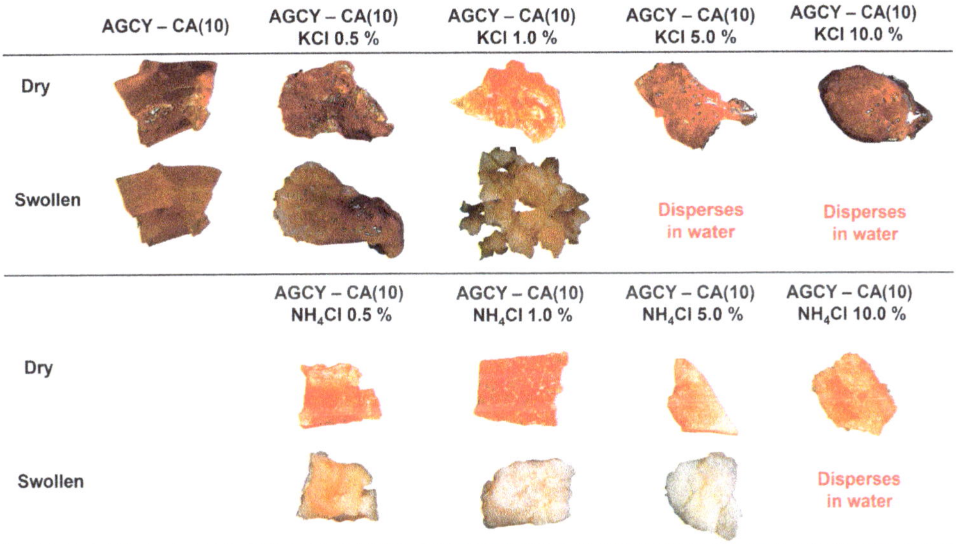


**Figure 3.** Digital photos of AGCY-CA(10) with and without several KCl and $NH_4Cl$ concentrations (0.5 to 10.0% *w/w*) before and after water-absorbing process.

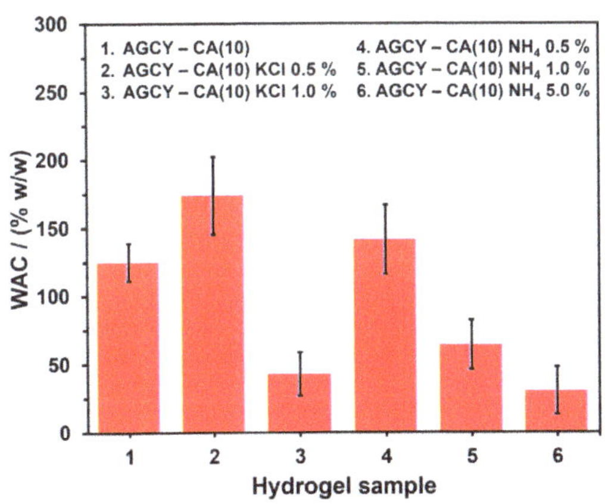


**Figure 4.** WAC of the AGCY-CA(10) hydrogel with and without KCl and $NH_4Cl$ in water.

ATR-FTIR spectra of hydrogels loaded with fertilizers are shown in Figures 5 and 6. The spectrum of $NH_4Cl$ exhibited three principal bands at 3122, 3032, and 1398 cm$^{-1}$ attributed to the $NH_4^+$ stretching vibrations [30]. The intensity of the band at 1398 cm$^{-1}$ increased with a higher $NH_4Cl$ concentration in the hydrogels. Furthermore, the peak associated with ester formation, reaching its maximum intensity between 1710 and 1720 cm$^{-1}$, suggests that the presence of low concentrations of fertilizer did not significantly impact the crosslinking of cassava starch. However, based on both digital photos of the materials and ATR-FTIR results, it is evident that despite observing a band at 1712 cm$^{-1}$ in the material containing 10% $NH_4Cl$, the degree of esterification was not substantial. This is evidenced by the material dispersing in water after 24 h.

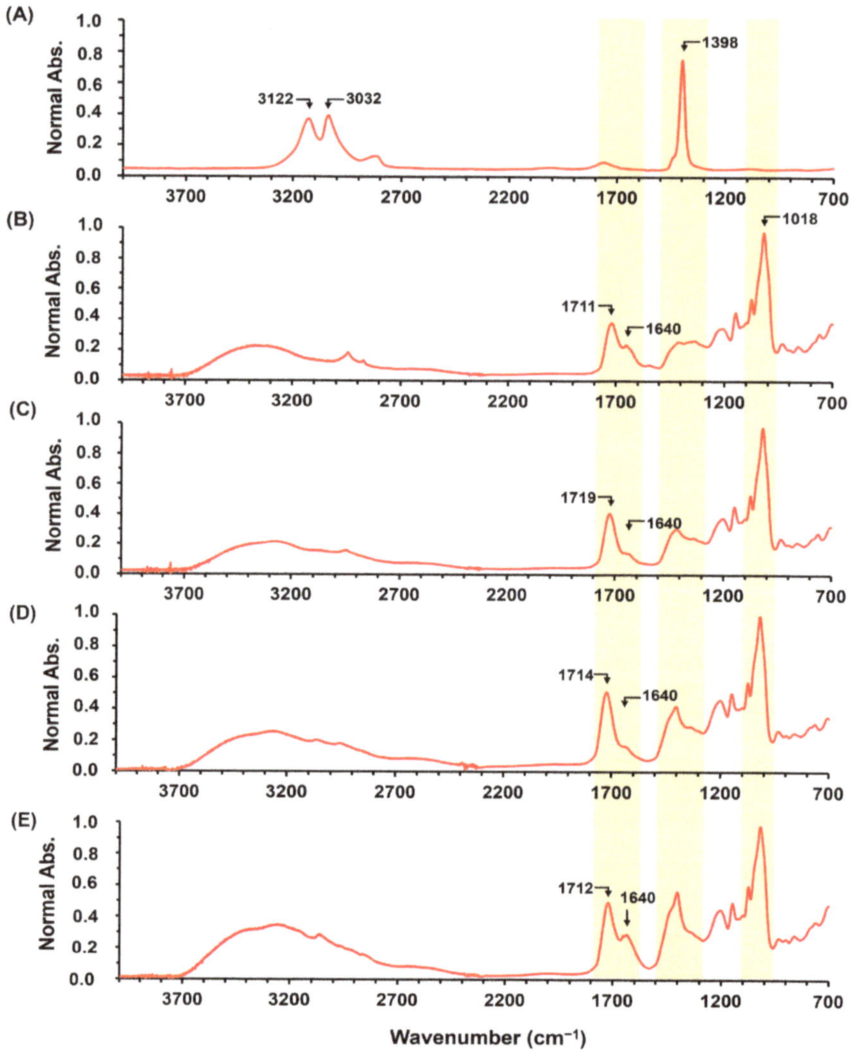


**Figure 5.** ATR-FITR spectra of fertilizer and hydrogels: (**A**) $NH_4Cl$ and AGCY-CA(10) with several concentrations of $NH_4Cl$: (**B**) 0.5%, (**C**) 1.0%, (**D**) 5.0%, and (**E**) 10.0%.

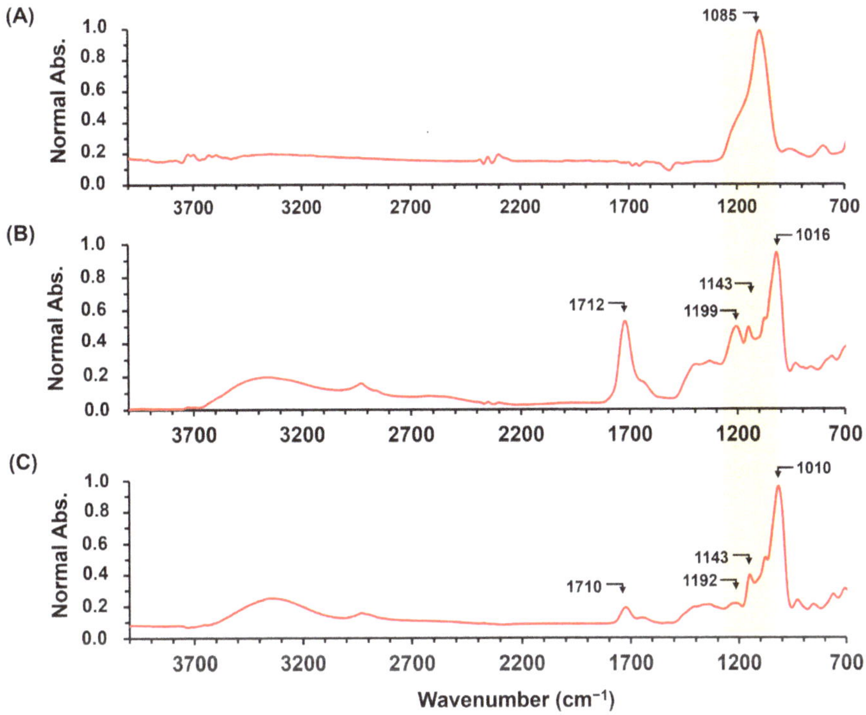


**Figure 6.** ATR-FITR spectra of (**A**) KCl and hydrogel AGCY-CA(10) with several concentrations of KCl: (**B**) 0.5% and (**C**) 1.0%.

Furthermore, all AGCY-CA materials with fertilizer showed the same band profiles of glucose units in the starch, demonstrating that there were no significant modifications in the structure of amylose and amylopectin during material formation. On the other hand, KCl exhibits an intense band at 1085 $cm^{-1}$, which is likely attributed to the Si–O stretching of impurities in the fertilizer used. In the hydrogel AGCY-CA(10) with KCl 0.5%, a band around 1712 $cm^{-1}$ is observed, indicating AGCY-CA esterification. However, this band notably decreases with 1.0% KCl, suggesting that the fertilizer reduces the efficiency of the crosslinking reaction. This finding is consistent with the analysis using digital photos of the dry and swollen material, where an increase in KCl concentration leads to the reduced physical stability of the material. Even at concentrations higher than 5.0%, the material does not swell and disperses in water.

During the material formation, three processes are involved: (i) gelatinization, (ii) crosslinking reaction, and (iii) retrogradation. In the system, the fertilizer is added after cassava starch gelatinization; therefore, the fertilizer is expected to affect steps (ii) and (iii) of the material formation process, limiting crosslinking and/or the association and hydrogen bonding between the linear chains of AM and AP. It is known that cations can significantly influence starch retrogradation, particularly bivalent cations such as $Ca^{2+}$ and $Mg^{2+}$, as these ions affect the hydrogen bonds in water–starch and starch–starch systems [31,32]. The negative effect on retrogradation occurs because the interaction between the cation and starch hydroxyl groups causes the starch to become negatively charged, resulting in a positive charge in the water phase. This creates a potential difference known as the Donnan potential [33]. Therefore, it is likely that potassium ions induce a negative charge on the starch, leading to intramolecular repulsive forces within the starch and affecting the starch–starch association. This explains why, at high concentrations of salt, the material does not form a rigid structure. Interestingly, this effect was also observed with ammonium salts, although it only occurred at fertilizer salt concentrations higher than 5.0%.

### 2.2. Release Studies In Vitro of Potassium and Nitrogen

The potassium and nitrogen release profiles from hydrogels AGCY-CA(10) were evaluated using materials with 0.5 and 1.0% KCl, and the material with 5.0% $NH_4Cl$ (Figure 7A,B). The in vitro release profile of potassium from the cassava starch hydrogel showed an initial burst release, with approximately 80% and 60% of the entrapped nutrient released in the first 6 h for the materials formed with 0.5% and 1.0% KCl, respectively. Subsequently, for the systems with low KCl concentrations, the nutrient release increased gradually, reaching almost 90% of the entrapped potassium after approximately 60 h. After that, the released potassium levels remained relatively constant with increasing time. This behavior may be attributed to the rapid dissolution and diffusion of potassium at the beginning of the swelling process. However, for the material with 1.0% KCl, the release did not exceed 64% of the entrapped potassium. This suggests that the higher KCl concentration led to multiple desorption–adsorption steps, which contributed to maintaining a significant proportion of entrapped potassium (approximately 34%). On the other hand, when nitrogen was entrapped, the percentage of released nutrients increased rapidly from 27.4% to 96.2% as time increased from 30 min to 24 h, indicating its ability to gradually release the nutrient within the first 24 h. The rapid release of ammonium ions could be attributed to their instantaneous migration to the release medium when the sample is immersed. A similar release behavior was observed by León et al. (2019), who found that starch oxidate hydrogel released 90% of the entrapped ammonium within the first 4 h of the release experiment [34].

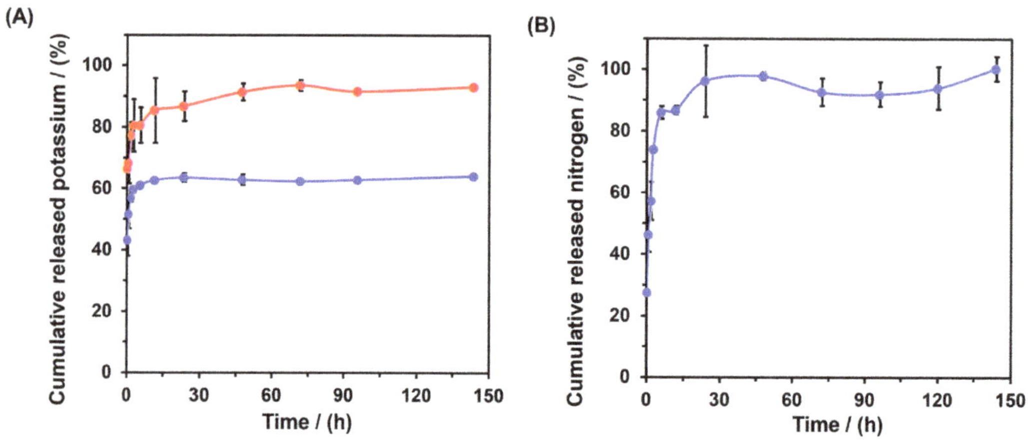


**Figure 7.** Cumulative nutrient release profiles from hydrogel AGCY-CA(10) with (**A**) KCl 0.5% (•) and 1.0% (•) and (**B**) $NH_4Cl$ 1.0% (•).

Four kinetic models (zero-order, first-order, Higuchi, and power-law) were considered to evaluate the mechanism of nutrient release from the materials formed. Table 1 presents the calculated kinetic parameters obtained by fitting the experimental data. Among the models applied to the release data, the power-law model showed the highest regression coefficients ($R^2$) for both nutrients.

**Table 1.** Release kinetics parameters for potassium and nitrogen release from the AGCY-CA(10) hydrogel.

| Fertilizer/Concentration | Mathematical Models | | | | | | | | |
|---|---|---|---|---|---|---|---|---|---|
| | Zero-Order | | First-Order | | Higuchi | | Power-Law | | |
| | $k_0$ ($1 \times 10^{-2}\ h^{-1}$) | $R^2$ | $k_1$ ($1 \times 10^{-1}\ h^{-1}$) | $R^2$ | $k_H$ ($1 \times 10^{-1}\ h^{-1}$) | $R^2$ | K ($1 \times 10^{-1}\ h^{-1}$) | $R^2$ | n |
| KCl/0.5% | 3.4 | 0.7747 | 7.0 | 0.8157 | 1.0 | 0.9005 | 7.81 | 0.9447 | 0.13 |
| KCl/1.0% | 1.7 | 0.2154 | 1.5 | 0.2662 | 0.7 | 0.3958 | 1.49 | 0.995 | 0.14 |
| $NH_4Cl$/5% | 1.2 | 0.8564 | 2.7 | 0.8907 | 3.92 | 0.8986 | 4.21 | 0.9791 | 0.55 |

$k_0$, $k_1$ "$k_H$", and "k" represent the apparent release rate constants of the respective mathematical models, and n is the release exponent of the power-law model.

This mathematical model describes solute diffusion from a matrix with different geometric dimensions and porous systems, allowing for inference of the mechanisms associated with nutrient release based on the release exponent (n) value: n = 0.45 corresponds to a Fickian diffusion mechanism or solute release that is diffusion-controlled; n = 0.45–0.89 corresponds to non-Fickian transport (anomalous diffusion) or solute release that is both diffusion-controlled and erosion-controlled; n = 0.89 relates to a case II mechanism (polymer relaxation or swelling-controlled); and n > 0.89 relates to a super case II mechanism (transport or solute release that is erosion-controlled) [35–37]. According to the power-law model, the release of nutrients occurs through the diffusion of nutrients, which are rapidly delivered upon swelling of the hydrogel AGCY-CA(10). When the polymeric material swells, it induces macromolecular relaxation of the biopolymer, resulting in a rubbery state. In this state, solute diffusion into the aqueous medium generally occurs. However, in the case where the water penetration rate is much slower than the polymer chain relaxation rate, the value of "n" can be less than 0.5, as observed in the tested materials. This phenomenon is still considered Fickian diffusion and is referred to as "Less-Fickian". For the hydrogels studied, the release of potassium (n = 0.5) and nitrogen (n = 0.55) follows a Less-Fickian and anomalous diffusion pattern, respectively [36,38]. This means that the release process of potassium is primarily dependent on the rate of water penetration, where the biopolymers have high mobility, and water easily penetrates the rubbery network, promoting rapid diffusion of the nutrient out of the matrix. Conversely, for nitrogen, the material exhibits high mobility, allowing for the rapid penetration of water into the biopolymer matrix [34].

### 2.3. *Production and Application of Economic Starch—CA Hydrogel*

CA is an economical raw material and is considered a green chemical compared to other crosslinking agents reported in the literature for starch, such as glutaraldehyde, epichlorohydrin, boric acid, and sodium hypophosphite. In order to decrease production costs, cassava with peel (AYD) was milled to replace the cassava starch used in the previous experiments and applied in the entrapment of N and K macronutrients. Figure 8A shows the ATR-FTIR spectra of AYD, AYD-CA(10) with $NH_4Cl$ 1.0% (Figure 8B), and KCl 1.0% (Figure 8C). The AYD material exhibits similar bands to those reported for cassava starch, especially in the fingerprint region (600–1400 cm$^{-1}$), indicating the presence of a high quantity of amylose and amylopectin units in the material. This allows for its application in forming hydrogels, eliminating the need for starch extraction. Additionally, both spectra of the formed material show an intense band around 1700 cm$^{-1}$, indicating material crosslinking and the successful production of the economical hydrogel. These results are important because they demonstrate that it is not necessary to extract cassava starch in order to obtain a hydrogel with the capacity to reduce nutrient release in soils.

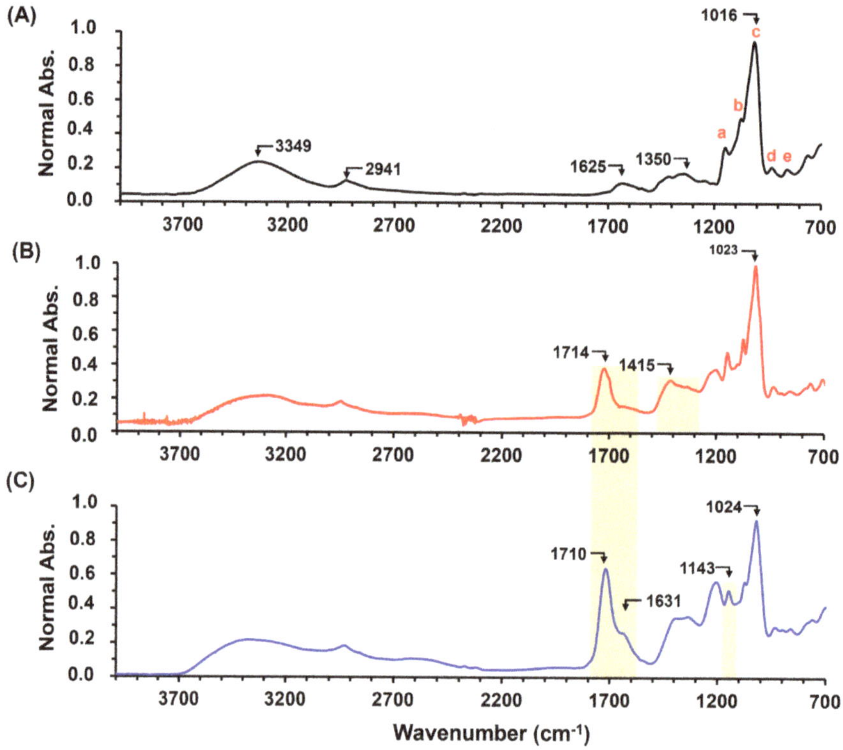


**Figure 8.** ATR-FITR spectra of (**A**) AYD, (**B**) AYD-CA(10)-$NH_4Cl$ 1.0%, and (**C**) AYD-CA(10)-KCl 1%. In spectrum A, bands related to monomeric units are identified with a, b, c, d and e (see details in Section 2.1).

The material AYD-CA(10), with and without fertilizer, exhibits a rigid structure both before and after immersion in water for 24 h, maintaining its physical integrity without dispersion in the swelling medium (Figure 9A). However, a heterogeneous surface can be observed due to the presence of cassava peel. Additionally, the materials have the ability to absorb water, with AYD-CA(10) showing a WAC of 183.4 $\pm$ 0.6%, which is higher than that of AGCY-CA(10). This effect is likely caused by the heterogeneity of the cassava sample, which reduces the crosslinking points in the material and allows for greater penetration of water into the biopolymer network. Furthermore, similar to the AGCY-CA(10) materials, the addition of nutrients decreases the WAC of the material AYD (Figure 9B), resulting in a WAC of 117.9 $\pm$ 3.7% for the material with $NH_4Cl$ and 157.4 $\pm$ 25.0% for the material with KCl. This is likely because the fertilizers reduce the efficiency of crosslinking, decreasing the formation of cavities in the material for water entrapment. Moreover, the application of the material for potassium release shows similar behavior to AGCY-CA(10)-KCl 1.0%, delivering approximately 40% of the entrapped nutrient in 48 h (Figure 9C). However, for nitrogen, the material exhibits improved nutrient retention, delivering approximately 60% of the entrapped nutrient. This indicates that the material AYD-CA(10) is economical and enhances the capacity to retain nutrients compared to AGCY-CA(10). The above is a result of raw materials and methodology; for instance, hydrogels are produced from cassava starch with peel, which eliminates the need for AGCY and reduces the cost of hydrogel production, as AGCY is obtained from AYD and requires different physical processes. Additionally, after 48 h of nutrient release, AYD-CA(10) retains nearly 60% K and 40% N. These values are higher than those for AGYD-CA(10), which retains 40% K and 3% N.

Thus, it has the potential to facilitate nutrient applications in various agricultural practices, reducing the environmental impacts of conventional application methods.

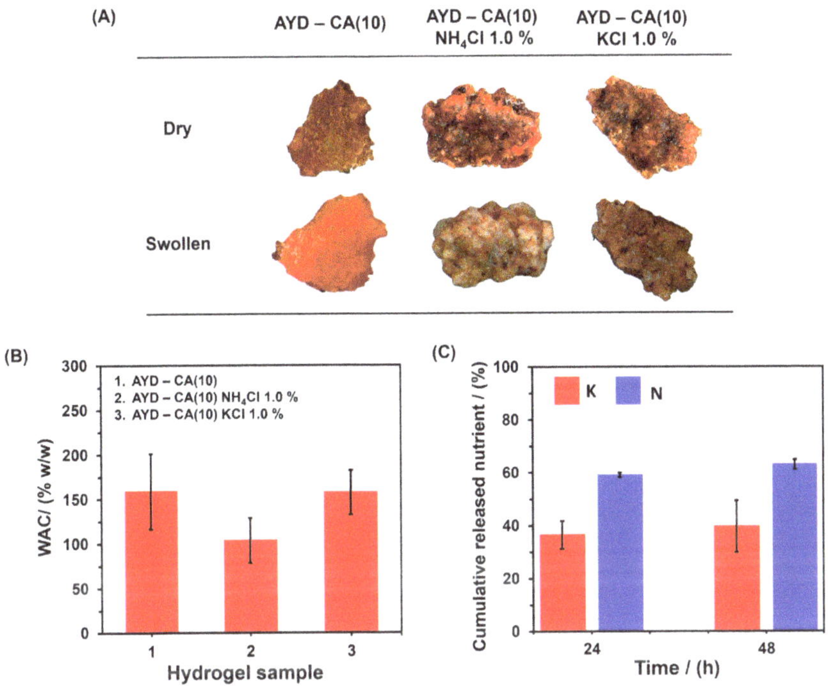


**Figure 9.** AYD-CA(10) with and without KCl (1.0%) and $NH_4Cl$ (1.0%): (**A**) Digital photos before and after water-absorbing process, (**B**) WAC in water, and (**C**) Potassium and nitrogen cumulative nutrient release during 24 h and 48 h.

## 3. Conclusions

Cassava starch hydrogels were successfully prepared through a crosslinking reaction between starch and CA in an aqueous solution. This methodology is considered simple and economical. The hydrogel material was then utilized to entrap potassium and nitrogen, facilitating the controlled release of these macronutrients. The ATR-FTIR analysis demonstrated that the incorporation of nutrient sources, such as KCl and $NH_4Cl$, influenced the formation of the hydrogel material. Higher concentrations of these salts were found to reduce the efficiency of crosslinking. Additionally, the presence of nutrients resulted in a decrease in the WAC of the hydrogel materials. The cassava starch hydrogel exhibited the ability to effectively entrap and release the tested macronutrients. The release mechanism followed a Less-Fickian diffusion for potassium and anomalous diffusion for nitrogen. This indicates that the release of potassium is mainly dependent on the water penetration rate, allowing for rapid diffusion outside the rubbery network of the hydrogel. On the other hand, the release of nitrogen is influenced by the high mobility of the hydrogel, which facilitates the penetration of water into the biopolymer matrix. Furthermore, the production of hydrogels using cassava with peel material, which was previously milled, was successfully demonstrated. This approach aimed to reduce production costs and eliminate the need for starch extraction. The resulting materials exhibited improved nitrogen retention compared to cassava starch hydrogels. This economic hydrogel material offers the advantage of being suitable for production in farms with limited infrastructure and can be applied in the agroindustry to minimize the negative environmental impacts of nutrient application.

## 4. Materials and Methods

### 4.1. Materials

Cassava and cassava starch (AGCY) were acquired from a local market in Cali, Colombia. Citric acid (CA) ($C_6H_8O_7$), sodium hydroxide (NaOH), formaldehyde ($CH_2O$), and potassium hydrogen phthalate ($KHC_8H_4O_4$) were purchased from Sigma-Aldrich (St. Louis, MA, USA). Potassium chloride (KCl) and ammonium chloride ($NH_4Cl$) were obtained from a fertilizer warehouse in Cali, Colombia.

### 4.2. Preparation of Starch-CA Hydrogels

The AGCY-CA hydrogels were prepared using the following procedure: A specific amount of AGCY that would achieve a starch concentration of 15% $w/w$ was mixed with distilled water and stirred. The mixture was then heated at 75 °C for 30 min to facilitate the gelatinization of starch. After that, a predetermined quantity of CA was added to obtain a CA concentration of 10% $w/w$. The mixture was continuously stirred for 1 h at 75 °C, and the resulting material was dried at 50 °C for 72 h. On the other hand, to form cost-effective hydrogels, the cassava was washed and milled to obtain a cassava flour (AYD). The AYD-CA hydrogels were then formed using the same methodology described above.

### 4.3. Swelling Studies

The WAC was evaluated using the tea bag method described by Palencia et al. (2017) [39]. A known mass of dried hydrogel was placed inside a pre-weighed tea bag and submerged in distilled water for 24 h. Subsequently, tea bag containing the hydrogel was suspended for 15 min to allow the excess water to drain off. The tea bag, along with the sample, was weighed to determine the WAC using Equation 1. As a control, a blank experiment was conducted using only the empty tea bag.

$$\mathrm{WAC} = \left(\frac{\mathrm{w}_2 - (\mathrm{w}_1 + \mathrm{w}_0)}{\mathrm{w}_1}\right) - \mathrm{WAC}_0 \tag{1}$$

where $w_0$, $w_1$, and $w_2$ are the masses of the tea bag, the sample, and the sample tea-bag system, respectively, and $WAC_0$ is the correction of WAC given the results of blank experiment. These experiments were conducted in triplicate to ensure accuracy and reproducibility of the results.

### 4.4. Physicochemical and Morphological Characterizations of the Hydrogels

Fourier-transformed infrared spectroscopy measurements using attenuated total reflectance (ATR-FTIR) were conducted on the hydrogels to verify the esterification reaction. The measurements were performed using an IR-Affinity-1 instrument (Shimadzu, Kyoto, Japan). Data acquisition was carried out in the range of 4000 to 700 $cm^{-1}$ at a rate of 2 $cm^{-1}$ per data point, with 20 scans recorded for each sample. The measurement was performed in triplicate. In addition to spectroscopy analysis, digital photographs were taken of the hydrogel materials before and after immersion in water. These photographs were captured using a Flexible Digital Electronic Microscope (Suzhou Jingtong Instrument Co., Ltd., Suzhou, China) to visually assess the changes in the samples upon water exposure.

### 4.5. Preparation of AGCY-CA with Potassium and Nitrogen Nutrients Entrapped

AGCY-CA and AYD-CA materials were prepared as described earlier in the "Preparation of starch-CA hydrogels" section. In addition, KCl and $NH_4Cl$ were independently added to form Y-CA-KCl-X and Y-CA-$NH_4Cl$-X materials, where Y represents the starch source (Y = AGCY or AYD) and X represents the salt concentration (0.5% or 1.0% or 5.0% or 10.0% $w/w$). For AYD-CA, both KCl and $NH_4Cl$ were incorporated at a concentration of 1.0% $w/w$. To determine the concentration of potassium and nitrogen, the following methodology was employed: Potassium quantification was carried out using an atomic absorption spectrophotometer (AA-6300, Shimadzu, Kyoto, Japan) following the procedure outlined in the Colombian Institute of Technical Standards and Certification (NTC 202) [40]. The sample was first digested in a mixture of $HCl/HNO_3$ with a 2:1 ratio, followed by filtration using quantitative filter paper. A potassium curve was prepared using concentrations ranging from 0.2 to 1.0 ppm in deionized water with the addition of 1.0% lanthanum chloride ($LaCl_3$).

On the other hand, nitrogen concentration was determined by titration of ammonium in the fertilizer sample, using the methodology outlined in the Colombian Institute of Technical Standards and Certification (NTC 5268) [41]. First, the sample was dissolved in deionized water, followed by the addition of 40% $v/v$ formaldehyde and 30 mL of water. The sample was titrated with NaOH in the presence of phenolphthalein. NaOH solution was standardized using $KHC_8H_4O_4$, which was previously dried at 98 °C for 4 h. The amount of NaOH used in the titration provided the ammonium content in the sample, with a 1:1 molar ratio, allowing the determination of the nitrogen concentration.

### 4.6. Nutrients Release Kinetics—In Vitro Study

To evaluate the release of nutrients from AGCY-CA(10%) and AYD-CA(10%) hydrogels, approximately 1.0 g of each hydrogel with entrapped nutrients was placed in tea bags. Release experiments were conducted by continuously agitating the samples at 100 rpm in deionized water as the release medium. At specified time intervals over a period of six days, 2.0 mL aliquots of the resulting solution were withdrawn and immediately replaced with 2.0 mL of fresh medium. The experiments were performed in triplicate to ensure accuracy and reproducibility. The concentrations of potassium and nitrogen in the withdrawn solution samples were determined using atomic absorption spectroscopy and titration methods, respectively, as described in the previous sections. For the nitrogen quantification, a titration blank was performed using 1.0 mL of the sample solution with NaOH, employing phenolphthalein as an indicator, without the addition of formaldehyde. To analyze the release mechanism of the nutrients, four kinetic models were employed: zero-order, first-order, Higuchi, and power-law. These models can be represented by the following equations (Equations (2)–(5), respectively):

$$\text{Zero-order model}: M_t/M_\infty = k_0 t \tag{2}$$

$$\text{First-order model}: M_t/M_\infty = 1 - e^{(-k_1 t)} \tag{3}$$

$$\text{Higuchi model}: M_t/M_\infty = k_H t^{\frac{1}{2}} \tag{4}$$

$$\text{Power-law model}: M_t/M_\infty = kt^n \tag{5}$$

where $M_t/M_\infty$ represents the fraction of nutrient released at time t. $k_0$, $k_1$, $k_H$, and k represent the zero-order release constant, first-order release constant, Higuchi constant, and power-law constant, respectively. The release constants and model parameters were determined by fitting the experimental release data [42,43].

**Author Contributions:** A.F.C. investigation, formal analysis, and writing—review and editing: M.P. conceptualization, investigation, formal analysis, and writing—review and editing: Á.A.A. conceptualization, investigation, formal analysis, and writing—review and editing. All authors have read and agreed to the published version of the manuscript.

**Funding:** Authors acknowledge the Universidad del Valle, Universidad de Córdoba, Universidad de Sucre, and National Planning Department of Colombia, specifically, for the general royalty system (Sistema General de Regalías, SGR) for the project BPIN 2020000100027, and this research has been funded by Dirección General de Investigaciones of Universidad Santiago de Cali under call No. 01-2024 and grant number No. 939-621122-030.

**Institutional Review Board Statement:** Not applicable.

**Informed Consent Statement:** Not applicable.

**Data Availability Statement:** Data are contained within the article.

**Conflicts of Interest:** The authors declare no conflicts of interest. The funders had no role in the design of the study; in the collection, analysis, and interpretation of data; in the writing of the manuscript; or in the decision to publish the results.

## References


1. United Nations, World Population Prospects 2019: Highlights. 2019. Available online: https://population.un.org/wpp/Publications/Files/WPP2019_10KeyFindings.pdf (accessed on 22 May 2024).
2. Chakraborty, R.; Mukhopadhyay, A.; Sarkar, P.; Mukhopadhyay, R. Nanocomposite-based smart fertilizers: A boon to agricultural and environmental sustainability. *Sci. Total Environ.* **2023**, *863*, 160859. [CrossRef] [PubMed]
3. Choudhury, A.T.; Kennedy, I.R. Nitrogen Fertilizer Losses from Rice Soils and Control of Environmental Pollution Problems. *Commun. Soil Sci. Plant Anal.* **2005**, *36*, 1625. [CrossRef]
4. Li, M.; Gao, L.; White, J.C.; Haynes, C.L.; O'Keefe, T.L.; Rui, Y.; Zhang, P. Nano-enabled strategies to enhance biological nitrogen fixation. *Nat. Nanotechnol.* **2023**, *18*, 688–691. [CrossRef] [PubMed]
5. Amodeo, G.; Giacometti, R.; Spagnoletti, F.; Santagapita, P.R.; Perullini, M. Eco-friendly routes for obtaining nanoparticles and their application in agro-industry. In *Nano-Enabled Agrochemicals in Agriculture*; Ghorbanpour, M., Muhammad Shahid, M.A., Eds.; Elsevier: London, UK, 2022; pp. 49–62. [CrossRef]
6. Narjary, B.; Aggarwal, P.; Singh, A.; Chakraborty, D.; Singh, R. Water availability in different soils in relation to hydrogel application. *Geoderma* **2012**, *187*, 94. [CrossRef]
7. Guilherme, M.R.; Aouada, F.A.; Fajardo, A.R.; Martins, A.F.; Paulino, A.T.; Davi, M.F.; Muniz, E.C. Superabsorbent hydrogels based on polysaccharides for application in agriculture as soil conditioner and nutrient carrier: A review. *Eur. Polym. J.* **2015**, *72*, 365. [CrossRef]
8. Liu, M.; Liang, R.; Zhan, F.; Liu, Z.; Niu, A. Preparation of superabsorbent slow-release nitrogen fertilizer by inverse suspension polymerization. *Polym. Int.* **2007**, *56*, 729. [CrossRef]
9. Tyliszczak, B.; Polaczek, J.; Pielichowski, J.; Pielichowski, K. Preparation and Properties of Biodegradable Slow-Release PAA Superabsorbent Matrixes for Phosphorus Fertilizers. *Macromol. Symp.* **2009**, *279*, 236. [CrossRef]
10. Ramli, R.A. Slow-release fertilizer hydrogels: A review. *Polym. Chem.* **2019**, *10*, 6073. [CrossRef]
11. Zhao, L.Z.; Zhou, C.H.; Wang, J.; Tong, D.S.; Yu, W.H.; Wang, H. Recent advances in clay mineral-containing nanocomposite hydrogels. *J. Soft Matter* **2015**, *11*, 9229. [CrossRef]
12. Michalik, R.; Wandzik, I.A. Mini-Review on Chitosan-Based Hydrogels with Potential for Sustainable Agricultural Applications. *Polymers* **2020**, *12*, 2425. [CrossRef]
13. Klein, M.; Poverenov, E. Natural biopolymer-based hydrogels for use in food and agriculture. *J. Sci. Food Agric.* **2020**, *100*, 2337. [CrossRef]
14. Alharbi, K.; Ghoneim, A.; Ebid, A.; El-Hamshary, H.; El-Newehy, M.H. Controlled release of phosphorous fertilizer bound to carboxymethyl starch-g-polyacrylamide and maintaining a hydration level for the plant. *Int. J. Biol. Macromol.* **2018**, *116*, 224. [CrossRef]
15. Ismail, H.; Irani, M.; Ahmad, Z. Starch-Based Hydrogels: Present Status and Applications. *Int. J. Polym. Mater.* **2013**, *62*, 411. [CrossRef]
16. Qamruzzaman, M.; Ahmed, F.; Mondal, M.I.H. An Overview on Starch-Based Sustainable Hydrogels: Potential Applications and Aspects. *J. Environ. Polym. Degrad.* **2022**, *30*, 190. [CrossRef]
17. Garces, V.; García-Quintero, A.; Lerma, T.; Palencia, M.; Combatt, E. Characterization of Cassava Starch and Its Structural Changes Resulting of Thermal Stress by Functionally-Enhanced Derivative Spectroscopy (FEDS). *Polysaccharides* **2021**, *2*, 866–877. [CrossRef]
18. Nikonenko, N.; Buslov, D.; Sushiko, N.; Zhbankov, R. Investigation of stretching vibrations of glycosidic linkages in disaccharides and polysaccarides with use of IR spectra deconvolution. *Biopolymers* **2000**, *57*, 257. [CrossRef]
19. Ghorpade, V.S.; Yadav, A.V.; Dias, R.J. Citric acid crosslinked cyclodextrin/hydroxypropylmethylcellulose hydrogel films for hydrophobic drug delivery. *Int. J. Biol. Macromol.* **2016**, *93*, 75. [CrossRef] [PubMed]
20. Uliniuc, A.; Hamaide, T.; Popa, M.; Băcăiță, S. Modified Starch-Based Hydrogels Cross-Linked with Citric Acid and their use as Drug Delivery Systems for Levofloxacin. *Soft Mater.* **2013**, *11*, 483–493. [CrossRef]
21. Hassan, M.M.; Tucker, N.; Le Guen, M.J. Thermal, mechanical and viscoelastic properties of citric acid-crosslinked starch/cellulose composite foams. *Carbohydr. Polym.* **2020**, *230*, 115675. [CrossRef]
22. Seligra, P.G.; Jaramillo, C.M.; Famá, L.; Goyanes, S. Biodegradable and non-retrogradable eco-films based on starch–glycerol with citric acid as crosslinking agent. *Carbohydr. Polym.* **2016**, *138*, 66. [CrossRef]
23. Zhou, J.; Tong, J.; Su, X.; Ren, L. Hydrophobic starch nanocrystals preparations through crosslinking modification using citric acid. *Int. J. Biol. Macromol.* **2016**, *91*, 1186. [CrossRef]
24. Shen, L.; Xu, H.; Kong, L.; Yang, Y. Non-Toxic Crosslinking of Starch Using Polycarboxylic Acids: Kinetic Study and Quantitative Correlation of Mechanical Properties and Crosslinking Degrees. *J. Environ. Polym. Degrad.* **2015**, *23*, 588. [CrossRef]
25. Golachowski, A.; Drożdż, W.; Golachowska, M.; Kapelko-Żeberska, M.; Raszewski, B. Production and Properties of Starch Citrates—Current Research. *Foods* **2020**, *9*, 1311. [CrossRef]

26. Xie, X.; Liu, Q. Development and Physicochemical Characterization of New Resistant Citrate Starch from Different Corn Starches. *Starch-Stärke* **2004**, *56*, 364. [CrossRef]
27. Nusrat, Z.; Tahira, M.A.; Abid, H. Comparative study on citric acid modified instant starches (alcoholic alkaline treated) isolated from white sorghum and corn grains. *Int. J. Biol. Macromol.* **2020**, *150*, 1331. [CrossRef]
28. Kapelko-Żeberska, M.; Zięba, T.; Pietrzak, W.; Gryszkin, A. Effect of citric acid esterification conditions on the properties of the obtained resistant starch. *Int. J. Food Sci. Technol.* **2016**, *51*, 1647. [CrossRef]
29. Mei, J.Q.; Zhou, D.N.; Jin, Z.Y.; Xu, X.M.; Chen, H.Q. Effects of citric acid esterification on digestibility, structural and physicochemical properties of cassava starch. *Food Chem.* **2015**, *187*, 378. [CrossRef]
30. Pajarito, B.B.; Llorens, C.; Tsuzuki, T. Effects of ammonium chloride on the yield of carbon nanofiber aerogels derived from cellulose nanofibrils. *Cellulose* **2019**, *26*, 7727. [CrossRef]
31. Fu, Z.; Chen, J.; Luo, S.J.; Liu, C.M.; Liu, W. Effect of food additives on starch retrogradation: A review. *Starch-Stärke* **2015**, *67*, 69. [CrossRef]
32. Beck, B.M.; Jekle, M.; Becker, T. Starch re-crystallization kinetics as a function of various cations. *Starch-Stärke* **2011**, *63*, 792. [CrossRef]
33. Oosten, J. Tentative Hypothesis to Explain How Electrolytes Affect the Gelatinization Temperature of Starches in Water. *Starch-Stärke* **1982**, *34*, 233. [CrossRef]
34. León, O.; Soto, D.; Antúnez, A.; Fernández, R.; González, J.; Piña, C.; Fernandez-García, M. Hydrogels based on oxidized starches from different botanical sources for release of fertilizers. *Int. J. Biol. Macromol.* **2019**, *136*, 813. [CrossRef]
35. Ritger, P.L.; Peppas, N.A. A simple equation for description of solute release I. Fickian and non-fickian release from non-swellable devices in the form of slabs, spheres, cylinders or discs. *JCR* **1987**, *5*, 23. [CrossRef]
36. Ganji, F.; Vasheghani, F.S.; Vasheghani, F.E. Theoretical Description of Hydrogel Swelling: A Review. *Iran. Polym. J.* **2010**, *19*, 375.
37. Langer, R.S.; Peppas, N.A. Present and future applications of biomaterials in controlled drug delivery systems. *Biomaterials* **1981**, *2*, 201. [CrossRef]
38. Wang, J.; Wu, W.; Lin, Z.J. Kinetics and thermodynamics of the water sorption of 2-hydroxyethyl methacrylate/styrene copolymer hydrogels. *Appl. Polym. Sci.* **2008**, *109*, 3018. [CrossRef]
39. Palencia, M.; Mora, M.; Palencia, S. Biodegradable Polymer Hydrogels Based in Sorbitol and Citric Acid for Controlled Release of Bioactive Substances from Plants (Polyphenols). *Curr. Chem. Biol.* **2017**, *11*, 36. [CrossRef]
40. The Colombian Institute of Technical Standards and Certification, ICONTEC, Soil Quality. Quantitative Methods to Determine Water Soluble Potassium, in Fertilizers and Material Sources for Fertilizers Production, NTC 202. 2001. Available online: https://tienda.icontec.org/gp-metodos-cuantitativos-para-la-determinacion-de-potasio-soluble-en-agua-en-abonos-o-fertilizantes-y-fuentes-de-materias-para-su-fabricacion-ntc202-2001.html (accessed on 23 May 2024).
41. The Colombian Institute of Technical Standards and Certification, ICONTEC, Soil Quality. Determination of Cationic Exchange Capacity, NTC 5268. 2014. Available online: https://tienda.icontec.org/ (accessed on 23 May 2024).
42. Bauli, C.R.; Lima, G.F.; de Souza, A.G.; Ferreira, R.R.; Rosa, D.S. Eco-friendly carboxymethyl cellulose hydrogels filled with nanocellulose or nanoclays for agriculture applications as soil conditioning and nutrient carrier and their impact on cucumber growing. *Coll. Surf. A* **2021**, *623*, 126771. [CrossRef]
43. Lakshani, N.; Wijerathne, H.S.; Sandaruwan, C.; Kottegoda, N.; Karunarathne, V. Release Kinetic Models and Release Mechanisms of Controlled-Release and Slow-Release Fertilizers. *ACS Agric. Sci. Technol.* **2023**, *3*, 939–956. [CrossRef]

gels

MDPI

Article

# Biodegradable Cassava Starch/Phosphorite/Citric Acid Based Hydrogel for Slow Release of Phosphorus: *In Vitro* Study

**Andrés F. Chamorro [1,*], Manuel Palencia [2,*] and Enrique M. Combatt [3]**

1 Research Group of Electrochemistry and Environment (GIEMA), Faculty of Basic Sciences, Universidad Santiago de Cali, Cali 760035, Colombia
2 Research Group in Science with Technological Applications (GICAT), Department of Chemistry, Faculty of Natural and Exact Science, Universidad del Valle, Cali 760032, Colombia
3 Department of Agricultural and Rural Development, Faculty of Agricultural Sciences, Universidad de Córdoba, Monteria 230002, Colombia; ecombatt@fca.edu.co
* Correspondence: andres.chamorro03@usc.edu.co (A.F.C.); manuel.palencia@correounivalle.edu.co (M.P.); Tel.: +57-2-5183000 (A.F.C.); +57-3205271934 (M.P.)

**Abstract:** Phosphorous (P) is one the most important elements in several biological cycles, and is a fundamental component of soil, plants and living organisms. P has a low mobility and is quickly adsorbed on clayey soils, limiting its availability and absorption by plants. Here, biodegradable hydrogels based on Cassava starch crosslinked with citric acid (CA) were made and loaded with $KH_2PO_4$ and phosphorite to promote the slow release of phosphorus, the storing of water, and the reduction in P requirements during fertilization operations. Crosslinking as a function of CA concentrations was investigated by ATR-FTIR and TGA. The water absorption capacity (WAC) and P release, under different humic acid concentration regimens, were studied by in vitro tests. It is concluded that hydrogel formed from 10% *w*/*w* of CA showed the lowest WAC because of a high crosslinking degree. Hydrogel containing 10% *w*/*w* of phosphorite was shown to be useful to encouraging the slow release of P, its release behavior being fitted to the Higuchi kinetics model. In addition, P release increased as humic acid contents were increased. These findings suggest that these hydrogels could be used for encouraging P slow release during crop production.

**Keywords:** hydrogels; Cassava starch; P slow release; phosphorite

## 1. Introduction

The rapidly growing world population demands an increase in food production to ensure food and nutritional security. In this context, in recent decades, there has been a growing demand for resources in the agricultural field, which has placed pressure on ecosystem resources such as water and soil. In the first case, as the agricultural frontier increases, water requirements are increasing as well, which competes with the supply needs of the community, fauna, and other productive sectors. On the other hand, due to over-exploitation, the productive capacity of soils is diminished, experiencing a marked decrease in their agricultural suitability. The above is usually addressed through the increasing incorporation of fertilizers, mainly nitrogen (N), phosphorus (P) and potassium (K), which in turn, when used excessively and in an uncontrolled manner, negatively impact the environment by disturbing ecosystem balances, through processes such as eutrophication, and the progressive degradation of the physical, chemical, and biological properties of soils (i.e., salinization, erosion, loss of structure, disintegration and loss of the clay fraction, decrease in the population of microorganisms, etc.). For example, chemical fertilizers can cause soil acidification, increasing the availability of heavy metals such as Cd found in phosphate fertilizers derived from phosphate rock [1]. Another example is the salinization and disintegration of the soil structure due to the increase in salts, such as sodium and potassium, which increase the soil pH and promote erosion; but also, water availability for plants is reduced [2]. In addition, hazardous emissions resulting from the hydrolysis of nutrients and the action of microorganisms (i.e., $NH_3$, $N_2O$, etc.), residue accumulation, and pesticide resistance in insects and pathogens pose risks to human health, among other concerns [3–6].

As consequence, agricultural-use products including fungicides, herbicides, nematicides, fertilizer, and other substances are used in large quantities [7]. Particularly, though fertilizers improve the crop production, 60–90% of them are lost due to evaporation, adsorption, degradation, and environmental runoff [8,9]; consequently, the costs of crop production are increased and transferred to the final consumer, restricting its accessibility in the most impoverished communities. In this way, accessibility to food is restricted for the population with lower purchasing power. As a result of the above, new technologies focused on improving the supply of fertilizers, increasing their use, and reducing their losses are highly desired.

Polymer-based controlled-release systems are a technological alternative set out to overcome the disadvantages and side effects of the uncontrolled (or inadequate) use of fertilizers. Recently, the utilization of nano- and micromaterials has been a strong alternative set out to promote the slow delivery of fertilizer to plants [10]. However, whereas slow-release fertilizers are usually prepared by the encapsulating of urea, used as nitrogen source, slow-release systems based on the encapsulation of phosphorus and potassium are limited. Furthermore, the agricultural application and industrialization of nanomaterials are still limited due to limited information on the human and environmental health risks of them; but also, there are technical and scientific reports that show the regulatory difficulties resulting from the uncertainty of the dynamics of nanomaterials when they are released into uncontrolled environments [11]. For instance, ZnO–urea nanoparticles reduce N leaching in the soil because they can be taken up by plant roots, decreasing fertilizer loss. However, they produce oxidative stress in soybean roots, affecting hormone metabolism [12]. Additionally, ZnO nanoparticles cause death in aquatic organism such as *Daphnia magna*, *Lymnaea stagnalis*, *Caenorhabditis elegans* [13], and *P. lividus embryos* [14,15]. This negative effect has also been observed in human cells, where ZnO nanoparticles decrease cell viability, produce DNA damage, and cause carcinogenesis [16]. In humans, medical studies on ZnO nanoparticles are limited, but there are reports of severe pulmonary inflammatory diseases [17]. Another limitation is the intrinsic complexity of the characterization and standardization of its properties during the large-scale manufacturing process, which implies the use of technology the costs of which are transferred to the product, leading to a higher relative cost of these compared to other technologies. On the contrary, materials on the microscale are easier to manufacture, control, characterize and standardize, and consequently both their properties as well as their effects have less uncertainty, which implies lower risks.

On the other hand, eco-friendly hydrogels are highly hydrophilic polymeric materials with the ability to absorb a large amount of water. Interest in these materials for application in agriculture has increased because of their ability to store and release both water and the solutes dissolved in it; therefore, they can be used to release fertilizers and reduce their losses simultaneously with the storage and release of water. In addition, hydrogels can be synthesized from natural raw materials to obtain biodegradable materials with a low or null residuality [18]. Examples of raw materials used for obtaining biodegradable-matrix hydrogels include starch, proteins, and cellulosic materials, among others. In particular, starch is one of the most abundant polysaccharides in nature, together with cellulose and chitin. Starch-based hydrogels have shown many advantages such as nontoxicity, biocompatibility, high biodegradability, efficient encapsulation, and prolonged release [6]. These have been classified into two main groups: (i) starch-based hydrogels resulting from the graft copolymerization of vinyl monomers onto starch in the presence of crosslinker, and (ii) starch-based hydrogels obtained by the direct crosslinking of starch chains among themselves or with other polymer chains [19]. The first group requires the use of petroleum-derivate polymers (i.e., acrylic acid); however, their use decreases the material's biodegradability, limiting its application and causing the generation of persistent solid wastes. On the other hand, direct crosslinking can be induced using biodegradable species with both low and high molecular weight. Some examples are cellulose, collagen, and phosphorous compounds such as phosphorus oxychloride, sodium trimetaphosphate, and sodium tripolyphosphate; also, polyacids such as succinic acid and citric acid (CA) can be used [20–23].

Here, to promote the slow release of phosphorus, the storing of water, and the reduction in P requirements during fertilization operations, starch-based biodegradable hydrogels (SBHGs) were obtained from Cassava starch using CA as crosslinker, and these were characterized and evaluated to be applied in the controlled release of phosphorous, using two phosphorous sources: $KH_2PO_4$ and phosphate rock (phosphorite).

## 2. Results and Discussion

### 2.1. Hydrogel Formation and Characterization

SBHG(7.0%) was obtained from an aqueous solution by the gelatinization of Cassava starch. The purpose of gelatinization was to open the starch granule and thus ease the reaction among the Cassava starch hydroxyl groups and carboxylic acid groups of CA. Thus, steric repulsion is reduced, and the crosslinking reaction is promoted. The spectra of FTIR of Cassava starch, CA and SBHG(7.0%) are shown in Figure 1. For Cassava starch, the bands in the range between 1700 $cm^{-1}$ and 900 $cm^{-1}$ are attributed to glucose (monomer units of amylose and amylopectin) [24,25]. Bands at 997, 925 and 860 $cm^{-1}$ are attributed to the vibration of C–O–C bonds from carbohydrates, and the bands around 1075 $cm^{-1}$ are assigned to C–O stretching. At 1408 $cm^{-1}$, it is possible to see the symmetric deformation and scissoring of $CH_2$ in the Cassava starch. Above 1700 $cm^{-1}$, the stretching band from hydroxyl groups (3323 $cm^{-1}$) and the stretching from C–H groups in the polymer chain (2931 $cm^{-1}$) are identified. In addition, due to the high hydrophilicity, at 1640 $cm^{-1}$, a band attributed to water molecules absorbed into the starch is observed [25–27].

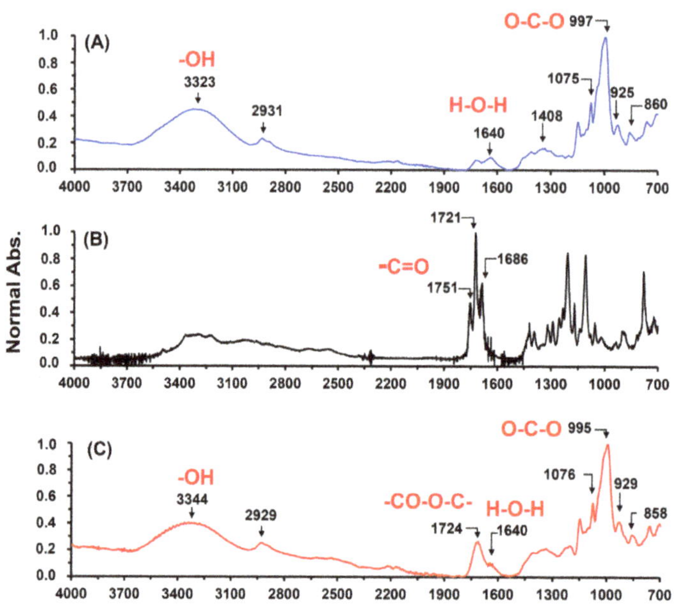


**Figure 1.** ATR-FITR spectra of (**A**) Cassava starch, (**B**) CA, and (**C**) SBHG(7%).

For CA, characteristic bands of carbonyl group (C=O) from carboxylic groups are observed at approximately 1751, 1721, and 1686 cm$^{-1}$; these correspond to stretching vibrations related to three acid groups [28,29]. On the other hand, as shown in Figure 1, for SBHG(7%), it is possible to identify the absorption band of the carbonyl group at 1724 cm$^{-1}$, corresponding to the ester groups formed by the crosslinking reaction between the CA and Cassava starch. Similar bands in the region of 1720–1750 cm$^{-1}$ were related to the stretching vibration of the carbonyl group in the spectra of starch crosslinked with CA by others researchers [20,30,31]. Therefore, the band at 1724 cm$^{-1}$ was considered as evidence of the successful derivation of SBHG. In addition, in the fingerprint region, the bands at 1100–700 cm$^{-1}$ did not show significant changes in their intensities or shifts in their wavenumbers, demonstrating that glucose units in Cassava starch maintain their chemical structure during the formation of the SBHG(7.0%). Additionally, the band around 1640 cm$^{-1}$ evidences water entrapped inside the polymeric matrix, even after drying the material, which is consistent with the highly hydrophilic nature of the material. Similar results were obtained for the other compositions.

The Fisher esterification reaction between carboxyl groups of CA and the hydroxyl of Cassava starch is shown in Figure 2. The reaction is catalyzed by a low pH or by Lewis acids and drops in alkaline pH values due to the formation of carboxylates, which limit the reaction and decrease the esterification yield [29]. Thus, CA catalyzes the esterification, and, at the same time, it allows the crosslinking of starch chains (pH between 2 and 3). In this experiment, the use of another acid was not needed since the CA was sufficient to catalyze the reaction and form the SBHG. Thus, the making of a hydrogel is easy when using only three components: water, Cassava starch, and CA.

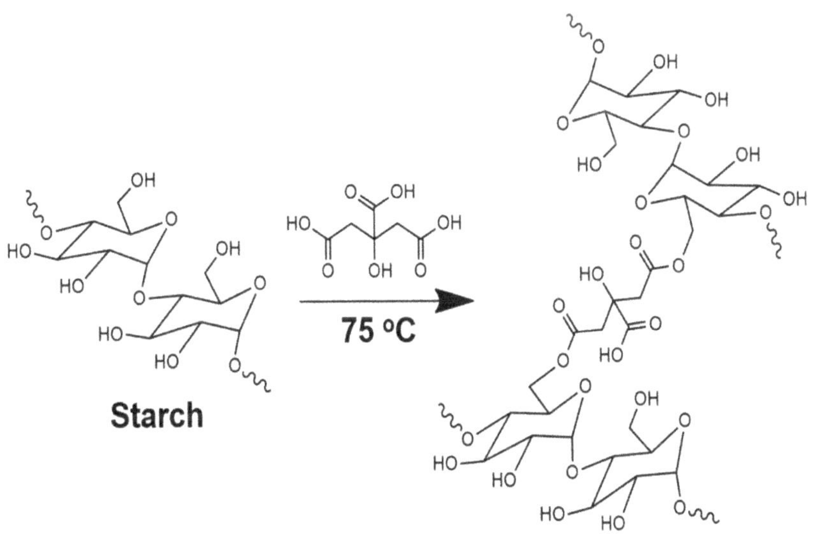


**Figure 2.** Illustration of the crosslinking of starch by Fischer esterification reaction with CA.

To further understand the formation process of SBHG(x), the effect of CA concentration (0.5 to 40.0% *w*/*w*) on esterification was monitored by ATR-FTIR. Figure 3 shows that the main bands change in the ATR infrared spectra of dry SBHG(x) as the CA concentration is increased. In all hydrogels, we observed a signal at 1724 cm$^{-1}$ demonstrating the successful esterification reaction. In addition, as the CA concentration is increased, the intensity of the signal at 1724 cm$^{-1}$ is increased, whereas the band related to hydroxyl groups is decreased (3344 cm$^{-1}$); this behavior is associated with the conversion of hydroxyl groups to ester groups during esterification. The above is also evidenced by the increase in the intensity of asymmetric stretching vibration in ester groups (-CO–O–C-) at 1392 cm$^{-1}$. A similar behavior was described by reference [32] during the esterification of amylose from corn starch and propionic anhydride. In addition, a new band at 1206 $cm^{-1}$, corresponding to the C–H deformation vibration of CA, appears in the fingerprint region, which implies an increase in intensity as CA concentration is increased. From the previously described results, it can be concluded that crosslinking is promoted by the contents of CA, what allows the formation of a three-dimensional network of starch chains.

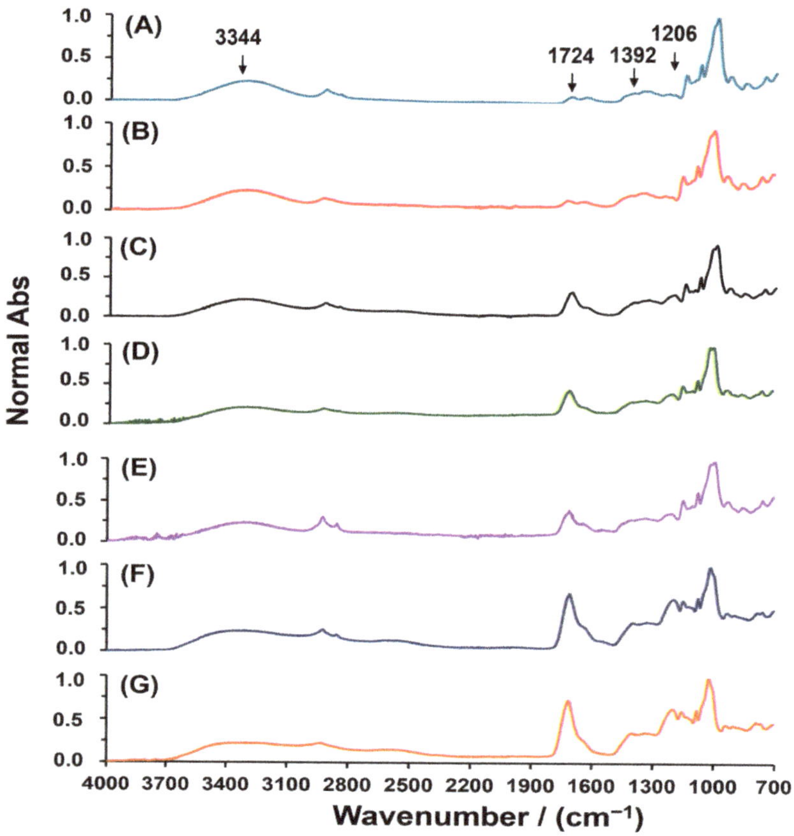


**Figure 3.** ATR-FITR spectra of dry hydrogels of Cassava starch crosslinked with various CA concentrations: 0.5 (**A**), 1.0 (**B**), 5.0 (**C**), 7.0 (**D**), 10.0 (**E**), 20.0 (**F**) and 40.0% *w*/*w* (**G**).

On the other hand, the gel fraction can be understood as an indirect measurement of the percentage of polymer chain that is covalently linked and does not leach out from the hydrogel matrix. Figure 4A shows the effects of CA concentration on gel fraction percentage. The gel fraction increased from 69.1 ± 2.3% to 75.0 ± 0.09% as the CA concentration was increased from 0.5 to 10.0% of CA; after that, the gel fraction decreased to 39.1 ± 0.2 for hydrogel with 40.0% of CA. The results show that for all SBHG, the use of up to 20.0% of CA allows for a high degree of crosslinking between CA and Cassava starch. However, at higher concentration of CA (40.0%), a decrease in gel fraction was observed, which is probably due to the depolymerization reaction promoted by acid hydrolysis, which is a competitive reaction of esterification.

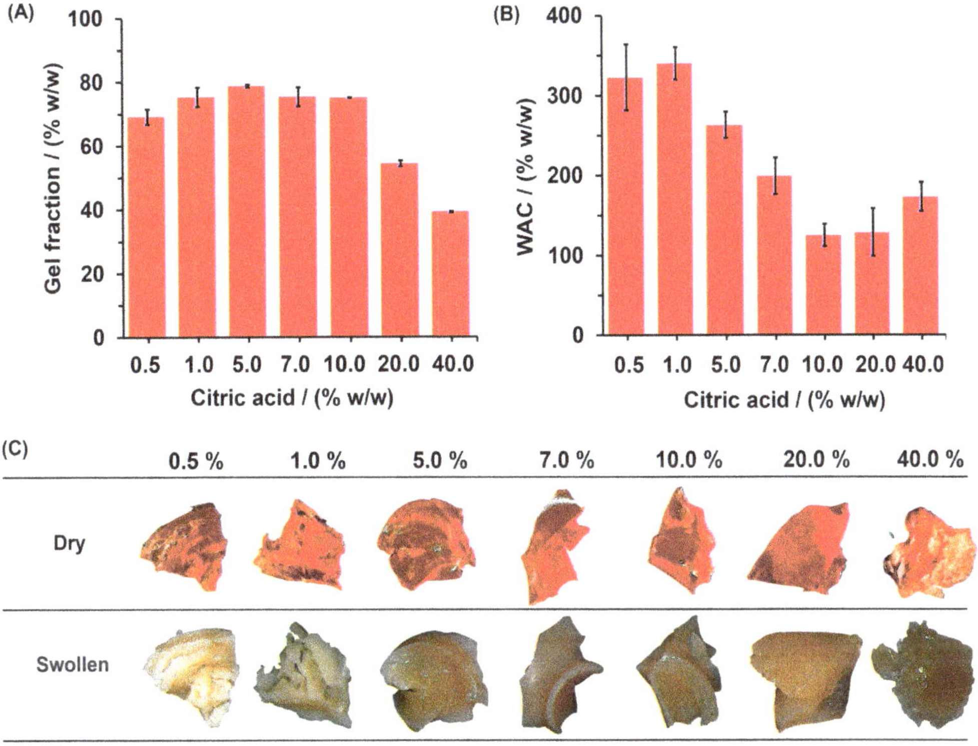


**Figure 4.** Effect of CA concentration on: (**A**) fraction gel and (**B**) WAC of SBHG. In addition, (**C**) digital photos of SBHG with various CA concentrations before (dry) and after the water-absorbing process (swollen).

On the other hand, WAC decreased from 322.9 ± 41.0% to 125.1 ± 13.9% as CA concentration increased from 0.5 to 1.0%. For higher values of CA contents, WAC increased up to 172.9 ± 18.2%, corresponding to SBHG(40.0%) (Figure 4B). The decrease in WAC with the increase in the concentration of CA can be explained by the increase in crosslinikng and the subsequent decreasing of pore size and cavities in the polymer structure, which limits the diffusion of water towards the inside of the hydrogel. The above is consistent with previous studies, wherein a relationship between CA content and the degree of swelling was identified [30,31]. This behavior could be related to the material's appearance as dry or swollen (Figure 4C). When the hydrogel is dry, hydrogels are hard to the touch and have a red-brown hue; however, when they are swollen, hydrogels are softened, increase in volume, and characterized by a whitish hue. After immersion in distilled water, swollen SBHG(7.0%) and SBHG(10.0%) showed less swelling compared to other hydrogels. This is an important characteristic of hydrogels for agricultural use, since it allows a reduction in the fertilizer release time. Thus, a lower rate of water diffusion will be related to lower fertilizer transference from the loading phase (hydrogel) to the surrounding environment, and from there to the site of action (plant).

Though the swelling capacity of a hydrogel is expected to be influenced by medium composition, ionic strength, and pH as a result of changes in chemical and intermolecular interactions inside the hydrogel network, the results shows that there are no significant differences ($p \geq 0.05$—paired student $t$ test) when distilled water is used as the swelling medium in comparison with salt solutions (NaCl, KCl, and $CaCl_2$); therefore, it is concluded that, at the tested concentration, the saline enviroment does not alter intermolecular interactions and does not produce significant changes in the hydrogel's osmotic pressure, which is the driving force of hydrogel swelling. On the other hand, NaOH and HCl had opposing effects on WAC. Thus, whereas NaOH decreased the WAC, HCl increased it (see Figure 5). In both cases, significant differences in WAC were observed with respect to distilled water ($p \leq 0.05$). In an alkaline medium, ester groups are more easily hydrolyzed, producing a decrease in crosslinking, and as a consequence, a part of the water contained in the hydrogel is drained by gravity. On the contrary, although hydrolysis can occur in acidic environments, its kinetic activity is relatively slower compared to alkaline hydrolysis, but also, the acid load inside the hydrogel can be decreased by the protonation of the carboxylic acid groups remaining in the polymer matrix. According to Le Chatelier's principle, the esterefication reaction is controlled by chemical equilibrium, the acid enviroment being one of the factors that promote a shift in equilibrium towards the formation of the ester, with the subsequent decrease in protons in the surrounding environment.

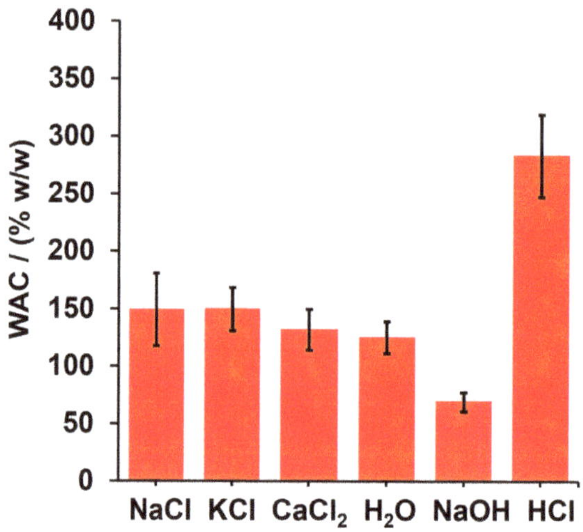


**Figure 5.** WAC of SBHG(10.0%) in different media containing 0.5% $w/v$ of solute (NaCl, KCl, $CaCl_2$, NaOH and HCl).

To determine the effects of the CA concentration on the thermal stability of hydrogels, measurements of TGA were performed on starch, SBHG(1.0%), SBHG(5.0%) and SBHG(10.0%). The results are shown in Figure 6A,B. The starch showed two notable weight losses caused by thermal decomposition. The first one, which occered between 60 and 110 °C and invovled an 8% loss of mass, is attributed to the water absorbed and entrapped inside the starch sample, whereas the second one, at between 250 and 400 °C, corresponds to the thermal degradation of the starch's backbone. The maximum mass loss was identified at 315 °C, according to the derivate curves (Figure 6B).

The SBHGs showed behaviors similar to Cassava starch; however, the first stage, which ranged between 60 and 230 °C, was characterized by an almost linear trend of mass loss as the temperature increased, the mass loss being approximately 20% with respect to the mass of each samples. This stage is related to the elimination of water from the polymer matrix. The second stage started at 250 °C and finished at 400 °C, with its maximum mass loss decreasing as the CA concentration increased in the order of SBHG(1.0%) (at 311 °C) > SBHG(5.0%) (at 275 °C) > SBHG(10.0%) (at 274 °C). After that, at temperatures greater than 274 °C, the mass loss rate in hydrogels decreased, with a residue of about 25% $w/w$ being obtained in the tested samples, ths being greater than that obtained from starch (~12.0%), which indicates a very much improved thermal resistance for SBHG compared to Cassava starch. A similar behavior was observed by reference [33], who concluded that CA increases the resistance to thermal degradation in hydrogels made from CA and carboxymethyl sago starch.

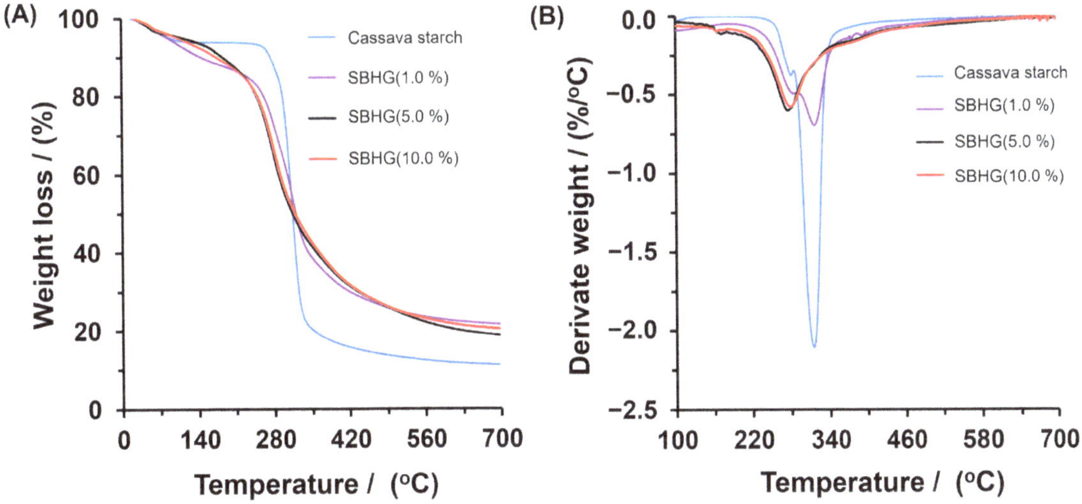


**Figure 6.** (**A**) TGA and (**B**) weight loss derivate curves of Cassava starch, SBHG(1.0%), SBHG(5.0%), and SBHG(10.0%).

## *2.2. In Vitro Studies of Phosphorus Release*

SBHG(10.0%) showed the lowest WAC, indicating that hydrogels with 10% *w*/*w* of CA are characterized by the greatest crosslinking degree within the range of concentrations evaluated. Thus, SBHG(10.0%) was loaded with phosphorus in order to evaluate its release. Thus, the content of phosphorous comprised the sum of phosphorus derived from $KH_2PO_4$ (82.5 $\pm$ 1.6 mg $g^{-1}$) and from phosphorite (68.6 $\pm$ 0.4 mg $g^{-1}$). The SBHG(10.0%)-P showed a visual appearance similar to that of SBHG(10.0%) (Figures 4C and 7A). However, hydrogels containing phosphorite showed a grayish hue. Phosphorite (Figure 7A) is a P-rich rock, normally composed of carbonate hydroxyl fluorapatite ($Ca_{10}(PO_4CO_3)_6F_{2\text{-}3}$) [34], which provides $PO_4^{3-}$ ions to the crops when it is used as a fertilizer. The phosphorite used in this research contained 169.5 $\pm$ 0.32 mg $g^{-1}$ of phosphorus.

The phosphorus release profile from SBHG(10.0%)-P showed a rapid and progressive release, reaching approximately 80% in the first 15 h (Figure 7B), followed by a slight increase up to 88% in 144 h (i.e., a variation of only 8 percentage units over almost ten times as long). When phosphorite was used, the phosphorus released from hydrogel showed a similar behavior to that observed for SBHG(10.0%)-P, releasing almost 73% in the first 15 h until reaching a constant value independent of time. On the other hand, the phosphorus release time decreased for the phosphorite contained in SBHG(10.0%), which delivering around 48% of its phosphorus in the first 15 h, a rate that gradually increased until approximately 50% in 144 h. The results indicate that the phosphorite entrapped in SBHG(10.0%) was delivered slower than the rate observed for other tested materials, which is desirable, since it will aid in creating phosphorus slow-release systems. In addition, according to Figure 7C, phosphorus concentration deacreses as time is increased for all the materials; however, SBHG(10.0%)-phosphorite showed the greatest level of nutrient release after 144 h.

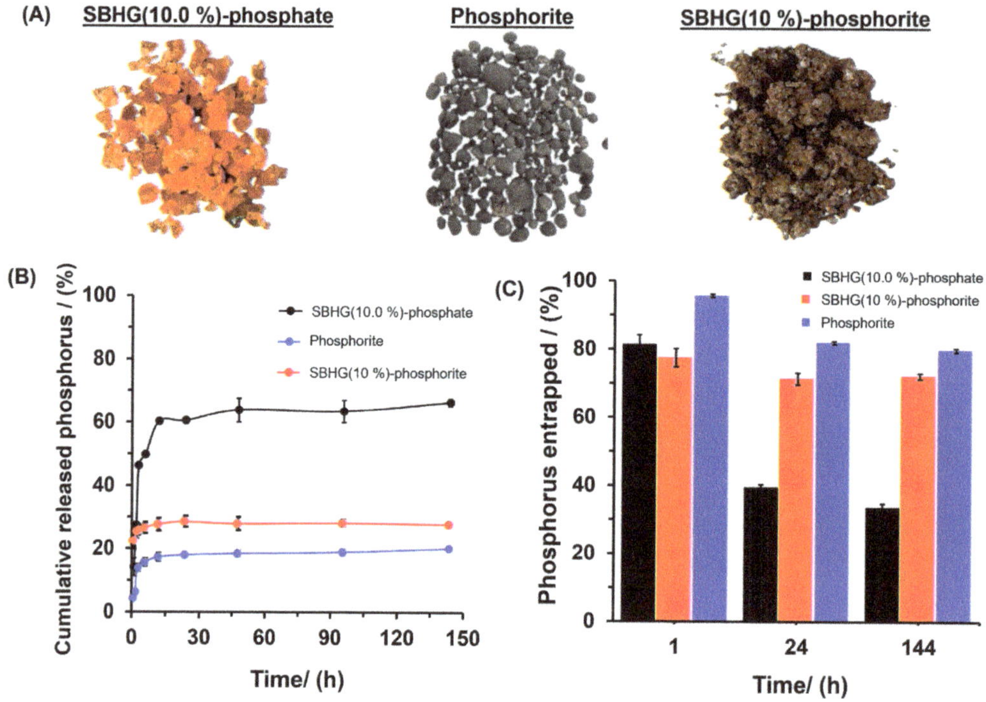


**Figure 7.** (**A**) Digital photos of materials containing P, and (**B**) in vitro release profiles of phosphorus and (**C**) phosphorus entrapped after 1, 24, and 144 h in the release experiment on SBHG(10.0%)-phosphate, phosphorite (without ground), and SBHG(10%)-phosphorite.

Four kinetic models (zero-order, first-order, Higuchi, and Power law) were used to evaluate the mechanism of phosphorus release from hydrogels (see Table 1). The Higuchi model showed higher correlation coefficients ($R^2$) for both SBHG(10.0%)-P and SBHG(10.0%)-phosphorite. This model describes the diffusion of a solute from a matrix with different geometric dimensions and porous systems, wherein the diffusion process occurs via matrix swelling, promoting the diffusion of solute at a constant rate [35,36]. Thus, in both cases, it is possible to identify three stages of the absorption–release processes (loading–unloading): (i) When the hydrogel is introduced into an aqueous solution, due to the difference in water potential inside the hydrogel with respect to the surrounding medium, the solvent and the solutes flow into the interior of the hydrogel, leading to the incorporation of water and solutes, and subsequent matrix swelling. Although this first stage corresponds to the loading of the material and not to its release, it is important because it suggests that after swelling the hydrogel is not in equilibrium, but rather that an adsorption processes has taken place inside it. Flow-mediated incorporation is the result of the dry condition of the material, so diffusion and adsorption are not significant. However, inside the hydrogel, the forces of interaction of the matrix with the medium are different; in principle, the medium is more viscous, which makes diffusive processes difficult, but they are not eliminated. (ii) During the release experiments, the swollen hydrogels are introduced into water, creating a heterogeneous system with two characteristics. The first relates to its two zones, an external zone (water) and an internal zone (a gelled fertilizer solution), separated by a common interface. In this conext, the change in hydric potential between the different states is zero; therefore, it is expected that water flow does not occur in any direction. However, the phosphate ions contained in the polymer matrix can diffuse from the interior of the hydrogel to the surrounding medium (i.e., the chemical potential of the phosphate ion is higher in the hydrogel compared to pure water). Consequently, the first stage is expected to be controlled by diffusion. (iii) In addition to diffusion, dissolution appears to be a determinative process when phosphorite is present. Unlike phosphate ions, which are found in the solution, phosphorite is dispersed in particulate form. Therefore, the viscosity of the surrounding medium reduces the availability of water to promote dissolution. Inside the hydrogel, water molecules interact strongly with the matrix, their mobility is reduced, and the hydration of the phosphotite particles is strongly limited. Consequently, an appreciable fraction of the phosphorite remains retained inside the hydrogel matrix.

**Table 1.** Kinetics parameters of phosphorus release from the SBHG(10.0%) loaded with phosphate and phosphorite.

| Parameters of Model | | Load | |
|---|---|---|---|
| | | Phosphate | Phosphorite |
| Zero-order | $k_0$ $(1 \times 10^{-2} h^{-1})$ | 9.71 | 6.89 |
| | $R^2$ | 0.844 | 0.922 |
| First-order | $k_1$ $(1 \times 10^{-2} h^{-1})$ | 0.19 | 0.15 |
| | $R^2$ | 0.728 | 0.705 |
| Higuchi | $k_H$ $(1 \times 10^{-2} h^{-1})$ | 0.36 | 0.25 |
| | $R^2$ | 0.929 | 0.973 |
| Power law | k $(1 \times 10^{-2} h^{-1})$ | 2.76 | 1.85 |
| | $R^2$ | 0.928 | 0.911 |
| | n | 0.62 | 0.50 |

$k_0$, $k_1$ "$k_H$" and "k" represent the apparent rate constants of respective models, and n is the release exponent of the Power law model.

The above is consistent with Higuchi's model. It can be seen that for SBHG(10.0%), the Higuchi's rate constant was higher when hydrogel was loaded with phosphate than with phosphorite (see Figure 7A). This result shows that SBHG(10.0%)-phosphorite is a promising material that can be used to provide phosphorus gradually in crop production; according to the literature, phosphorus showed a faster release profile from polysaccharide hydrogel. For instance, reference [37] reported a release profile lasting around 40 min for phosphorus derived from cellulose hydrogel. Similar results were reported for monoammonium phosphate fertilizer composed of a sodium carboxymethyl cellulose/hydroxyethyl cellulose blend filled with 5% spherical regenerated cellulose particles, where ~90% of the fertilizer encapsulated was released over 3 h [38].

Previous studies have demonstrated that the addition of organic matter such as humic acid (HA) and fulvic acid produces a positive effect on soils, due to increasing phosphorus mobility and enhanced crop yield [39,40]. Therefore, to evaluate the effect of HA concentration on the delivery behavior of phosphorus from SBHG(10.0%)-phosphorite, the phosphorus release kinetic was determined up to 144 h (Figure 8). In HA 0.05% and 0.10%, after 144 h, the SBHG(10.0%)-phosphorite delivered between 35% and 40% of phosphorus, which is lower than the level of phosphorus released from distilled water (~50%). However, with 0.2% HA, the hydrogel delivered approximately 55% of the phosphorus at 144 h. Thus, the results suggest that a low HA concentration (0.05 and 0.10%) decreases the capacity for phosphorus delivery, whereas at higher concentrations (0.2% of HA), we see a significant enhancement in the release of phosphorus. These results can be attributed to the chemical interaction between the HA molecules and phosphate ions. In an alkaline medium (used to ensure solubility of HA), HA molecules present electrostatic repulsion produced by the deprotonated phenolic and carboxylic groups [41].

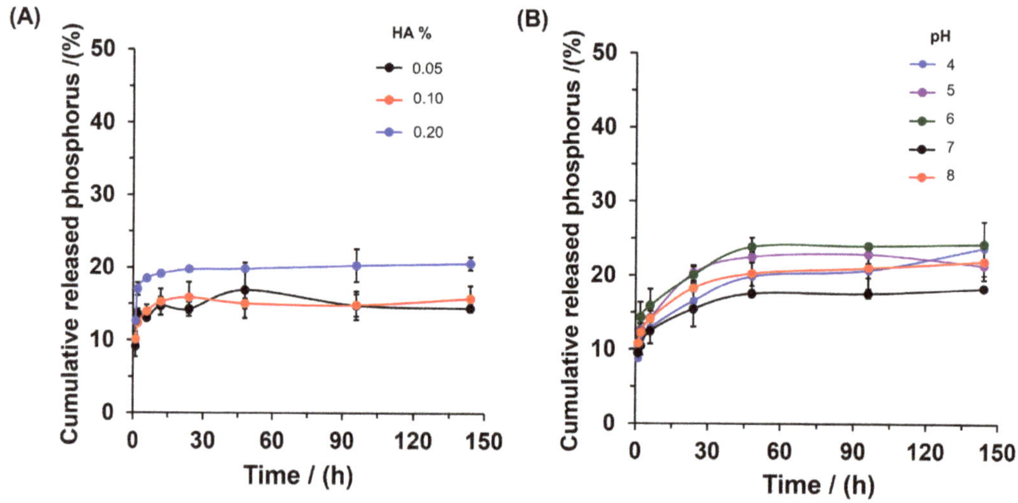


**Figure 8.** In vitro release profiles of phosphorus from SBHG(10.0%)-phosphorite at different concentrations of HA (**A**) and different pH values (**B**). the NaOH 1.0 M solution in deionized water was used as the release medium for HA, and the results are represented as the mean SD of triplicate experiments.

Thus, at low HA concentrations, an unfavorable chemical interaction between $PO_4^{3-}$ ions and HA is expected; as a consequence, the $PO_4^{3-}$ diffusion rate is decreased. However, the aggregation of HA in solution could decrease the electrostatic repulsion between $PO_4^{3-}$ ions and HA; thus, by increasing the favorable interaction between $PO_4^{3-}$ ions and HA, we see a transfer of $PO_4^{3-}$ from the solid phase to the alkaline aqueous solution. These results are in agreement with those reported by reference [42], who found that HA affects the adsorption of $PO_4^{3-}$ on goethite-coated silica sand via batch isotherms and column experiments under steady-state flow conditions. However, the presence of HA could produce a positive effect due to the increase in $PO_4^{3-}$ release, favoring phosphorus' contribution to plant nutrition. On the other hand, the effectof pH on phosphorus delivery was evaluated for SBHG(10.0%)-phosphorite (Figure 8B). It was found that the hydrogel exhibited a tendency to increase phosphorus release in an acidic medium (pH 4–6). As was demonstrated, in an acidic medium, the material showed higher WAC (Figure 5), increasing the content of medium released from the hydrogel. Consequently, under this condition, the solubility of the phosphorite rock entrapped in the hydrogel increased, thereby increasing the nutrient diffusion rate.

The slow release of P from SBHG(10.0%) described above offers an alternative means to provide the nutritional requirements of crops while reducing the negative impact of conventional P fertilizers. Additionally, SBHG could offer an alternative way to overcome the limitations of the large-scale production and commercialization of hydrogel nutrient carriers. For example, PUSA hydrogel, a cellulose-graft anionic polyacrylate superabsorbent polymer applied to encourage water release during dry periods (without fertilizer), showed a higher cost (USD 14–18/kg) with application levels of 2.5–3.0 kg/ha. The high cost of production and manufacturing limited its application in crop production. ALSTA hydrogel, on the other hand, is a more economical alternative (USD 8–9/kg) for use in nutrient release, and is based on potassium polyacrylate, which can reduce fertilizer use [43]. Thus, the low-cost production of SBHG(10.0%), owed to it being composed of only Cassava starch and CA, presents advantages related to commercial viability. Additionally, SBHG is a green, biodegradable material, thus overcoming the biodegradability issue of hydrogels made from synthetic polymers. However, even in this work, the mechanical properties of SBHG were not measured, and its mechanical strength could be limited upon swelling, characteristic of biopolymer-based hydrogels [44]. New experiments should be performed to understand the behavior of SBHG in different soils and its effects on nutrient release to promote plant growth.

## 3. Conclusions

An SBHG was synthesized and applied to encourage the slow release of phosphorus through a low-cost and simple method; thus, from the results, different aspects can be inferred. First, the crosslinking of starch with CA is easily assessed by ATR-FTIR and swelling tests. This is occurs at different concentrations of CA from 0.5 to 40%. In addition, it was shown that the degree of starch crosslinking reduces the WAC and improves the thermal resistance of hydrogels. On the other hand, the release of phosphorus from SBHG(10.0%)-phosphorite was initially rapid, followed by a slow, sustained and continuous release stage until 144 h of immersion. In addition, it was found that the HA concentration promotes the delivery of the nutrient from a hydrogel (SBHG(10.0%) loaded with phosphorite) via the favorable chemical interaction between HA aggregates and $PO_4^{3-}$ ions. In summary, this study has demonstrated that the combined use of SBHG and HA can be useful in promoting the steady and sustained release of phosphorus, which could have high potential utility in the slow release of nutrients in crops; however, the potential use of this hydrogel in plant growth, the effects of field conditions, and its residual time in the soil must be evaluated.

## 4. Materials and Methods

### 4.1. Materials

Cassava starch and phosphorite were obtained from a local market in Cali, Colombia. CA ($C_6H_8O_7$), potassium dihydrogen phosphate ($KH_2PO_4$), sodium chloride (NaCl), potassium chloride (KCl), sodium hydroxide (NaOH), calcium chloride ($CaCl_2$), and hydrochloric acid (HCl) were purchased from Sigma-Aldrich (St. Louis, MA, USA).

### 4.2. Preparation of SBHGs

SBHGs were formed as follows: A mixture of 6.0 g of starch and 50 g of distilled water was combined, stirred, and heated at 75 °C for 30 min until starch gelatinization was achieved. A predetermined quantity of CA was added to obtain a CA concentration of 0.5% *w*/*w*. The mixture was continuously stirred for 1 h at 75 °C. Later, the material was dried at 60 °C for 24 h, washed with distilled water to remove the unreacted CA, and dried again at 60 °C for 72 h. The same procedure was used to form several SBHGs named SBHG(x), where x is the concentration of CA used (x = 0.5, 1.0, 5.0, 7.0, 10.0, 20.0 and 40.0%).

### 4.3. Gel Fraction

The gel fraction was determined following the research reported by reference [45], with some modifications. To determine the gel fraction, the pre-weighed SBHG(x) was taken in a beaker and immersed in a sufficient amount of distilled water for 24 h. After this time, the water was removed, and the material was dried in an oven at 50 °C until the weight was constant. The gel fraction was calculated using Equation (1):

$$\text{Gel fraction}(\%) = \left(\frac{W_f}{W_0}\right) \tag{1}$$

where $W_f$ and $W_0$ are the final dried weight and the initial dried weight, respectively. Note that the gel fraction is a measure of the degree of effectiveness of the starch crosslinking process, performed through esterification with CA.

### 4.4. Swelling Studies

The water absorption capacity (WAC) was determined by the tea bag method [46]. A mass of SBHG(x) was added to a tea bag that had previously been weighed. Later, the tea bag containing the hydrogel was immersed in water for 24 h. After that, the bag with the hydrogel was hung up for 15 min to remove the excess water via gravity, and weighed.

The WAC was calculated using Equation (2). A blank experiment was performed using only tea bag.

$$\mathrm{WAC} = \frac{w_2 - (w_1 + w_0)}{w_1} \quad (2)$$

where $w_0$, $w_1$ and $w_2$ are the masses of the tea bag, the SBHG(x) sample, and the tea bag containing SBHG(x), respectively. These experiments were performed in triplicate.

### 4.5. Physicochemical and Morphological Characterization of SBHG(x)

The SBHG(x) was characterized by Fourier-transformed infrared spectroscopy using the technique of attenuated total reflectance (ATR-FTIR, IR-Affinity-1, Shimadzu, Kyoto, Japan). The spectra were recorded to verify the esterification reaction. The data were acquired from 4000 to 700 $cm^{-1}$ at a rate of 2 $cm^{-1}$ per point, with 20 scans per sample. Thermal analysis was performed using a thermogravimetric analyzer (TGA, SDT-Q600, TA instrument, New Castle, DE, USA). Thus, powdered starch and SBHG(x) were weighed (5–10 mg) and heated from room temperature to 650 °C at a constant heating rate of 10 °C/min under a nitrogen flow of 20 mL per min. The results were evaluated by considering the first-order derivative of the raw weight loss thermogram.

### 4.6. Preparation of Phosphorous-Loaded SBHG(x)

SBHG(10.0%) was produced via the methodology described in Section 4.2, with the addition of $KH_2PO_4$ and phosphorite (previously ground and sieved to 250 μm) to form SBHG with phosphate (10% *w*/*w*) and SBHG with phosphorite (10.0% *w*/*w*). Samples of both materials were digested with concentrated HCl, and the phosphorus concentrations were quantified by UV-Vis spectroscopy (DU-8800DS, Chongqing Drawell Instrument Co., Ltd., Chongqing, China) using a calibration curve via the molybdovanadate method reported by NTC 6259 [47].

### 4.7. Phosphorous Release Kinetics—In Vitro Study

In separate experiments, 0.5 g samples of SBHG(10.0%) loaded with phosphate, and SBHG(10.0%) loaded with phosphorite and phosphorite, were placed in a tea bag. The phosphorous concentration was evaluated in the solution as a function of time. The samples were kept under conditions of constant stirring at 100 rpm, using deionized water as the release medium. Aliquots (0.5 mL) of the resulting solution were taken at specified time intervals over six days, and immediately replaced with fresh medium. Additionally, following the methodology described above, an in vitro release study was performed using SBHG(10.0%)-phosphorite at different HA concentrations (0.05, 0.1, and 0.2% *w*/*v*) and pH values (4, 5, 6, 7, and 8). To determine the effect of HA, NaOH 1.0 M solution in deionized water was used as the release medium. The experiments were carried out in triplicate.

The release mechanism was evaluated using four kinetic models: zero-order, first-order, Higuchi, and Power law. These can be represented via the following Equations (3)–(6):

$$\text{Zero-order model}: \ M_t/M_\infty = k_0 t \quad (3)$$

$$\text{First-order model}: \ M_t/M_\infty = 1 - e^{(-k_1 t)} \quad (4)$$

$$\text{Higuchi model}: \ M_t/M_\infty = k_H t^{\frac{1}{2}} \quad (5)$$

$$\text{Power law model}: \ M_t/M_\infty = k t^n \quad (6)$$

where $M_t/M_\infty$ is the drug fraction released at time t; $k_0$, $k_1$, $k_H$ and k represent the zero-order release constant, first-order release constant, Higuchi constant and Power law constant, respectively [48]. The release constants and the parameters of the models were derived from the fitting of the experimental release data.

**Author Contributions:** A.F.C., investigation, formal analysis, writing—review and editing. M.P., conceptualization, investigation, formal analysis, writing—review and editing. E.M.C., conceptualization, investigation, formal analysis, writing—review and editing. All authors have read and agreed to the published version of the manuscript.

**Funding:** Authors acknowledge Universidad del Valle, Universidad de Córdoba, Universidad de Sucre, and National Planning Department of Colombia, specifically, and the general royalty system (Sistema General de Regalías, SGR) for project BPIN 2020000100027. This research was funded by Dirección General de Investigaciones of Universidad Santiago de Cali under call No. DGI-01-2024.

**Institutional Review Board Statement:** Not applicable.

**Informed Consent Statement:** Not applicable.

**Data Availability Statement:** Data are contained within the article.

**Conflicts of Interest:** The authors declare that they have no known competing financial interests or personal relationships that could have appeared to influence the work reported in this paper.

## References

1. Rashid, A.; Schutte, B.J.; Ulery, A.; Deyholos, M.K.; Sanogo, S.; Lehnhoff, E.A.; Beck, L. Heavy Metal Contamination in Agricultural Soil: Environmental Pollutants Affecting Crop Health. *Agronomy* **2023**, *13*, 1521. [CrossRef]
2. Savci, S. An agricultural pollutant: Chemical fertilizer. *Int. J. Environ. Sci. Dev.* **2012**, *3*, 73. [CrossRef]
3. Schoumans, O.F.; Chardon, W.J.; Bechmann, M.E.; Gascuel-Odoux, C.; Hofman, G.; Kronvang, B.; Dorioz, J.M. Mitigation options to reduce phosphorus losses from the agricultural sector and improve surface water quality: A review. *Sci. Total Environ.* **2014**, *468*, 1255–1266. [CrossRef]
4. Sigurdarson, J.J.; Svane, S.; Karring, H. The molecular processes of urea hydrolysis in relation to ammonia emissions from agriculture. *Rev. Environ. Sci. Biotechnol.* **2018**, *17*, 241–258. [CrossRef]
5. Skrzypczak, D.; Witek-Krowiak, A.; Dawiec-Liśniewska, A.; Podstawczyk, D.; Mikula, K.; Chojnacka, K. Immobilization of biosorbent in hydrogel as a new environmentally friendly fertilizer for micronutrients delivery. *J. Clean. Prod.* **2019**, *241*, 118387. [CrossRef]
6. Mujtaba, M.; Khawar, K.M.; Camara, M.C.; Carvalho, L.B.; Fraceto, L.F.; Morsi, R.E.; Wang, D. Chitosan-based delivery systems for plants: A brief overview of recent advances and future directions. *Int. J. Biol. Macromol.* **2020**, *154*, 683–697. [CrossRef]
7. Qiao, D.; Liu, H.; Yu, L.; Bao, X.; Simon, G.P.; Petinakis, E.; Chen, L. Preparation and characterization of slow-release fertilizer encapsulated by starch-based superabsorbent polymer. *Carbohydr. Polym.* **2016**, *147*, 146–154. [CrossRef] [PubMed]
8. Ghormade, V.; Deshpande, M.V.; Paknikar, K.M. Perspectives for nano-biotechnology enabled protection and nutrition of plants. *Biotechnol. Adv.* **2011**, *29*, 792–803. [CrossRef]
9. Xu, T.; Ma, C.; Aytac, Z.; Hu, X.; Ng, K.W.; White, J.C.; Demokritou, P. Enhancing agrichemical delivery and seedling development with biodegradable, tunable, biopolymer-based nanofiber seed coatings. *ACS Sustain. Chem. Eng.* **2020**, *8*, 9537–9548. [CrossRef]
10. Subramanian, K.S.; Manikandan, A.; Thirunavukkarasu, M.; Rahale, C.S. Nano-fertilizers for Balanced Crop Nutrition. In *Nanotechnologies in Food and Agriculture*; Rai, M., Ribeiro, C., Mattoso, L., Duran, N., Eds.; Springer: Cham, Switzerland; London, UK, 2015; pp. 69–80. [CrossRef]
11. An, C.; Sun, C.; Li, N.; Huang, B.; Jiang, J.; Shen, Y.; Wang, Y. Nanomaterials and nanotechnology for the delivery of agrochemicals: Strategies towards sustainable agriculture. *J. Nanobiotechnol.* **2022**, *20*, 11. [CrossRef]
12. Hossain, Z.; Mustafa, G.; Sakata, K.; Komatsu, S. Insights into the proteomic response of soybean towards $Al_2O_3$, ZnO, and Ag nanoparticles stress. *J. Hazard. Mater.* **2016**, *304*, 291–305. [CrossRef] [PubMed]
13. Wong, S.W.; Leung, P.T.; Djurisic, A.B.; Leung, K.M. Toxicities of nano zinc oxide to five marine organisms: Influences of aggregate size and ion solubility. *Anal. Bioanal. Chem.* **2010**, *396*, 609–618. [CrossRef] [PubMed]
14. Radenac, G.; Fichet, D.; Miramand, P. Bioaccumulation and toxicity of four dissolved metals in *Paracentrotus lividus* sea-urchin embryo. *Mar. Environ. Res.* **2001**, *51*, 151–166. [CrossRef]
15. Falugi, C.; Aluigi, M.G.; Chiantore, M.C.; Privitera, D.; Ramoino, P. Toxicity of metal oxide nanoparticles in immune cells of the sea urchin. *Mar. Environ. Res.* **2012**, *76*, 114–121. [CrossRef] [PubMed]
16. Medici, S.; Peana, M.; Pelucelli, A.; Zoroddu, M.A. An updated overview on metal nanoparticles toxicity. *Semin. Cancer Biol.* **2021**, *76*, 17–26. [CrossRef] [PubMed]
17. Hadrup, N.; Rahmani, F.; Jacobsen, N.R.; Saber, A.T.; Jackson, P. Acute phase response and inflammation following pulmonary exposure to low doses of zinc oxide nanoparticles in mice. *Nanotoxicology* **2019**, *13*, 1275–1292. [CrossRef] [PubMed]
18. Palencia, M.; Lerma, T.A.; Garcés, V.; Mora, M.A.; Martínez, J.M.; Palencia, S.L. Eco-friendly hydrogels. In *Eco-Friendly Functional Polymers: An Approach from Application-Targeted Green Chemistry*; Elsevier: Amsterdam, The Netherlands, 2021; pp. 141–153. [CrossRef]

19. Ismail, H.; Irani, M.; Ahmad, Z. Starch-based hydrogels: Present status and applications. *Int. J. Polym. Mater.* **2013**, *62*, 411–420. [CrossRef]
20. Reddy, N.; Yang, Y. Citric acid cross-linking of starch films. *Food Chem.* **2010**, *118*, 702–711. [CrossRef]
21. Ergün, H.; Ergün, M.E. Modeling Xanthan Gum Foam's Material Properties Using Machine Learning Methods. *Polymers* **2024**, *16*, 740. [CrossRef]
22. Cecone, C.; Fiume, V.; Bracco, P.; Zanetti, M. Maltodextrin-Based Cross-Linked Electrospun Mats as Sustainable Sorbents for the Removal of Atenolol from Water. *Polymers* **2024**, *16*, 752. [CrossRef]
23. Ojstršek, A.; Gorgieva, S. Cellulose Nanofibrils-Reinforced Pectin Membranes for the Adsorption of Cationic Dyes from a Model Solution. *Polymers* **2024**, *16*, 724. [CrossRef] [PubMed]
24. Nybacka, L. FTIR Spectroscopy of Glucose. Bachelor's Thesis, Uppsala Universitet, Uppsala, Sweden, 2016.
25. Garces, V.; García-Quintero, A.; Lerma, T.A.; Palencia, M.; Combatt, E.M. Characterization of cassava starch and its structural changes resulting of thermal stress by functionally-enhanced derivative spectroscopy (FEDS). *Polysaccharides* **2021**, *2*, 866–877. [CrossRef]
26. Nikonenko, N.A.; Buslov, D.K.; Sushiko, N.I.; Zhbankov, R.G. Investigation of Stretching Vibrations of Glycosidic Linkages in Disaccharides and Polysaccarides with Use of IR Spectra Deconvolution. *Biopolymers* **2000**, *57*, 257–262. [CrossRef] [PubMed]
27. Mina, J.; Valadez-González, A.; Herrera-Franco, P.; Zuluaga, F.; Delvasto, S. Physicochemical characterization of natural and acetylated thermoplastic cassava starch. *Dyna* **2011**, *78*, 166–173.
28. Chandrasiri, K.K.; Nguyen, C.C.; Parimalam, B.S.; Jurng, S.; Lucht, B.L. Citric acid based pre-sei for improvement of silicon electrodes in lithium-ion batteries. *J. Electrochem. Soc.* **2018**, *165*, A1991. [CrossRef]
29. Wilpiszewska, K.; Antosik, A.K.; Zdanowicz, M. The effect of citric acid on physicochemical properties of hydrophilic carboxymethyl starch-based films. *J. Environ. Polym. Degrad.* **2019**, *27*, 1379–1387. [CrossRef]
30. Lipatova, I.M.; Yusova, A.A. Effect of mechanical activation on starch crosslinking with citric acid. *Int. J. Biol. Macromol.* **2021**, *185*, 688–695. [CrossRef] [PubMed]
31. Uliniuc, A.; Hamaide, T.; Popa, M.; Băcăiță, S. Modified starch-based hydrogels cross-linked with citric acid and their use as drug delivery systems for levofloxacin. *Soft Mater.* **2013**, *11*, 483–493. [CrossRef]
32. Zhang, Z.; Macquarrie, D.J.; Clark, J.H.; Matharu, A.S. Chemical modification of starch and the application of expanded starch and its esters in hot melt adhesive. *RSC Adv.* **2014**, *4*, 41947–41955. [CrossRef]
33. Keirudin, A.A.; Zainuddin, N.; Yusof, N.A. Crosslinked carboxymethyl sago starch/citric acid hydrogel for sorption of $Pb^{2+}$, $Cu^{2+}$, $Ni^{2+}$ and $Zn^{2+}$ from aqueous solution. *Polymers* **2020**, *12*, 2465. [CrossRef]
34. Douglas, S.; Beveridge, T.J. Mineral formation by bacteria in natural microbial communities. *FEMS Microbiol. Ecol.* **1998**, *26*, 79–88. [CrossRef]
35. Dash, S.; Murthy, P.N.; Nath, L.; Chowdhury, P. Kinetic modeling on drug release from controlled drug delivery systems. *Acta Pol. Pharm.* **2010**, *67*, 217–223. [PubMed]
36. Siepmann, N.A. Peppas, Higuchi equation: Derivation, applications, use and misuse. *Int. J. Pharm.* **2011**, *418*, 6–12. [CrossRef] [PubMed]
37. Li, X.; Li, Q.; Xu, X.; Su, Y.; Yue, Q.; Gao, B. Characterization, swelling and slow-release properties of a new controlled release fertilizer based on wheat straw cellulose hydrogel. *J. Taiwan Inst. Chem. Eng.* **2016**, *60*, 564–572. [CrossRef]
38. Kassem, I.; Ablouh, E.H.; El Bouchtaoui, F.Z.; Kassab, Z.; Hannache, H.; Sehaqui, H.; El Achaby, M. Biodegradable all-cellulose composite hydrogel as eco-friendly and efficient coating material for slow-release MAP fertilizer. *J. Prog. Org. Coat.* **2022**, *162*, 106575. [CrossRef]
39. Delgado, A.; Madrid, A.; Kassem, S.; Andreu, L.; del Carmen del Campillo, M. Phosphorus fertilizer recovery from calcareous soils amended with humic and fulvic acids. *Plant Soil* **2002**, *245*, 277–286. [CrossRef]
40. Yoon, H.Y.; Lee, J.G.; Esposti, L.D.; Iafisco, M.; Kim, P.J.; Shin, S.G.; Adamiano, A. Synergistic release of crop nutrients and stimulants from hydroxyapatite nanoparticles functionalized with humic substances: Toward a multifunctional nanofertilizer. *ACS Omega* **2020**, *5*, 6598–6610. [CrossRef]
41. de Melo, B.A.G.; Motta, F.L.; Santana, M.H.A. Humic acids: Structural properties and multiple functionalities for novel technological developments. *Mat. Sci. Eng. C* **2016**, *62*, 967–974. [CrossRef]
42. Wu, Y.; Li, S.; Chen, G. Impact of humic acids on phosphorus retention and transport. *J. Soil Sci. Plant Nutr.* **2020**, *20*, 2431–2439. [CrossRef]
43. Kaur, P.; Agrawal, R.; Pfeffer, F.M.; Williams, R.; Bohidar, H.B. Hydrogels in agriculture: Prospects and challenges. *J. Environ. Polym. Degrad.* **2023**, *31*, 3701–3718. [CrossRef]
44. Elhaouzi, F.; Mdarhri, A.; Brosseau, C.; El Aboudi, I.; Almaggoussi, A. Effects of swelling on the effective mechanical and electrical properties of a carbon black–filled polymer. *Polym. Bull.* **2018**, *76*, 2765–2776. [CrossRef]
45. Gull, N.; Khan, S.M.; Butt, M.T.Z.; Khalid, S.; Shafiq, M.; Islam, A.; Khan, R.U. In vitro study of chitosan-based multi-responsive hydrogels as drug release vehicles: A preclinical study. *RSC Adv.* **2019**, *9*, 31078–31091. [CrossRef] [PubMed]
46. Palencia, M.; Mora, M.; Palencia, S. Biodegradable polymer hydrogels based in sorbitol and citric acid for controlled release of bioactive substances from plants (polyphenols). *Curr. Chem. Biol.* **2017**, *11*, 36–43. [CrossRef]

47. The Colombian Institute of Technical Standards and Certification, ICONTEC. Soil Quality. Determination of Total Phosphorus. (NTC 6259). 2018. Available online: https://tienda.icontec.org/ (accessed on 15 May 2024).
48. Hou, Y.; Deng, B.; Wang, S.; Ma, Y.; Long, X.; Wang, F.; Qin, C.; Liang, C.; Yao, S. High-Strength, High-Water-Retention Hemicellulose-Based Hydrogel and Its Application in Urea Slow Release. *Int. J. Mol. Sci.* **2023**, *24*, 9208. [CrossRef]

*gels*

MDPI

*Article*

# Synthesis and Characterization of Nanocomposite Hydrogels Based on Poly(Sodium 4-Styrene Sulfonate) under Very-High Concentration Regimen of Clays (Bentonite and Kaolinite)

**Tulio A. Lerma [1,2,*], Enrique M. Combatt [3] and Manuel Palencia [1,*]**

1 Research Group in Science with Technological Applications (GI-CAT), Department of Chemistry, Faculty of Natural and Exact Sciences, Universidad del Valle, Cali 760042, Colombia
2 Mindtech Research Group (Mindtech-RG), Mindtech s.a.s., Monteria 230003, Colombia
3 Department of Agricultural Engineering and Rural Development, Universidad de Córdoba, Monteria 230002, Colombia; ecombatt@fca.edu.co
* Correspondence: t.lerma@mindtech.com.co (T.A.L.); manuel.palencia@correounivalle.edu.co (M.P.); Tel.: +57-3166316951 (T.A.L.); +57-3205271934 (M.P.)

**Abstract:** The aim of this work was to synthesize and study the functional properties of polymer-clay nanocomposite (PCNCs) based on poly(sodium 4-styrene sulfonate) (NaPSS) and two types of clay in the dispersed phase: bentonite and kaolinite, in order to advance in the development of new geomimetic materials for agricultural and environmental applications. In this study, the effect of adding high concentrations of clay (10–20 wt. %) on the structural and functional properties of a polymer–clay nanocomposite was evaluated. The characterization by infrared spectroscopy made it possible to show that the PCNCs had a hybrid nature structure through the identification of typical vibration bands of the clay matrix and NaPSS. In addition, scanning electron microscopy allowed us to verify its hybrid composition and an amorphous particle-like morphology. The thermal characterization showed degradation temperatures higher than ~300 °C with Tg values higher than 100 °C and variables depending on the clay contents. In addition, the PCNCs showed a high water-retention capacity (>2900%) and cation exchange capacity (>112 meq/100 g). Finally, the results demonstrated the ability of geomimetic conditioners to mimic the structure and functional properties of soils, suggesting their potential application in improving soil quality for plant growth.

**Keywords:** soil polymer conditioner; organoclay; hybrid composite; geomimetic

## 1. Introduction

Polymer–clay nanocomposites (PCNCs) are hybrid organo-inorganic systems formed by dispersion of one clay, into relatively small quantities, in a polymer-nature continuous phase. In these materials, the dispersed phase has a marked effect on the mechanical, barrier and thermal properties, which usually increase, reach a maximum value, and subsequently experience a decrease as clay content is increased. The maximum value is generally identified between 3 and 5 wt. %, therefore, studies for the development of PCNCs at higher clay concentrations are very scarce (i.e., 5–10 wt. %), but also, these are practically non-existent for very high clay concentration regimens (upper to 10 wt. %). The above is mainly due to the rise of structural materials for applications where mechanical, barrier and thermal properties are desired; however, within the framework of functional polymers, the areas of application, processability and properties of interest go beyond those properties previously mentioned. Some examples are hydrogels, ionic exchangers, membranes, sensors, among others [1,2].

On the other hand, soil plays a vital role in supporting terrestrial ecosystems and sustaining life on Earth. According to the Food and Agriculture Organization of the United Nations (FAO), it is estimated that by 2050, agriculture will have to double the production of food, livestock fodder and biofuels; in addition, to transform the way food is processed and consumed to meet the needs of approximately ten billion people in the world [3]. However, to date, excessive and indiscriminate practice of anthropogenic activities on soils, including industrialization and agriculture, has led to the alteration of their properties and their undeniable degradation, which currently poses important environmental challenges [4]. Furthermore, in places where the degree of soil degradation is very high, the expansion of the agricultural frontier becomes relevant, generating deforestation, soil erosion, nutrient depletion, and alteration of environmental biodiversity [5].

In recent decades, practices to improve physicochemical properties, increase soil fertility and promote healthy plant growth have been developed and promoted, which are based on crop rotation practices, the use of crops of coverage, the application of no-till agriculture, precision agriculture, among others [6]. In fact, practices based on the use of agrochemicals are widely used to increase crop yields; however, excessive, and indiscriminate use generates harmful effects on the health of soils and surrounding water bodies [7]. Consequently, soil conditioners have acquired relevance through being used as natural or synthetic materials, which modulate different characteristics of the soil, such as its structure, texture, water retention capacity, pH, and nutrient content. Among the best-known conditioners, a wide variety of materials stand out, such as compost, manure, peat, clays, polymers, and various types of organic and inorganic materials [8].

Particularly, the use of PCNCs has gained attention as a promising approach to address problems related to the restoration of degraded soils [9,10]. Furthermore, the progress in the development of geomimetic soil conditioners (materials whose design and function are inspired by natural systems of geological origin, e.g., the soil, and allow improvement in the physicochemical properties of degraded soils or soils with low agricultural suitability) has allowed progress in the construction of materials based on PCNCs capable of imitating the structures, biological/physicochemical properties, or functions of soils [11–13]. Thus, among the countless combinations that have not yet been explored, the synergistic interaction between NaPSS and clays has attracted significant attention towards the development of advanced materials [14,15].

Particularly, in soil science and agriculture Poly(sodium 4-styrene sulfonate) (NaPSS) has been gaining great interest for its ability to improve soil structure, water retention, and nutrient availability [16]. In addition, NaPSS is widely used as a model polyelectrolyte (polymers containing ionic and/or ionizable functional groups) system [17], as a flocculating agent to eliminate heavy metals and other contaminants in water treatment processes [18,19], as raw material for manufacturing of ion exchange membranes [20], personal care products and medicines [21]. The NaPSS is characterized by having an alkyl main chain with aryl sulfonate groups. Thus, sulfonate groups give it exceptional properties including high solubility, biocompatibility, and ion exchange capacity, among others. By virtue of these groups, NaPSS exhibits ion exchange properties allowing it to interact with soil particles and humified organic matter, promoting flocculation of colloidal soil particles and subsequent stabilization of aggregates, and finally influencing soil properties and ecosystemic functions. Therefore, the use of NaPSS in soils promises to be an alternative to mitigate soil erosion, improve fertility and promote plant growth [22].

On the other hand, clays are particles of aluminosilicates minerals that are very abundant in nature with a diameter less than 0.002 mm. Clays encompass a diverse group of natural minerals with layered structures, including montmorillonite, kaolinite, smectite, among others. The unique arrangement of layers provides clays with exceptional properties, such as high surface area, swelling capacity and ion exchange capacity [23,24]. Specifically, bentonite and kaolinite are two types of clays with significant differences in their chemical composition, crystal structure, and physicochemical properties, making them useful in the petroleum and gasification industry for the purposes of ceramic manufacturing and wastewater treatment [25,26]. Thus, bentonite is a clay rich in montmorillonite minerals, with cation exchange capacity due to its 2:1 laminar structure, composed of an octahedral layer of alumina between two tetrahedral layers of silica with the presence of intercalated ions [27]. Additionally, it has a high capacity to absorb water, making it valuable in oil and gas drilling applications. In contrast, kaolinite, known for its whiteness and fineness, has a 1:1 lamellar structure, a tetrahedral layer of silica and another octahedral layer of alumina; however, it has low adsorbent properties due to its low specific surface area and minimal isomorphic substitution, which gives it high molecular stability, low shrinkage, plasticity, swelling and cohesion, making it valuable in the manufacture of ceramics, paper, and ceramic products [28].

Based on the above, the combination of NaPSS with clay emerges as an opportunity to create PCNCs with synergistic properties that exceed those of the individual components, providing improvements in their mechanical resistance, thermal stability, and functional properties; but also, these can be used in the improvement and restoration of degraded soils. Thus, the objective of this work was to synthesize and study the functional properties of clay–polymer composites based on NaPSS and two types of clay in the dispersed phase: bentonite and kaolinite, to advance the development of new geomimetic materials for agricultural and environmental applications. For this, clays were modified with organosilanes for the incorporation of vinyl groups in its structure, which act as anchor points for the growth of NaPSS polymer chains. Thus, here it is proposed that PCNCs can be used to imitate the clay-humin-humic acid structures present in soil particles and improve their cationic exchange properties, water retention and aggregate formation.

## 2. Results and Discussion

### 2.1. Organoclays Synthesis

Figure 1A,B show IR-ATR spectra of unmodified bentonite and kaolinite (dashed lines) and modified with tClVS (solid line), respectively. Thus, in the analysis of the region below 2000 $cm^{-1}$ in all of the cases spectra, at ~988 $cm^{-1}$ and ~908 $cm^{-1}$ the presence of two characteristic bands of clay minerals can be seen; these signals were associated with the tension vibration of Si-O-Si groups and deformation vibration of Si-OH groups, respectively. Likewise, at ~1632 $cm^{-1}$ the deformation vibration signal of the interlaminar water molecules contained by the clay and organoclays was identified, this being of lower intensity for the kaolinite spectra [29]. Finally, only in the spectra of clays modified with tClVS (solid line), a weak signal was identified at ~1404 $cm^{-1}$, linked to the in-plane deformation vibrations of the vinyl group, $=CH_2$, anchored on the surface of the clays. Now, for the analysis of the spectral region between 2500 and 3800 $cm^{-1}$, FEDS analysis was used for adequate identification and assignment of signals in areas of high overlap [30]. The IR-ATR-FEDS spectra of bentonite–tClVS and kaolinite–tClVS in the selected analysis region are shown in Figures 1C and 1D, respectively. In general, in both spectra the presence of typical signals of aluminosilicates were observed, which differ in intensity because of the structural differences between bentonite and kaolinite, which alter their ability to interact and absorb water molecules [31]. At ~3682 and ~3621 $cm^{-1}$, the tension vibration signals of hydroxyls from silanol groups and clays were identified, respectively. Likewise, at ~3436 $cm^{-1}$, a signal associated with tension vibrations of absorbed water molecules was observed for both spectra. In the spectral region below 3100 $cm^{-1}$, characteristic signals of organic vinyl functional groups incorporated into bentonite and kaolinite were observed. Thus, the tension vibration signal from the $=CH_2$ group was observed at ~3060 $cm^{-1}$, whereas a symmetric and asymmetric tension vibration signal from the =CH group was identified at ~2960 $cm^{-1}$.

Based on the above, the identification of typical spectral bands of the C-H bonds belonging to the vinyl group (organic nature fragment) in the spectra of the organoclays allowed the verification of their adequate modification. Thus, the signals are consistent with the procedure for modification of clays with tClVS to obtain organoclays with the ability to react by free radical polymerization. Note that tClVS is characterized by having chlorine atoms in its structure in Si-Cl bonds, which can be anchored covalently by nucleophilic substitution reactions on clay surface; but also, tClVS has a vinyl group in its structure, which can intervene in free radical polymerization reactions. Thus, tClVS acts as a covalent bonding molecule or a coupling agent between the inorganic–organic fractions of composites and allows the obtaining of materials with greater stability in contrast to materials obtained through non-covalent interactions [32,33].

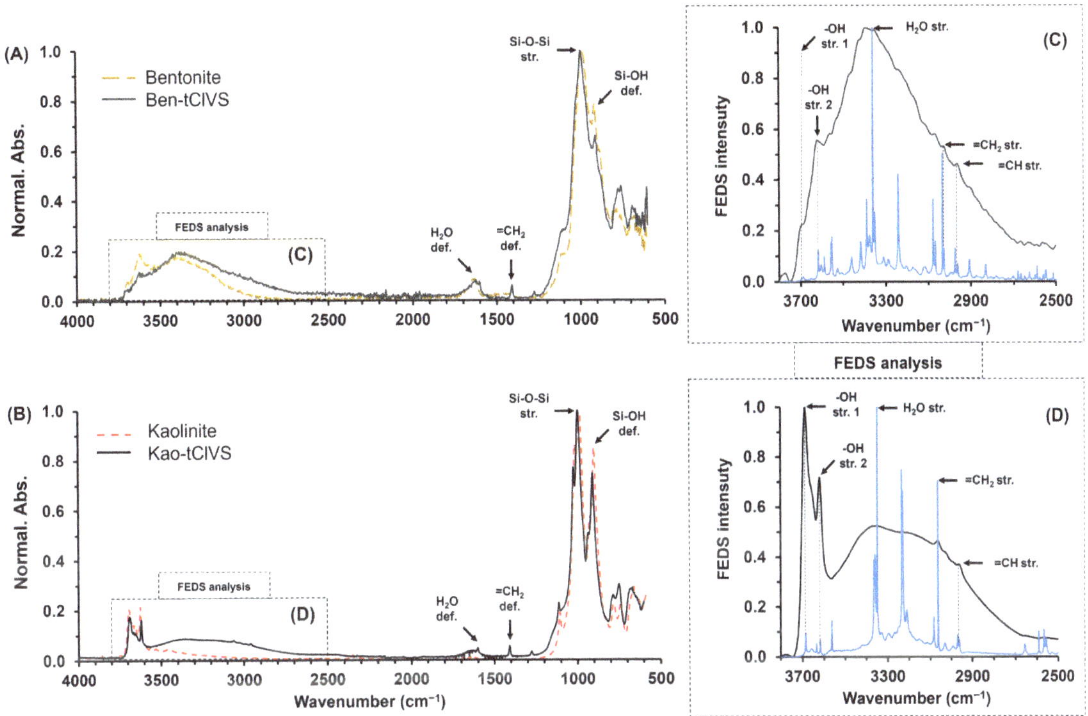


**Figure 1.** IR-ATR spectra of bentonite (dashed line) and bentonite–tClVS (solid line) (**A**) and kaolinite (dashed line) and kaolinite–tClVS (solid line) (**B**). IR-ATR-FEDS spectra of bentonite–tClVS (**C**) and kaolinite–tClVS (**D**) between 2500 and 3800 $cm^{-1}$.

On the other hand, Figure 2 shows the results of morphological analysis by SEM (image left), the EDS spectrum for determination of elemental composition and a digital image (top right image) for both kaolinite–tClVS and bentonite–tClVS. The EDS analysis zone corresponds to the cross enclosed in the red circle in the SEM image. Thus, at the macroscopic level, from the digital images in Figure 2, it can be seen that tClVS-modified clays did not show significant changes in their appearance after modification. Only the bentonite–tClVS organoclay showed a slight lightening in its color after synthesis; coloration caused by the presence of iron ions in its structure and their loss during washing processes [34]. At the microscopic level in the SEM images, a wide distribution in sizes and the absence of a defined shape were observed in kaolinite–tClVS and bentonite–tClVS particles (Figures 2A and 2B, respectively). Regarding the elemental characterization by EDS of the tClVS-based organoclays, the results are shown in Table 1. In general, it was observed that the kaolinite–tClVS and bentonite–tClVS organoclays presented a higher content of Si and O in contrast to the elements Al and Fe, all of them being typical constituents of aluminosilicates [35]. Furthermore, it was possible to determine by EDS analysis the presence of characteristic elements of tClVS, i.e., C and Cl, in the structure of the organoclays. Thus, C and Cl contents for kaolinite–tClVS were 8.65% and 5.74%, respectively; for its part, in the same order, the values for bentonite–tClVS were 7.72% and 2.46%.

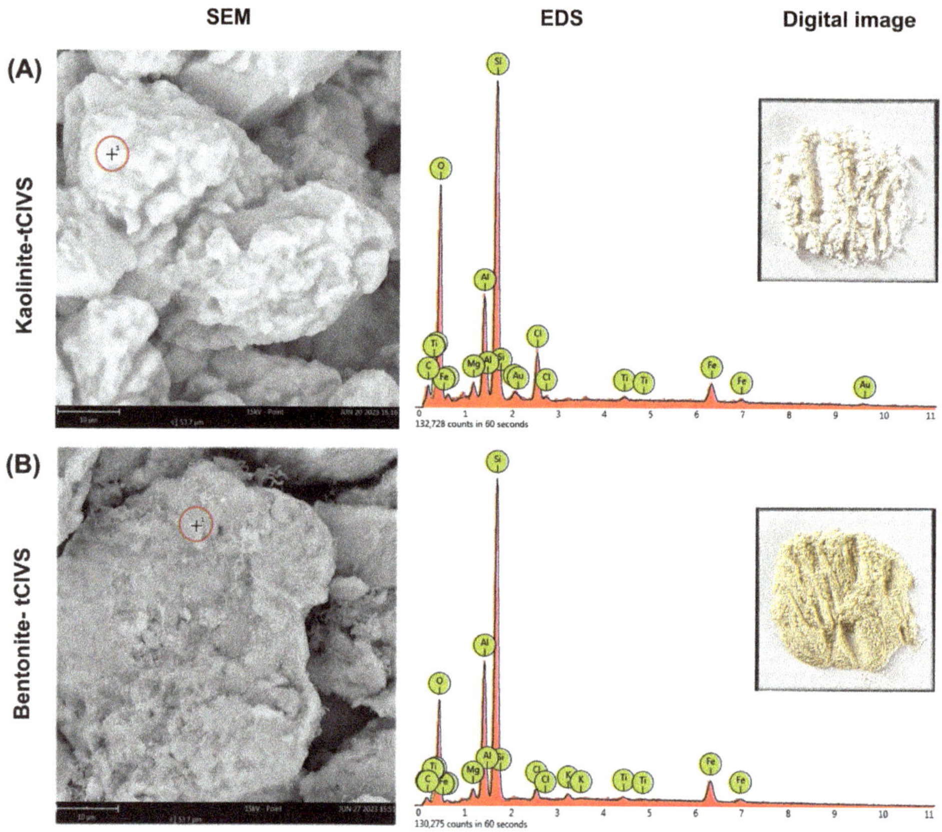


**Figure 2.** SEM images (left imagen), EDS spectrum for the determination of elemental composition (bottom right image, EDS analysis zone corresponds to the cross enclosed in the red circle in the SEM image) and digital images (top right image) of bentonite and kaolinite modified with tClVS: kaolinite–tClVS (**A**) and bentonite–tClVS (**B**).

**Table 1.** Elemental contents determined by EDS, size, zeta potential, WSC and CEC of clays (kaolinite and bentonite) and modified clays (kaolinite–tClVS and bentonite–tClVS).

| Sample | Elemental Composition (% *w*/*w*) | | | | | | Particle Size (nm) | Zeta Potential (mV) | WSC (% *w*/*w*) | CEC (meq/100 g) |
|---|---|---|---|---|---|---|---|---|---|---|
| | O | Si | Al | Fe | C | Cl | | | | |
| Bentonite | --- | --- | --- | --- | --- | --- | 268 ± 16 | −38.3 ± 0.2 | 621 ± 19 | 104 ± 3 |
| Kaolinite | --- | --- | --- | --- | --- | --- | 555 ± 20 | −38.8 ± 1.3 | 210 ± 90 | 57 ± 4 |
| Bentonite–tClVS | 31.55 | 31.75 | 13.02 | 11.90 | 7.72 | 2.46 | 1144 ± 333 | −33.3 ± 4.7 | 300 ± 38 | 43 ± 5 |
| Kaolinite–tClVS | 52.07 | 19.96 | 7.09 | 6.49 | 8.65 | 5.74 | 776 ± 34 | −16.6 ± 2.3 | 145 ± 43 | 38 ± 2 |

O: oxygen, Si: silicon, Al: aluminum, Fe: iron, C: carbon and Cl: chlorine. The elemental composition was determined by EDS, the particle size and Z potential were determined by DLS.

Finally, Table 1 shows the results of particle size and Z potential which were measured through DLS for both bentonite–tClVS and kaolinite–tClVS as their precursors. Initially, it was observed that bentonite and kaolinite had particle sizes of 268 ± 16 nm and 555 ± 20 nm, respectively. However, after modification with tClVS, it was observed that the obtained organoclays showed an increase in their size. Thus, bentonite–tClVS and kaolinite–tClVS presented particle sizes of 1144 ± 333 nm and 776 ± 34 nm, respectively. This variation and the increase in particle size are due to changes in the surface energies of clay particles caused by the incorporation of organosilane tClVS [36]. Covalent insertion of tClVS modifies the hydrophilic groups on the surface of the clay particles, decreasing their affinity for aqueous media and promoting their agglomeration [37,38]. On the other hand, the zeta potentials obtained at natural pH for bentonite–tClVS ($-33.3 \pm 4.7$ mV) and kaolinite–tClVS ($-16.6 \pm 2.3$ mV) showed a decrease in the measured values with respect to the unmodified clays: bentonite ($-38.3 \pm 0.2$ mV) and kaolinite ($-38.8 \pm 1.3$ mV). The decrease in the potentials is attributed to changes caused at the surface level of the clay minerals because of incorporating tClVS, which decreases the density of polar groups and increases the nonpolar fraction after the incorporation of vinyl groups [39,40]. Thus, the determination of the typical vibration bands of the vinyl organic groups through IR-ATR-FEDS, the identification of the C and Cl elements in the structure of the organoclays through EDS, and the identification of variations in the particle size and Z potential by DLS, allowed for verification of the adequate synthesis of organoclays from tClVS with bentonite or kaolinite.

### 2.2. Geomimetic Soil Conditioners Synthesis

After being obtained, bentonite–tClVS and kaolinite–tClVS were used together with NaSS for the synthesis of geomimetic conditioners. For this, the preparation of geomimetic conditioners was carried out through the formation of hybrid composites using the in situ polymerization technique, that is, the NaSS was polymerized via free radicals in the presence of kaolinite–tClVS or bentonite–tClVS; thus, the anchoring of NaPSS chains to the surface of the clays and their adequate dispersion in the material was performed [41,42]. In this work, clay–NaPSS hybrid composites were synthesized with clay proportions of 10.0 and 20.0% (i.e., under very high concentration regimen). The composition and name of the geomimetic conditioners are given in Table 2.

**Table 2.** Sample identification, composition and elemental analysis determined by EDS of PCNCs based on kaolinite–NaPSS and bentonite–NaPSS.

| **Composition (% *w/w*)** | **Sample** | | | |
|---|---|---|---|---|
| | **Kao-NaPSS-10** | **Kao-NaPSS-20** | **Bent-NaPSS-10** | **Bent-NaPSS-20** |
| NaSS | 90.0 | 80.0 | 90.0 | 80.0 |
| kaolinite–tClVS | 10.0 | 20.0 | --- | --- |
| bentonite–tClVS | --- | --- | 10.0 | 20.0 |
| O | 44.91 | 34.59 | 46.35 | 41.17 |
| C | 18.75 | 35.97 | 13.50 | 16.07 |
| S | 8.60 | 13.74 | 10.08 | 13.27 |
| Na | 17.54 | 11.14 | 12.10 | 11.44 |
| Si | 0.91 | 1.96 | 5.58 | 8.56 |
| Al | 1.43 | 2.59 | 3.32 | 4.40 |

O: oxygen, C: carbon, S: sulfur, Na: sodium, Si: silicon and Al: aluminum. The elemental composition was determined by EDS.

Figure 3 shows the IR-ATR-FEDS spectra of PCNCs Kao-NaPSS-10 (A), Kao-NaPSS-20 (B), Bent-NaPSS-10 (C), and Bent-NaPSS-20 (D) (being Kao = kaolinite, Bent = bentonite, and NaPSS = sodium poly(styrene sulfonate), where clay percentages are given by 10 and 20 in the end of each notation). In general, typical signals of the organic group's constituents of the NaPSS matrix were observed in all spectra; as well as characteristic signals of the bentonite–tClVS and kaolinite–tClVS. In both cases, it was possible to identify at ~3675 $cm^{-1}$ and ~3610 $cm^{-1}$ the stretching vibrations of silanol groups and interlaminar hydroxyl groups constituting the clays. In addition, two high-intensity signals related to tension vibrations of Si-O-Si groups (~1036 $cm^{-1}$) and deformation vibration of Si-OH groups (~988 $cm^{-1}$) were observed. All the above signals are characteristics of clay fraction in the composites [43].

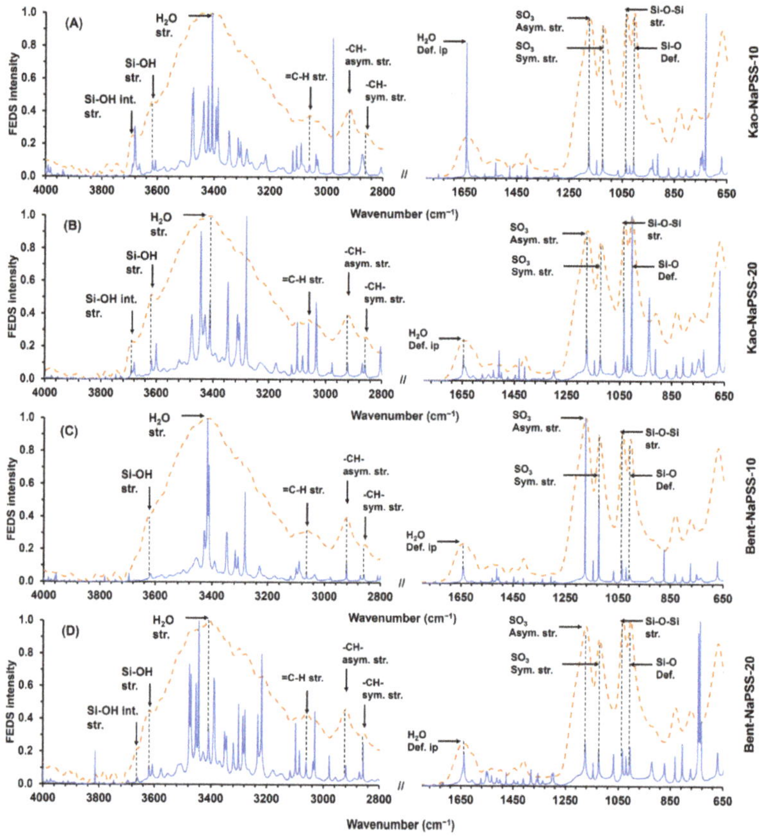


**Figure 3.** IR-ATR-FEDS spectra of PCNCs: Kao-NaPSS-10 (**A**), Kao-NaPSS-20 (**B**), Bent-NaPSS-10 (**C**) and Bent-NaPSS-20 (**D**). Dashed line IR-ATR spectra and solid line FEDS spectra. (being Kao = kaolinite, Bent = bentonite, and NaPSS = sodium poly(styrene sulfonate), where clay percentages are given by 10 and 20 in the end of each notation).

On the other hand, in all the IR-ATR-FEDS spectra of the all synthesized PCNCs shown in Figure 3, it was possible to identify at ~3050 $cm^{-1}$ the stretch vibration of the C-H bond from NaPSS aromatic ring. In addition, signals of asymmetric and symmetric stretch vibrations, at ~2940 $cm^{-1}$ and ~2860 $cm^{-1}$, respectively, related to C-H bonds of methyl and methylene groups were observed. Likewise, symmetric and asymmetric stretch vibrations of sulfonate groups were observed at ~1170 $cm^{-1}$ and ~1116 $cm^{-1}$, respectively [44]. From the above, it is possible to corroborate the inorganic-organic hybrid constitution of the synthesized materials. Finally, stress and strain vibration signals of water molecules constituting the clay fraction and are occluded in NaPSS matrix were observed at ~3430 $cm^{-1}$ and ~1650 $cm^{-1}$.

Continuing with morphological and elemental characterization of PCNCs, Figure 4 shows images obtained by digital photography (right image) and SEM (left image), EDS spectrum (bottom center image) and results of EDS mapping (upper center images) obtained from PCNCs: Kao-NaPSS-10 (A), Kao-NaPSS-20 (B), Bent-NaPSS-10 (C) and Bent-NaPSS-20 (D). In general, it was observed that materials obtained at a macroscopic level presented a white-cream hue for those based on kaolinite–NaPSS and a more intense white-yellow hue for bentonite–NaPSS composites. In both cases, the materials with clay contents of 20.0% $w/w$ were those that presented the highest intensity. Likewise, irregular particulates with a size less than 2.0 mm were observed for all materials. Now, at the microscopic level, the results of SEM analysis showed that PCNCs had fractions of bentonite and kaolinite in their structure. Furthermore, when distribution profiles of Si were analyzed, it was observed that it was distributed throughout the polymer matrix; however, some areas of greater intensity were characterized by higher Si contents, evidencing shortcomings related to the clay dispersion process in the synthesis of composites [45].

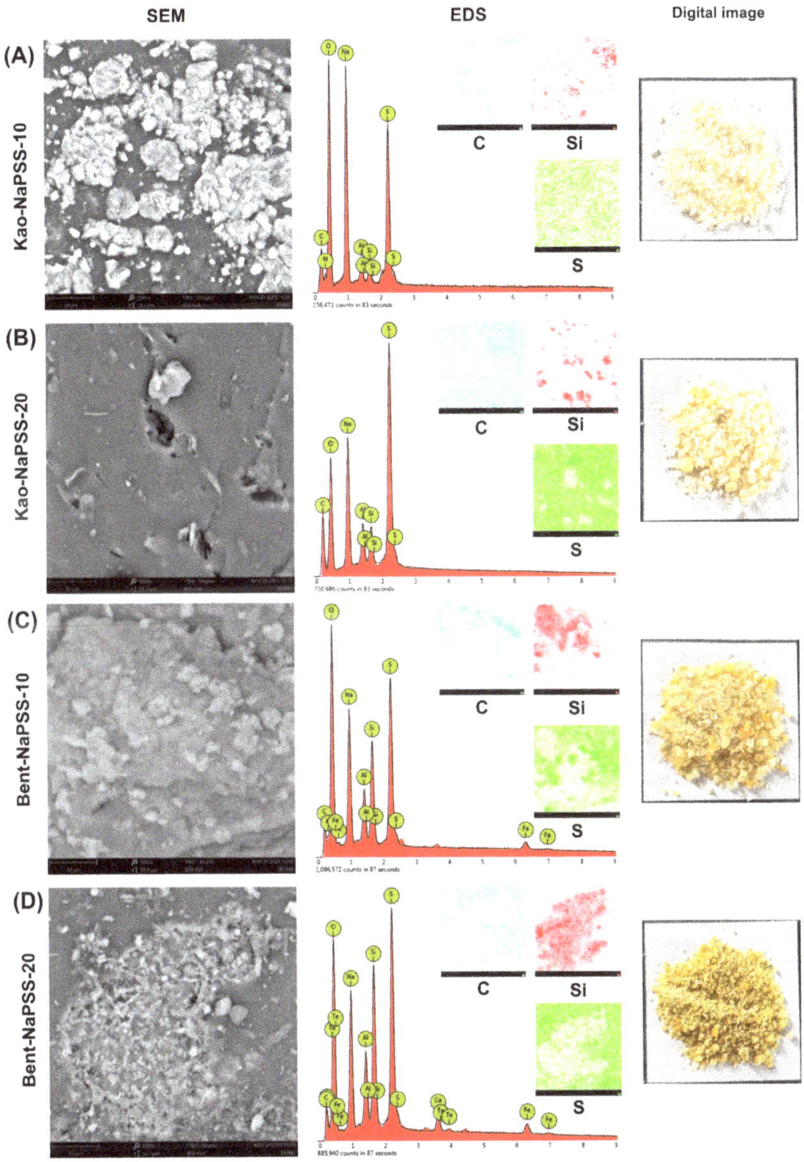


**Figure 4.** SEM images (left image), EDS spectrum for the determination of elemental composition (bottom center image, EDS analysis was carried out in mapping mode of the corresponding SEM image), results of EDS mapping (upper center images, each color represents an element, C (carbon), S (sulfur) and Si (silicon)) and digital images (right image) of PCNCs: Kao-NaPSS-10 (**A**), Kao-NaPSS-20 (**B**), Bent-NaPSS-10 (**C**) and Bent-NaPSS-20 (**D**).

Finally, results of compositional analysis of PCNCs are shown in Table 2. The results obtained show that the C, S, Na and O contents represent the majority in contrast to the Si and Al contents. Particularly, NaPSS is constituted by C, H, S, O and Na, and because of it is main component of PCNCs, a high compositional similarity between these materials and pure NaPSS is expected. Likewise, the presence of non-constituent elements of NaPSS provides evidence that the inclusion of bentonite and kaolinite into the polymer matrix has occurred.

*2.3. Geomimetic Soil Conditioners Thermal Characterization*

Figure 5 shows the results of thermal characterization by TGA (solid line) and derivative thermogravimetric curve (dashed line, mass/temperature derivation) of synthetized PCNCs: Kao-NaPSS-10 (A), Kao-NaPSS-20 (B), Bent-NaPSS-10 (C) and Bent-NaPSS-20 (D). The results showed that the hybrid materials had four stages of mass loss related to water loss and structural degradation. Thus, kaolinite-based composites, namely Kao-NaPSS-10 and Kao-NaPSS-20 (see Figures 5A and 5B, respectively), showed an initial stage of mass loss with a loss of ~16%. This loss was associated with the elimination of water molecules occluded in the polymer matrix and absorbed in clay minerals. Next, a second stage of mass loss was observed, that begins around ~340 °C and extends to ~600 °C, presenting losses of 24.06% and 26.06% for Kao-NaPSS-10 and Kao-NaPSS-20, respectively. Likewise, a third stage of mass loss was observed, in temperature ranges from ~600 °C to ~1000 °C, of 30.98% and 20.76%, for Kao-NaPSS-10 and Kao-NaPSS-20, respectively. According to previous studies, it has been reported that the thermal degradation of NaPSS occurs in two stages: the first occurs above 200 °C to below 600 °C and is characterized by the breaking of the C-S bonds of polymer chains, generating the formation of radicals and the release of sulfur dioxide. Next, a second stage of degradation occurs above 600 °C and extends up to 1000 °C; this is attributed to the degradation of the carbonated polymer chain of styrene, generating the release of gases rich in styrene, methyl styrene, toluene, and benzene [46]. Thus, according to the behavior observed for Kao-NaPSS-10 and Kao-NaPSS-20, degradation profiles are similar to those presented by NaPSS and those that overlap with the typical degradation processes of clays [28]. Finally, above 1000 °C, a residue of 29.09% and 37.18% was observed for Kao-NaPSS-10 and Kao-NaPSS-20, respectively, which is attributed to carbonaceous material and calcined kaolinite residues [47].

On the other hand, thermal characterizations of bentonite-based PCNCs, Bent-NaPSS-10 and Bent-NaPSS-20, Figures 5C and 5D, respectively, are summarized in Table 3. In general, the degradation profiles of composites obtained from bentonite showed a behavior like that observed in those obtained from kaolinite. These presented three regions of mass loss and generation of a residue above 1000 °C. However, it was observed that the amount of mass lost in the third stage was lower for these samples. In addition, a greater quantity of residue was obtained, with values of 40.90% and 45.57% for Bent-NaPSS-10 and Bent-NaPSS-20, respectively. Thus, since the thermal degradation processes of the NaPSS matrix and the bentonite clay fraction are overlaid; these variations are associated with the structural differences between bentonite and kaolinite that modulate their water loss and thermal dehydroxylation processes [48,49].

Finally, the results of DSC characterization of the PNCNs based on bentonite–NaPSS and kaolinite–NaPSS are shown in Figure 6. In general, an effect of the clay/NaPSS composition on the glass transition temperature (Tg) of synthesized composites was observed. Thus, for Kao-NaPSS-10 and Kao-NaPSS-20 composites, the Tg values were 112.1 °C and 113.4 °C, respectively (see Figures 6A and 6B, respectively). Likewise, an increase in Tg was observed between the bentonite-based composites, recording Tg values of 109.1 °C and 118.6 °C for Bent-NaPSS-10 and Bent-NaPSS-20, respectively (see Figures 6C and 6D, respectively). The increase in the Tg of composites can be explained by the insertion of bentonite and kaolinite particles into the NaPSS matrix. It has been reported that NaPSS without reinforcing materials in its structure presented Tg values of 69.5 °C [50]. Thus, the application of an in situ polymerization technique of NaSS and the insertion of organoclay with surface vinyl groups allowed for an increase in the degree of cross-linking and interaction among NaPSS polymer chains in the composite. This generated a decrease in the mobility of polymer chains and an increase in the temperature required to generate the thermal transition.

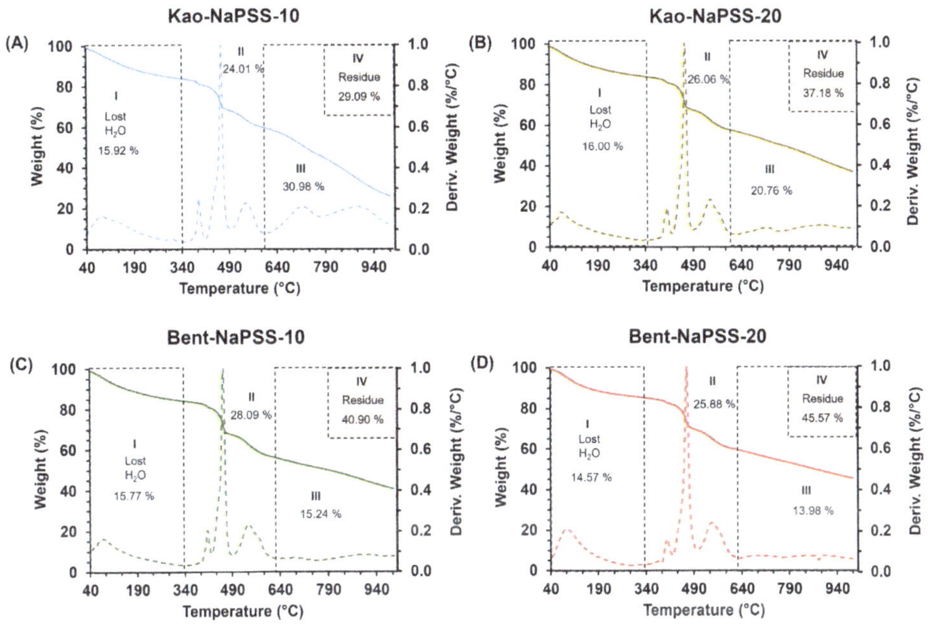


**Figure 5.** TGA results (solid line) and derivative thermogravimetric curve (dashed line, mass/temperature derivation) of PCNCs: Kao-NaPSS-10 (**A**), Kao-NaPSS-20 (**B**), Bent-NaPSS-10 (**C**) and Bent-NaPSS-20 (**D**).

**Table 3.** TGA and DSC results of PCNCs based on bentonite–NaPSS and kaolinite–NaPSS.

| Parameter | | Sample | | | |
|---|---|---|---|---|---|
| | | **Kao-NaPSS-10** | **Kao-NaPSS-20** | **Bent-NaPSS-10** | **Bent-NaPSS-20** |
| Stage 1 | Initial T (°C) | 40 | 40 | 40 | 40 |
| | Final T (°C) | 340 | 350 | 340 | 335 |
| | Δm (%) | 15.92 | 16.00 | 15.77 | 14.57 |
| Stage 1I | Initial T (°C) | 340 | 350 | 340 | 335 |
| | Final T (°C) | 600 | 600 | 630 | 630 |
| | Δm (%) | 24.01 | 26.06 | 28.09 | 25.88 |
| Stage 1II | Initial T (°C) | 600 | 600 | 630 | 630 |
| | Final T (°C) | 1000 | 100 | 1000 | 1000 |
| | Δm (%) | 30.98 | 20.76 | 15.24 | 13.98 |
| Residual mass | (%) | 29.09 | 37.18 | 40.90 | 45.57 |
| Tg | (°C) | 112.1 | 113.4 | 109.1 | 118.6 |

Tg: glass transition temperature.

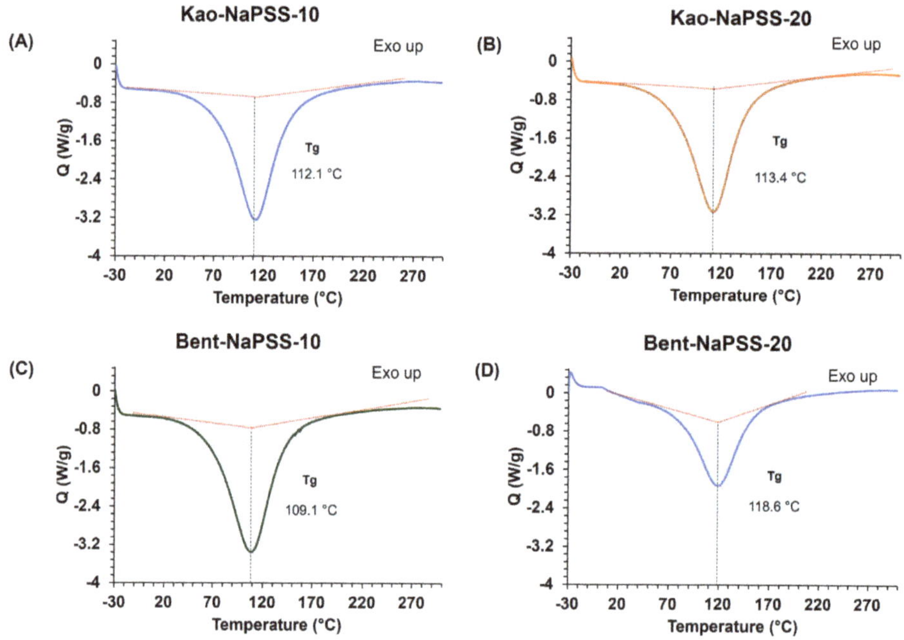


**Figure 6.** DSC results of PCNCs: Kao-NaPSS-10 (**A**), Kao-NaPSS-20 (**B**), Bent-NaPSS-10 (**C**) and Bent-NaPSS-20 (**D**). Analysis conditions: inert nitrogen atmosphere in a temperature range from room temperature to 300 °C and a heating ramp of 15 °C/min.

*2.4. Geomimetic Soil Conditioners Functional Characterization*

The results of functional characterization of WSC and CEC of PCNCs are shown in Figures 7A and 7B, respectively. The values of WSC from Kao-NaPSS-10, Kao-NaPSS-20, Bent-NaPSS-10, and Bent-NaPSS-20 were 3236 ± 337%, 2901 ± 43%, 3569 ± 400%, and 3179 ± 207%, respectively. When comparing the values obtained with those reported in Table 1 for the kaolinite–tClVS and bentonite–tClVS, the high hydrophilicity that NaPSS gives to the composites is evident, which allows it to retain more than 29 times its own weight. Thus, NaPSS contains sulfonate groups in its structure that allow it to interact and absorb large amounts of water [51]. Finally, values of CEC from PCNCs, shown in Figure 7B, were higher when the clay contents were lower (10%), being for each case the following: Kao-NaPSS-10 (159 ± 6 meq/100 g), Kao-NaPSS-20 (112 ± 6 meq/100 g), Bent-NaPSS-10 (203 ± 11 meq/100 g) and Bent-NaPSS-20 (146 ± 8 meq/100 g). Thus, when comparing the values obtained by clay–NaPSS composites with the values reported in Table 1 for the organoclays and their precursors, it can be observed that the insertion of NaPSS into materials produced a significant increase in the CEC. Therefore, the presence of sulfonate groups in the NaPSS polymer chains, with the capacity for ionization and development of a permanent negative charge, gives the composites the ability to interact electrostatically with cations and generate ion exchange processes [50]. Therefore, previous results demonstrate the potential of PCNCs based on bentonite–NaPSS and kaolinite–NaPSS to mimic cation exchange processes as well as the water storage capacity of soils, which can be understood as fundamental processes for the proper growth and development of plants [52].

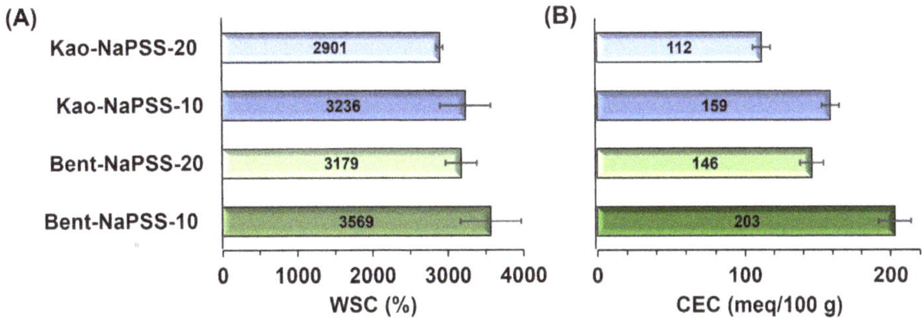


**Figure 7.** Water sorption capacity (WSC) (**A**) and cation exchange capacity (CEC) (**B**) of PCNCs: Kao-NaPSS-10, Kao-NaPSS-20, Bent-NaPSS-10 and Bent-NaPSS-20. The value in the center of the boxes and the error bars corresponds to the mean value and standard deviation of the measurements, respectively.

## 3. Conclusions

PCNCs are synthetic materials based on clay and functional polymers which are a promissory alternative for the making of geomimetic soil conditioners. These are synthetic geomaterials with hydrogel properties and CEC, they can be easily synthesized from NaPSS and surface-modified clays such as bentonite and kaolinite. The functionalization of clays is performed by direct reaction of clay and organosilanes compounds with vinyl groups in their structure. Thus, the presence of vinyl groups on the clay surface allows the anchoring of polymer chains and the formation of materials with greater stability due to the formation of covalent bonds. Characterization by IR-ATR spectroscopy and morphological analysis revealed the hybrid nature of the materials and an adequate dispersion of clays into the polymer matrix. Furthermore, the results of thermal characterization showed the influence of the clay/NaPSS ratio on the thermal properties of PCNCs, evidencing an increase in the Tg with the increase in clay content. Finally, the functional evaluation of PCNCs demonstrated the ability of geomimetic conditioners to retain high amounts of water and carry out cation exchange processes, which suggests their potential application in improving soil quality for plant growth.

## 4. Materials and Methods

### *4.1. Reagents and Materials*

Bentonite (BentoCol S.A.S, Bugalangrande, Colombia) and kaolinite (Caolines Superior Boyacá S.A.S, Tunja, Colombia) and trichlorovinylsilane ($H_2C{=}CHSiCl_3$, tClVS, 97 wt. %, Aldrich, Milwaukee, WI, USA) were used as precursors of organoclays. Sodium 4-styrenesulfonate (NaSS, >90 wt. %, Aldrich, Milwaukee, WI, USA) and ammonium persulfate ($(NH_4)_2S_2O_8$, wt. ≥98.0%, Merck, Darmstadt, Germany) were used for the synthesis of geomimetic soil conditioners. Ammonium acetate ($CH_3COONH_4$, ≥98%, Merck, Darmstadt, Germany), sodium chloride (NaCl, ≥99 wt. %, Merck, Darmstadt, Germany), aqueous formaldehyde (HCHO, 37 wt. %, Merck, Darmstadt, Germany), sodium hydroxide (NaOH, ≥98 wt. %, Merck, Darmstadt, Germany), and potassium phthalate monobasic ($HOOCC_6H_4COOK$, 99.95–100.05 wt. %, Aldrich, Milwaukee, WI, USA) were used to determine the cation exchange capacity (CIC). Toluene ($C_6H_5CH_3$, 99.8 wt %, Merck, Darmstadt, Germany), acetone ($CH_3COCH_3$, ≥99.5 wt. %, Merck, Darmstadt, Germany), absolute ethanol ($CH_3CH_2OH$, ≥99.5 wt. %, Merck, Darmstadt, Germany) and deionized water were used as solvents.

### *4.2. Organoclays Based on Bentonite and Kaolinite Synthesis*

Syntheses of organoclays based on bentonite and kaolinite were carried out through the covalent anchorage of tClVS on their surface. This procedure was previously published [13]. For this, 10.0 g of clay in toluene were added and subjected to ultrasound (M3800-E, 40 KHz, 130 watts, Branson Ultrasonics™, Danbury, CT, USA) for 1 h for its adequate dispersion. Next, dispersion was heated to 95 °C in a reflux unit and 1.0 g of tClVS (10% *w*/*w* with respect to the amount of clay used) was added and kept in continuous reflux for 24 h at 95 °C. Later, organoclay was filtered, washed with acetone, and dried at 60 °C for 24 h. This procedure was carried out for both bentonite and kaolinite, obtaining the bentonite–tClVS and kaolinite–tClVS organoclays, respectively.

### 4.3. Geomimetic Soil Conditioners Synthesis

Synthesis of geomimetic composites was carried out through the polymerization via free radicals of the NaPSS in the presence of organoclays (i.e., bentonite–ClVS and kaolinite–tClVS). For the above, 5.0 g of organo-clay was taken and dispersed in 200.0 mL of water under constant stirring at 1000 rpm for 4 h. Subsequently, the required amount of NaSS was added to prepare composites with 10.0 and 20.0 wt. % of organoclay. Next, ammonium persulfate was added at 1.0 mol % with respect to the amount of NaSS and left under constant stirring at 1000 rpm for 2 h. Mixture was heated continuous at 70 °C in at reflux equipment under stirring at 1000 rpm for 4 h. Composites were filtered, washed with distilled water, and dried in an oven at 105 °C for 24 h. Finally, the geomimetic composites were crushed and sieved through a 2.0 mm mesh. The composites obtained were identified as Kao-NaSS-10, Kao-NaPSS-20, Bent-NaPSS-10 and Bent-NaPSS-20. The composition and the identification of each material obtained are shown in Table 2.

### 4.4. Structural and Morphological Characterization

Structural characterization of organoclays was carried out by Fourier Transform Infrared Spectroscopy of Attenuated Total Reflectance (IR-ATR, IRAffinity-1, Shimadzu, Kyoto, Japan) and spectral analysis using Functionally Enhanced Derivative Spectroscopy (FEDS) [53]. Morphological characterization of organoclays was studied by scanning electron microscopy (SEM, Phenom Pro X, ThermoFisher Scientific, Waltham, MA, USA) at an acceleration voltage of 15 kV. For SEM observation, the particles were pasted on carbon tape as background and coated with gold using a sputtering technique. The analysis of its surface composition was performed by energy dispersive X-ray spectroscopy (EDS, Phenom Pro X, ThermoFisher Scientific, Waltham, MA, USA). Also, the dynamic light scattering technique (DLS, Zetasizer LAB, Malvern Panalytical, Malvern, UK) was used to study the size and charge of the organoclay particles. For DLS characterization, dispersions of the particulate material in water were prepared at concentrations of 0.1 mg/mL and measurements were performed at 25 °C. The Kao-NaPSS-10, Kao-NaPSS-20, Bent-NaPSS-10 and Bent-NaPSS-20 composites were characterized structurally by IR-ATR-FEDS, morphologically by SEM and their elemental composition was determined by EDS.

### 4.5. Thermal Characterization

Thermal properties of hybrid composites, based on kaolinite–NaPSS and bentonite–NaPSS, were studied by thermogravimetric analysis (TGA, TA Q50, TA Instruments, New Castle, DE, USA), using an inert nitrogen atmosphere in a temperature range from 40 to 1000 °C and a heating ramp of 15 °C/min. For the analysis by differential scanning calorimetry (DSC, DSC25, TA Instruments, New Castle, DE, USA), the material thermal memory was eliminated by heating from room temperature at 300 °C with a ramp of 15 °C/min. Next, isothermal heating at 300 °C for three minutes, and a 25 °C/min cooling ramp from 300 °C to −30 °C. Finally, Differential scanning calorimetry (DSC) analysis was performed by heating from −30 °C to 300 °C with a ramp of 15 °C/min.

### 4.6. Water Sorption Capacity Study

*WSC* evaluation of clays, organoclays, and geomimetic composites was performed by tea bag method [54]. For this, approximately 0.2 g of material (*M*) was placed in a tea bag (*Tb*) and sealed. Subsequently, the *M* + *Tb*, (*MTb*) was immersed in deionized water for 24 h.

Finally, the sample was removed from the water and *WSC* was determined gravimetrically by the following equation:

$$WSC\ (\%) = \frac{(MTb_{24} - Tb_{24} - M)}{M} * 100 \quad (1)$$

where $MTb_{24}$ is the weight of $MTb$ after 24 h submerged in water; $Tb_{24}$ is the blank, that is, the sorption of water by the tea bag after 24 h submersion. Tests were carried out in triplicate and a blank experiment was carried out to determine the amount of water absorbed by the tea bag.

*4.7. Cation Exchange Capacity Study*

Cation exchange capacity (CEC) of clays, organoclays and geomimetic composites were determined using 1.0 N ammonium acetate method defined by the Agustín Codazzi Geographic Institute [55]. For this, a defined quantity of the material (1.0 g) is placed in contact with 10.0 mL of 1.0 N ammonium acetate solution to promote cation exchange reactions. Subsequently, sample was filtered, and ammonium remaining in materials was exchanged with sodium ions. Finally, exchanged ammonium was quantified by formaldehyde titration with NaOH solution. The endpoint determination of the reaction was carried out using phenolphthalein in a 1% ethanolic solution as an indicator. The test was performed in triplicate for each sample and a control blank was included for the experiment.

**Author Contributions:** Conceptualization, T.A.L., E.M.C. and M.P.; methodology, T.A.L., E.M.C. and M.P.; formal analysis, T.A.L., E.M.C. and M.P.; investigation, T.A.L., E.M.C. and M.P.; resources, E.M.C. and M.P.; writing—original draft preparation, T.A.L., E.M.C. and M.P.; writing—review and editing, T.A.L., E.M.C. and M.P.; project administration, E.M.C. and M.P. All authors have read and agreed to the published version of the manuscript.

**Funding:** This research was funded by Universidad del Valle, grant number CI 71272 and National Planning Department of Colombia, specifically, to the general royalty system (Sistema General de Regalías, SGR) for project BPIN 2020000100027.

**Institutional Review Board Statement:** Not applicable.

**Informed Consent Statement:** Not applicable.

**Data Availability Statement:** The data presented in this study are openly available in article.

**Acknowledgments:** Authors acknowledge to Universidad del Valle, Universidad de Córdoba, Mindtech SAS, National Planning Department of Colombia and the general royalty system of Colombia.

**Conflicts of Interest:** The authors declare no conflicts of interest.

## References

1. Franco-Urquiza, E.A. Clay-Based Polymer Nanocomposites: Essential Work of Fracture. *Polymers* **2021**, *13*, 2399. [CrossRef] [PubMed]
2. Shunmugasamy, V.C.; Xiang, C.; Gupta, N. Clay/Polymer Nanocomposites: Processing, Properties, and Applications. In *Hybrid and Hierarchical Composite Materials*; Springer: Cham, Switzerland, 2015; pp. 161–200. [CrossRef]
3. FAO. *The State of the World's Land and Water Resources for Food and Agriculture 2021—Systems at Breaking Point*; Food and Agriculture Organization of the United Nations: Rome, Italy, 2022. [CrossRef]
4. Prăvălie, R. Exploring the Multiple Land Degradation Pathways across the Planet. *Earth Sci. Rev.* **2021**, *220*, 103689. [CrossRef]
5. Prakash, S.; Verma, A.K. Anthropogenic Activities and Biodiversity Threats. *Int. J. Biol. Innov.* **2022**, *04*, 94–103. [CrossRef]
6. Shah, F.; Wu, W. Soil and Crop Management Strategies to Ensure Higher Crop Productivity within Sustainable Environments. *Sustainability* **2019**, *11*, 1485. [CrossRef]
7. Raj, A.; Dubey, A.; Malla, M.A.; Kumar, A. Pesticide Pestilence: Global Scenario and Recent Advances in Detection and Degradation Methods. *J. Environ. Manag.* **2023**, *338*, 117680. [CrossRef] [PubMed]
8. Babla, M.; Katwal, U.; Yong, M.T.; Jahandari, S.; Rahme, M.; Chen, Z.H.; Tao, Z. Value-Added Products as Soil Conditioners for Sustainable Agriculture. *Resour. Conserv. Recycl.* **2022**, *178*, 106079. [CrossRef]

9. Ghobashy, M.M. The Application of Natural Polymer-Based Hydrogels for Agriculture. In *Hydrogels Based on Natural Polymers*; Elsevier: Amsterdam, The Netherlands, 2020; pp. 329–356. [CrossRef]
10. Huang, J.; Kogbara, R.B.; Hariharan, N.; Masad, E.A.; Little, D.N. A State-of-the-Art Review of Polymers Used in Soil Stabilization. *Constr. Build. Mater.* **2021**, *305*, 124685. [CrossRef]
11. Palencia, M.; Lerma, T.A.; Garcés, V.; Mora, M.A.; Martínez, J.M.; Palencia, S.L. Eco-Friendly Supramolecular Systems: Polymer Geo- and Biomimicry. In *Eco-Friendly Functional Polymers*; Elsevier: Amsterdam, The Netherlands, 2021; pp. 121–139. [CrossRef]
12. Lerma, T.A.; Garcés, V.; Palencia, M. Novel Multi- and Bio-Functional Hybrid Polymer Hydrogels Based on Bentonite-Poly(Acrylic Acid) Composites and Sorbitol Polyesters: Structural and Functional Characterization. *Eur. Polym. J.* **2020**, *128*, 109627. [CrossRef]
13. Lerma, T.A.; Combatt, E.M.; Palencia, M. Novel Multifuctional Geomimetic Soil Conditioner Based on Multilayer Hybrid Composites of Clay-Paa-Lignin: Synthesis and Functional Characterization. *Eur. Polym. J.* **2023**, *198*, 112376. [CrossRef]
14. Estrada, M.; Sepúlveda, F.; Nenen, A.; Bravo-Linares, C.; Nishide, H.; Suga, T.; Moreno-Villoslada, I. Novel Reusable Catalytic Poly(4-Styrenesulfonate-Co-Glycidylmethacrylate) Foams for Adsorption and Photodegradation of the Model Pollutant Dye Methylene Blue Based on Aromatic-Aromatic Interactions. *Chem. Eng. J.* **2023**, *459*, 141518. [CrossRef]
15. Liu, J.; Dai, Z.; Xu, K.; Yang, Y.; Lv, K.; Huang, X.; Sun, J. Water-Based Drilling Fluid Containing Bentonite/Poly(Sodium 4-Styrenesulfonate) Composite for Ultrahigh-Temperature Ultradeep Drilling and Its Field Performance. *SPE J.* **2020**, *25*, 1193–1203. [CrossRef]
16. Mercado, J.; Caballero, E.M.C.; Palencia, M. Adición de Polielectrolitos Sobre La Fase Acuosa de Un Suelo Sulfatado Ácido Interior de Cordoba, Colombia. *Temas Agrarios* **2015**, *20*, 60–70. [CrossRef]
17. Jokinen, M.; Manzanares, J.A.; Murtomäki, L. Thermodiffusion of Sodium Polystyrene Sulfonate in a Supporting Electrolyte. *Electrochim. Acta* **2019**, *317*, 542–550. [CrossRef]
18. Sen, A.K.; Roy, S.; Juvekar, V.A. On the Importance of Purification of Sodium Polystyrene Sulfonate. *ISRN Anal. Chem.* **2012**, *2012*, 514509. [CrossRef]
19. Palencia, M.; Córdoba, A.; Arrieta, Á. Stimuli-Sensitive Nanostructured Poly(Sodium 4-Styrene Sulfonate): Synthesis, Characterization, and Study of Metal Ion Retention Properties. *J. Appl. Polym. Sci.* **2018**, *135*, 46001. [CrossRef]
20. Cao, R.; Shi, S.; Li, Y.; Xu, B.; Zhao, Z.; Duan, F.; Cao, H.; Wang, Y. The Properties and Antifouling Performance of Anion Exchange Membranes Modified by Polydopamine and Poly (Sodium 4-Styrenesulfonate). *Colloids Surf. A Physicochem. Eng. Asp.* **2020**, *589*, 124429. [CrossRef]
21. Holleck, J.L.; Roberts, A.E.; Marhoffer, E.A.; Grimshaw, A.A.; Gunderson, C.G. Risk of Intestinal Necrosis With Sodium Polystyrene Sulfonate: A Systematic Review and Meta-Analysis. *J. Hosp. Med.* **2021**, *16*, 489–494. [CrossRef]
22. Huang, J.; Grajales, J.A.; Akula, P.; Little, D.N.; Kim, Y.-R.; Rushing, J.F. Strength and Fracture Properties of Sandy Subgrade Soil Treated with Sodium Polystyrene Sulfonate. *J. Mater. Civ. Eng.* **2024**, *36*, 04024065. [CrossRef]
23. Belghazdis, M.; Hachem, E.-K. Clay and Clay Minerals: A Detailed Review. *Int. J. Recent Technol. Appl. Sci. (IJORTAS)* **2022**, *4*, 54–75. [CrossRef]
24. Maj, I.; Matus, K. Aluminosilicate Clay Minerals: Kaolin, Bentonite, and Halloysite as Fuel Additives for Thermal Conversion of Biomass and Waste. *Energies* **2023**, *16*, 4359. [CrossRef]
25. Hubadillah, S.K.; Othman, M.H.D.; Matsuura, T.; Ismail, A.F.; Rahman, M.A.; Harun, Z.; Jaafar, J.; Nomura, M. Fabrications and Applications of Low Cost Ceramic Membrane from Kaolin: A Comprehensive Review. *Ceram. Int.* **2018**, *44*, 4538–4560. [CrossRef]
26. Corina, A.N.; Wollenweber, J.; Fischer, H.; van der Valk, K.; Castelein, K.; Moghadam, A.; Heerens, G.J. Evaluation of Bentonite Application for the Abandonment of Deep Geo-Energy Wells. *Rock Mech. Rock Eng.* **2023**, *56*, 301–317. [CrossRef]
27. Muhammad, N.; Siddiqua, S. Calcium Bentonite vs Sodium Bentonite: The Potential of Calcium Bentonite for Soil Foundation. *Mater. Today Proc.* **2022**, *48*, 822–827. [CrossRef]
28. Kgabi, D.P.; Ambushe, A.A. Characterization of South African Bentonite and Kaolin Clays. *Sustainability* **2023**, *15*, 12679. [CrossRef]
29. Kumar, A.; Lingfa, P. Sodium Bentonite and Kaolin Clays: Comparative Study on Their FT-IR, XRF, and XRD. *Mater. Today Proc.* **2020**, *22*, 737–742. [CrossRef]
30. Ramírez-Rincón, J.A.; Palencia, M.; Combatt, E.M. Fractionation of Optical Properties for Multicomponent Samples and Determination of Spectral Similarity Indices Based on Feds0 Algorithm. *SSRN Electron. J.* **2022**, *33*, 104528. [CrossRef]
31. Nazdracheva, T.; Morozov, A.; Yavna, V.; Kochur, A. Study of Hydration of Kaolinite and Montmorillonite Mixture by IR Spectroscopy. *J. Mol. Struct.* **2022**, *1250*, 131871. [CrossRef]
32. Bee, S.L.; Abdullah, M.A.A.; Bee, S.T.; Sin, L.T.; Rahmat, A.R. Polymer Nanocomposites Based on Silylated-Montmorillonite: A Review. *Prog. Polym. Sci.* **2018**, *85*, 57–82. [CrossRef]
33. Shi, G.; Araby, S.; Gibson, C.T.; Meng, Q.; Zhu, S.; Ma, J. Graphene Platelets and Their Polymer Composites: Fabrication, Structure, Properties, and Applications. *Adv. Funct. Mater.* **2018**, *28*, 1706705. [CrossRef]
34. Fontaine, F.; Christidis, G.E.; Yans, J.; Hollanders, S.; Hoffman, A.; Fagel, N. Characterization and Origin of Two Fe-Rich Bentonites from Westerwald (Germany). *Appl. Clay Sci.* **2020**, *187*, 105444. [CrossRef]
35. Muslim, W.A.; Al-Nasri, S.K.; Albayati, T.M. Evaluation of Bentonite, Attapulgite, and Kaolinite as Eco-Friendly Adsorbents in the Treatment of Real Radioactive Wastewater Containing Cs-137. *Prog. Nucl. Energy* **2023**, *162*, 104730. [CrossRef]
36. Musabekov, K.B.; Artykova, D.M.-K.; Tazhibayeva, S.M.; Oryntaeva, A.; Sugurbekova, G.K.; Kulichikhin, V. Surface modification of montmorillonite clay with organic molecules. *Rasayan J. Chem.* **2021**, *14*, 635–640. [CrossRef]

37. Raji, M.; Mekhzoum, M.E.M.; Rodrigue, D.; Qaiss, A.e.k.; Bouhfid, R. Effect of Silane Functionalization on Properties of Polypropylene/Clay Nanocomposites. *Compos. B Eng.* **2018**, *146*, 106–115. [CrossRef]
38. He, H.; Tao, Q.; Zhu, J.; Yuan, P.; Shen, W.; Yang, S. Silylation of Clay Mineral Surfaces. *Appl. Clay Sci.* **2013**, *71*, 15–20. [CrossRef]
39. Tsaffo Mbognou, M.H.; Lambert, S.D.; Mumbfu, E.M.; Caucheteux, J.; Farcy, A.; Fagel, N.; Woumfo, E.D.; Mahy, J.G. Silane Modified Clay for Enhanced Dye Pollution Adsorption in Water. *Results Surf. Interfaces* **2024**, *14*, 100183. [CrossRef]
40. Tang, N.; Yang, J.; Cen, W.; Pan, W.; Wu, L.; Xu, C. Preparation of Organic Montmorillonite Promoter for Improving the Adhesion between Bitumen and Acidic Aggregate. *Constr. Build. Mater.* **2021**, *274*, 121833. [CrossRef]
41. Adnan, M.M.; Dalod, A.R.M.; Balci, M.H.; Glaum, J.; Einarsrud, M.A. In Situ Synthesis of Hybrid Inorganic–Polymer Nanocomposites. *Polymers* **2018**, *10*, 1129. [CrossRef] [PubMed]
42. Palencia, M.; Lerma, T.A.; Garcés, V.; Mora, M.A.; Martínez, J.M.; Palencia, S.L. Eco-Friendly Composites and Nanocomposites. In *Eco-Friendly Functional Polymers*; Elsevier: Amsterdam, The Netherlands, 2021; pp. 105–120. [CrossRef]
43. Betiha, M.A.; Negm, N.A.; El-Sayed, E.M.; Mostafa, M.S.; Menoufy, M.F. Capability of Synthesized Sulfonated Aromatic Cross-Linked Polymer Covalently Bonded Montmorillonite Framework in Productivity Process of Biodiesel. *J. Clean. Prod.* **2020**, *261*, 120995. [CrossRef]
44. Chikkatti, B.S.; Sajjan, A.M.; Kalahal, P.B.; Banapurmath, N.R. Insight into the Performance of Valve-Regulated Lead-Acid Battery Using Sodium Salt of Poly(4-Styrene Sulfonic Acid-Co-Maleic Acid)-Poly(Vinyl Alcohol) Gel Electrolyte. *J. Energy Storage* **2023**, *72*, 108261. [CrossRef]
45. Zabihi, O.; Ahmadi, M.; Nikafshar, S.; Chandrakumar Preyeswary, K.; Naebe, M. A Technical Review on Epoxy-Clay Nanocomposites: Structure, Properties, and Their Applications in Fiber Reinforced Composites. *Compos. B Eng.* **2018**, *135*, 1–24. [CrossRef]
46. Hassam, C.; Lewis, D.A. Dispersion of Single and Multiwalled Nanotubes with Poly(Sodium Styrene Sulfonate)—Effect of PH and Ionic Strength on Dispersion Stability. *Aust. J. Chem.* **2013**, *67*, 66–70. [CrossRef]
47. Liu, X.; Liu, X.; Hu, Y. Investigation of the Thermal Behaviour and Decomposition Kinetics of Kaolinite. *Clay Miner.* **2015**, *50*, 199–209. [CrossRef]
48. Lerma H, T.A.; Combatt, E.M.; Palencia L, M.S. Effect of Temperature on Colloids of Agricultural Soils through Dynamic Light Scattering. *Rev. De Cienc. Agrícolas* **2015**, *32*, 94–103. [CrossRef]
49. Geng, J.; Sun, Q. Effects of High Temperature Treatment on Physical-Thermal Properties of Clay. *Thermochim Acta* **2018**, *666*, 148–155. [CrossRef]
50. Urbano, B.; Rivas, B.L. Poly(Sodium 4-Styrene Sulfonate) and Poly(2-Acrylamidoglycolic Acid) Nanocomposite Hydrogels: Montmorillonite Effect on Water Absorption, Thermal, and Rheological Properties. *Polym. Bull.* **2011**, *67*, 1823–1836. [CrossRef]
51. Wang, Y.; Wang, W.; Shi, X.; Wang, A. Enhanced Swelling and Responsive Properties of an Alginate-Based Superabsorbent Hydrogel by Sodium *p*-Styrenesulfonate and Attapulgite Nanorods. *Polym. Bull.* **2013**, *70*, 1181–1193. [CrossRef]
52. Bünemann, E.K.; Bongiorno, G.; Bai, Z.; Creamer, R.E.; De Deyn, G.; de Goede, R.; Fleskens, L.; Geissen, V.; Kuyper, T.W.; Mäder, P.; et al. Soil Quality—A Critical Review. *Soil Biol. Biochem.* **2018**, *120*, 105–125. [CrossRef]
53. Palencia, M. Functionally-Enhanced Derivative Spectroscopy (FEDS): A Methodological Approach. *J. Sci. Technol. Appl.* **2020**, *9*, 29–34. [CrossRef]
54. Otálora, A.; Lerma, T.A.; Palencia, M. Synthesis and Characterization of Polurea-Based Hydrogels by Multicomponent Polycondensation of 1,6-Hexamethylenediisocyanate, Sorbitol and Cysteine. *J. Sci. Technol. Appl.* **2019**, *7*, 5–16. [CrossRef]
55. Instituto Geografico Agustin Codazzi (IGAC); Zamudio Sánchez, A.M.; Carrascal Carrascal, M.L.; Pulido Roa, C.E.; Gallardo, J.F.; Gómez Guzmán, I.D. *Métodos Analíticos del Laboratorio de Suelos*; Departamento Administrativo Nacional de Estadística: Bogotá, Colombia, 2006.

*gels*

MDPI

*Review*

# Analysis of the Growth of Hydrogel Applications in Agriculture: A Review

**Carolina Buitrago-Arias [1], Piedad Gañán-Rojo [2], Mabel Torres-Taborda [2], Luisa Perdomo-Villar [2], Catalina Álvarez-López [1], Natalia Jaramillo-Quiceno [1] and Gustavo Adolfo Hincapié-Llanos [1,*]**

[1] Facultad de Ingeniería Agroindustrial, Universidad Pontificia Bolivariana, Circular 1 70-01, Medellín 050031, Colombia; carolina.buitrago@upb.edu.co (C.B.-A.); catalina.alvarezl@upb.edu.co (C.Á.-L.); natalia.jaramilloq@upb.edu.co (N.J.-Q.)

[2] Facultad de Ingeniería Química, Universidad Pontificia Bolivariana, Circular 1 70-01, Medellín 050031, Colombia; piedad.ganan@upb.edu.co (P.G.-R.); mabel.torres@upb.edu.co (M.T.-T.); luisa.perdomo@upb.edu.co (L.P.-V.)

* Correspondence: gustavo.hincapie@upb.edu.co

**Abstract:** Feeding a growing population under the pressures of climate change requires solutions that safeguard yields while strengthening agricultural resilience. Integrated Crop Management (ICM)—which combines precise fertilization, efficient water use, and targeted pest control—offers a promising framework. Hydrogels, with their water retention and controlled release properties, can enhance ICM by improving fertilizer efficiency, reducing water loss, and supporting soil health. Despite extensive research, their optimal use in agriculture remains unclear, and limitations continue to restrict large-scale adoption. To address this gap, this study applies the Preferred Reporting Items for Systematic Reviews and Meta-Analyses (PRISMA) methodology alongside bibliometric analysis to examine hydrogel applications in ICM from 2000 to 2024. Ninety Scopus-indexed publications were analyzed across four domains: pesticides, nutritional growth inputs, soil conditioners, and bioactive substances. The results reveal a marked increase in hydrogel structural complexity, greater diversity in characterization methods, ongoing reliance on high-impact pesticides despite advances in bio-based hydrogels, and persistent gaps in assessing environmental impacts and regulatory compliance. These findings underscore the need for stronger collaboration between academia and industry to translate hydrogel research into effective, sustainable agricultural practices under changing climatic conditions.

**Keywords:** hydrogel; integrated crop solutions; biostimulants; agriculture; fertilizer; herbicide; soil; PRISMA; bibliometric analysis; systematic review

## 1. Introduction

Plant diseases, soil salinity, drought, and extreme temperatures are increasing challenges to agricultural systems [1–3]. Biotic and abiotic stress considerably diminish agricultural output, with disease-related losses estimated to range from 20% to 40% [4]. In addition to diminishing productivity, they impair soil health, undermine crop quality, and jeopardize global food security [5]. Confronting these challenges needs novel agricultural practices that improve crop resilience and maximize resource efficiency.

Integrated Crop Management (ICM) is a holistic strategy that enhances agricultural productivity through the integration of precision fertilization, water management, and targeted pest control. The Food and Agriculture Organization (FAO) identifies ICM as a crucial strategy for sustainable agriculture, promoting soil health, optimizing water-use efficiency,

 https://doi.org/10.3390/gels11090731

and increasing crop yields by 10–14% compared to traditional practices [6]. Furthermore, ICM advocates for conservation agriculture by enhancing soil aggregation, improving water infiltration, and optimizing the controlled release of nutrients and agrochemicals [6].

ICM promotes a balanced and sustainable agroecosystem by integrating inputs such as fertilizers, plant protection agents (e.g., insecticides), and biostimulants derived from organic compounds, microorganisms including fungi [7], algae [7,8], inorganic compounds [7,9], and biopolymers like chitosan [10,11]. The significance of utilizing biostimulants pertains to their contribution to enhancing tolerance to abiotic stress, improving quality features, or facilitating the availability of limited nutrients in soil or the rhizosphere [12].

Persistent challenges remain regarding efficacy, particularly volatilization, degradation, and inadequate control over the release of fertilizers, plant protection agents, or biostimulants, necessitating the integration of these compounds into with other components or delivery systems [13–15]. In this context, polymer hydrogels emerge as a promising strategy, offering not only controlled release but also the capacity to alleviate soil-related problems such as soil compaction, erosion, and water runoff [3,16].

Hydrogels are versatile three-dimensional (3D) network structures formed by a monomer or polymer through a gelation process [17]. Generally, it can absorb substantial quantities of water without disintegrating or dissolving [18]. The 3D network can be created through physical crosslinking facilitated by non-covalent bonds, including hydrogen bonding, van der Waals forces, or physical interactions [18]; or through chemical crosslinking, where covalent bonds are essential [16,18]. Their agricultural relevance derives from their high water-holding capacity, which can alleviate drought stress and soil erosion while enabling the controlled release of agrochemicals, thereby enhancing plant water availability and nutrient uptake [19]. Hydrogel formulations have the potential to enhance soil aeration, microbial activity, and root development, which may lead to improved crop performance [16].

Hydrogels can be synthesized from various sources, including natural polymers such as chitosan and synthetic polymers like Polyvinyl Alcohol (PVA) [20]. Moreover, combinations of natural and synthetic polymers, such as PVA/sericin, are also feasible [21]. Additionally, polymer composites or nanocomposites, based on bacterial nanocellulose, can be recognized for these applications [22]. Resolving technical issues related to hydrogel structure, interactions with agricultural inputs, soil type, and environmental conditions is essential for the successful integration of hydrogel-based systems. This situation prompted extensive global research focused on the formulation and evaluation of hydrogels. Significant aspects of hydrogel structure have been extensively reviewed in the literature, including the sources of hydrogel materials, crosslinking processes that influence overall performance, and physical evaluations addressing absorption and desorption characteristics [18,23,24].

Nevertheless, these assessments have primarily concentrated on structural and physicochemical attributes, without providing enough connection to comprehensive agronomic frameworks such as Integrated Crop Management (ICM). The absence of integration hinders the optimization of hydrogels for field conditions and the exploitation of potential synergies with fertilizers, insecticides, and other agricultural inputs. Moreover, insufficient emphasis has been placed on identifying the variables that limit the widespread adoption of hydrogels, along with their ramifications for sustainability and the circular economy. Investigating the progression of this thematic domain enables the identification of trends and the specification of priority elements that should guide future research aimed at enhancing these applications. This work addresses this gap by integrating a systematic literature review, conducted in accordance with the Preferred Reporting Items for Systematic Reviews and Meta-Analyses (PRISMA) criteria, with bibliometric analysis. This dual methodology facilitates the identification of research trends, thematic clusters, and underexplored areas, establishing a systematic foundation for advancing hydrogel applications in agriculture. Although both PRISMA and bibliometric techniques are widely applied in materials research [25,26], their combined use in the context of hydrogel-based agricultural solutions has not been previously reported, representing a novel contribution. To operationalize this approach, this study examined 90 papers published between 2000 and 2024, indexed in the Scopus database, across four primary domains: pesticides, nutritional growth inputs, soil conditioners, and bioactive compounds. The search strategy targeted studies on chemical substances or agricultural inputs that enhance plant health or improve soil properties. The methodology used to select publications that met the search equation and the inclusion/exclusion criteria is detailed in Section 4. The results provide valuable insights for both early-stage and advanced investigations aiming to combine hydrogels with complementary materials for ICM or other sustainable agricultural practices. This research is relevant to large-scale production systems as well as small-scale contexts, including urban agriculture, where resource optimization and climate resilience are critical. Furthermore, it may support both emerging and established investigations seeking to explore the advantages of hydrogels combined with substances, inputs, or payloads—such as nutrients, agrochemicals, active agents, or biostimulants—applicable to strategies that promote ICM or similar approaches aimed at advancing sustainable agricultural practices.

## 2. Results and Discussion

### 2.1. Macro- and Micro-Level

Grounded in the study's scope—encompassing both large-scale production systems and small-scale contexts such as urban agriculture, and emphasizing hydrogels in association with nutrients, agrochemicals, active agents, and biostimulants within Integrated Crop Management (ICM) strategies—the results are structured to first provide a macro-level assessment of research evolution, followed by a micro-level analysis of individual contributions. This structure, enabled by the combined application of the Preferred Reporting Items for Systematic Reviews and Meta-Analyses (PRISMA) methodology and bibliometric analysis, provides a comprehensive framework for identifying publication trends, thematic clusters, and research gaps. The macro-level overview begins with an examination of publication growth patterns over time, offering context for subsequent in-depth analysis at the document level. As shown in Figure 1, the documents analyzed in this work cover the years 2000 to 2024. The earliest publications include to the works of the Saraydın et al. (2000) [27] which focused on the use of the hydrogel in pest control by evaluating the controlled release of the water-soluble agrochemical herbicide such as sodium 2,2 dichloropropionate (Dowpon). In the same year, Karadağ et al. (2000) [28] explored applications in crop nutrition by the analysis of the interaction between the hydrogel and fertilizers such as ammonium nitrate, potassium nitrate and ammonium sulfate. They also evaluated the potential for the pest control of the grasses and weeds using Dalapon. Both studies employed the same type of the pest control approach, involve the **same** active **compound**, and utilized hydrogels are based on acrylamide.

Figure 1 illustrates the steady increase in publications on this topic, with 2023 representing the most active year. The growth trend was modeled using a cubic polynomial regression ($y = 0.0081x^3 - 0.1078x^2 + 0.318x + 1.384$), which yielded an $R^2$ value of 0.8294. In the physical sciences and engineering disciplines, $R^2$ values above 0.70 are generally regarded as acceptable indicators of fit, providing meaningful descriptive insight into data trends [29]. Although this model is not intended as a predictive tool, it effectively summarizes the observed growth—particularly between 2020 and 2023—and suggests that publication activity in this field is likely to continue increasing over the next five years. This macro-level overview sets the stage for the detailed bibliometric analysis that follows, which integrates both macro- and micro-level perspectives to identify key institutions, leading authors, and thematic keyword patterns shaping research in this domain.

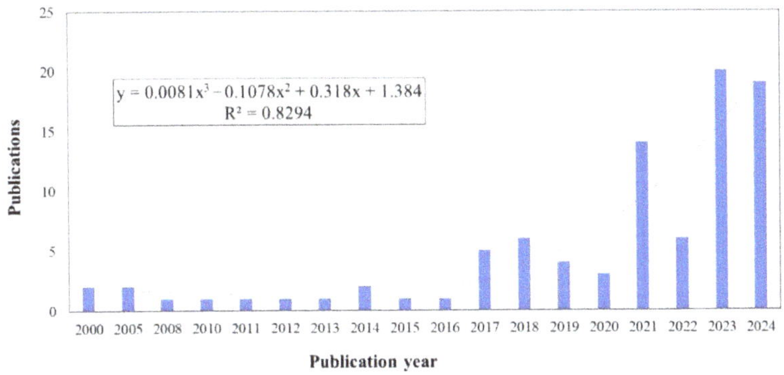


**Figure 1.** Annual number of publications on hydrogel applications in agriculture (2000–2024).

Following the upward publication trend shown in Figure 1, the macro-level analysis also examined the geographical distribution of research activity. As shown in Figure 2, contributions originate from 31 states across multiple regions, reflecting the widespread scientific interest in hydrogel applications for agriculture. This global distribution provides a basis for examining the relative research intensity of individual countries, offering insight into why certain regions demonstrate higher activity levels than others. The highest proportion of documents (29%) were authored by researchers based in China, followed by those in India (19%), and Brazil (7%). This point is noteworthy, given that these three emerging market economies have significantly expanded their agricultural activities over the past two decades and play a major role in agricultural trade policy issues [30,31]. In the case of China, the earliest publication identified corresponds to the work of Liu et al. (2005) [31], which explored the use of sodium carboxymethylcellulose hydrogel to improve water absorption However, until the continuous publications are observed since 2020, and the most productive year corresponds to 2024 with seven works. The trends observed in these documents correspond to more interest in the last year on delivery efficiency of compounds, especially on fertilizers [32,33]. Also, it is observed an interest on the water retention considering aspects such as soil moisture management [32,34], and in some cases are linked to the use of the humidic substance, especially considering soil fertility by use of the fertilizers [35]. and water retention [36].

In the case of India, the first documented identified in this work corresponds to Kumar and Kaith (2010) [37] that are focused on the develop of the hydrogel based on Psyllium/acrylic acid for the release of fungicide as copper sulfate that is useful in the control of the multiple fungi and bacteria that promote plant diseases. As similar in the case of the documents published by Chinese authors is observed interest by the compound control release, and the use of natural polymers or raw materials to develop the hydrogel. In the case of Brazil, the first document identified corresponds to Tanaka et al. (2021) [38] work where is observed again a concern related to the compound release, as the commercial product herbicide dibromide monohydrate considering a bio-based nanocomposite. In this work the trends observed are linked to the developed hydrogel based on natural resources supporting also the developed ecofriendly materials, as well as a more interest in advanced functionalization of the hydrogels. Tendency that is observed also in the case of the Chinese and Indian works.

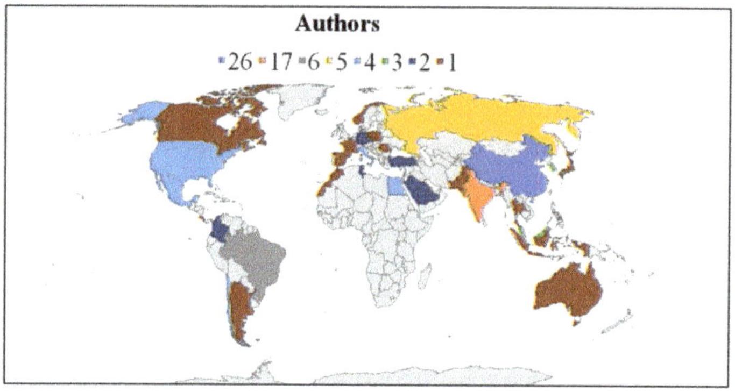


**Figure 2.** Geographical distribution of authors based on Scopus-indexed publications (2000–2024).

Additionally, it is important to note that certain states participating in these challenges are not solely middle-income but also serve as importers and exporters of agrochemicals, such the Russian Federation [39], Mexico [40], and Argentina [41]. This geographical profile of research activity provides the foundation for narrowing the focus to the micro-level, where the analysis examines the institutions and authors driving research in this domain.

In total, 160 authors were affiliated with 158 organizations, with 60% based in university departments, and 2% in industry; the remaining institutions comprised research and development institutes or centers supported by universities or national governmental entities. Notably, industry-affiliated authors consistently collaborated with academic institutions [2,42]. This represents an opportunity to further promote interinstitutional collaboration, a factor of particular relevance for advancing the integration of hydrogels to the ICM practices.

Building on the importance of inter-institutional collaboration for advancing the integration of hydrogels into ICM practices, the author network analysis provides further insight into the dynamics of research partnerships. As shown in Figure 3a, the non-gray colored clusters represent groups of authors collaborating across different documents, while the gray clusters indicate isolated author groups working independently. Overall, the observed level of collaboration is relatively low. This situation can be linked to the diversity of the focus on the topics, motivated also by the type of hydrogels produced and the necessities to evaluate their efficiency on ICM strategies. Strengthening cross-institutional and interdisciplinary networks, particularly those connecting research teams working on complementary aspects of hydrogel design, evaluation, and field implementation, could enhance knowledge exchange and accelerate the translation of laboratory findings into practical agricultural applications. This need for stronger collaboration is further reflected in the temporal pattern of existing partnerships. As shown in Figure 3b, collaborations among the most active clusters identified in Figure 3a have been recorded only since 2018, underscoring both the novelty of the topic and the opportunity to expand cooperative research efforts. In particular, several works involving authors from industry also included participation from academic institutions, highlighting the importance of cross-sector collaboration in advancing hydrogel research for agricultural applications. Complementing this collaboration analysis, citation data provide another perspective on the field's maturity and influence. The Scopus database yielded a total of 1,816 citations, with 91% of papers cited at least once—reinforcing the previously noted relevance of this research domain. Notably, 63% of the non-cited documents were published in 2024, reflecting the short time available for citation accumulation. The most referenced document, "Guar gum-crosslinked-soya lecithin nano-hydrogel sheets as effective adsorbent for the removal of thiophanate methyl fungicide" [43], has received 118 citations. This study offered novel insights into the physical interaction mechanisms between thiophanate-methyl fungicide and the synthesized hydrogel, highlighting potential applications in water detoxification methods to address pesticide contamination.

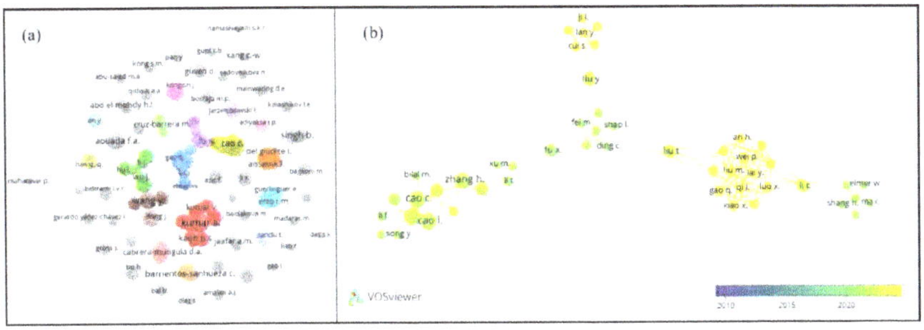


**Figure 3.** Authorship analysis: (**a**) distribution of authors; (**b**) collaboration networks among authors. Elaborated using VOSviewer.

With regard to the journals preferred by the authors, a total of 63 journals were identified, spanning 15 subject areas, as shown in Figure 4. These results suggest a notable degree of interdisciplinarity within the topic. However, a clear preference is observed for topics related to materials and polymer science. For example, the journal with the highest number of documents is Carbohydrate Polymers with 6 documents, followed by International Journal of Biological Macromolecules and Journal of Polymers and the Environment, both with 4 documents.

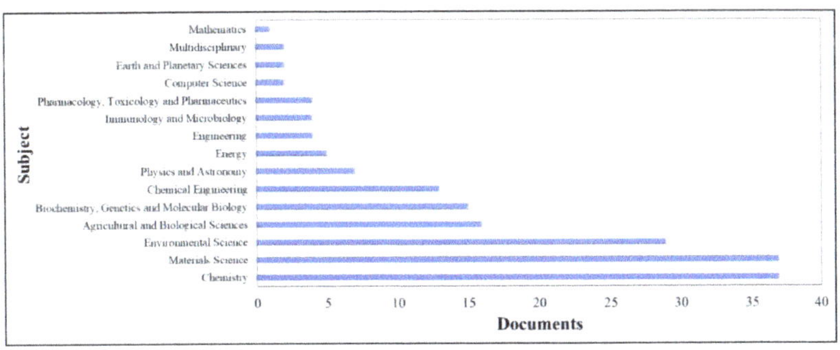


**Figure 4.** Distribution of documents across subject areas (2000–2024).

From the distribution of journals, the analysis then shifts to the thematic content of the publications, as reflected in the authors' chosen keywords. A total of 242 author keywords were detected across the studied articles, as shown in Figure 5a. As expected, the most frequently mentioned author keyword is "hydrogel." However, as in other cases, the studies also reference specific hydrogel types, including bio-based substances [44–46] or nanocomposites [38,46–49], as well as variations in crosslinking processes, ranging from beam-based methods such as gamma radiation [50] to more conventional chemical approaches using the crosslinker N,N′-methylenebisacrylamide (MBA) and ammonium persulfate (APS) as the initiator [51].

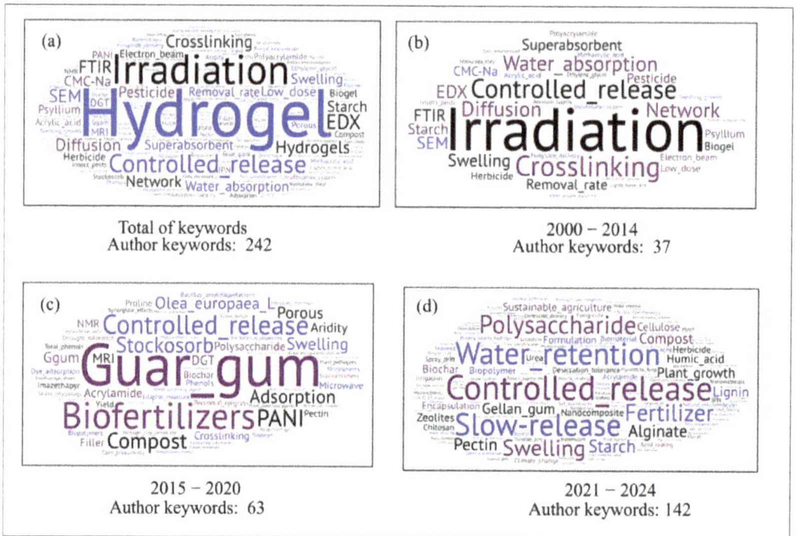


**Figure 5.** Temporal variation in author keywords illustrating the evolution of research themes, where (**a**) presents the overall frequency of keywords, (**b**) highlights terms used between 2000 and 2014, (**c**) depicts those from 2015 to 2020, and (**d**) reflects keywords from 2021–2024.

Figure 5b–d depict the temporal evolution of the contextual focus. To improve the interpretability of this analysis, the term hydrogel was excluded. Although the number of words increased, it was possible to observe the inclusion of more bio-based components during the period 2015–2020, such as Guar gum (Figure 5c), or the starch or other polysaccharides (Figure 5d). As shown Figure 5b–d, there is clear evidence of an evolution in applications. For instance, in the most recent works (Figure 5d) the keywords include terms such as "sustainable agriculture" [52]. In this same period the use of natural polymers is remarkable in order to produce more eco-friendly materials [53–55].

With regard to controlled release, an increase in publications has been observed, particularly those addressing fertilizer [34] or pesticides [56]. One of the most notable aspects concerns the advancement of the precision input delivery [57]. However, no specific use of the term associated with IMC was observed. This may be attributed to only 26 works (29%) in which there is a clear notice of collaboration between authors from departments in materials or chemical science and those in agricultural fields. This highlights an excellent opportunity to foster more interdisciplinary research.

From Figure 5, author keywords highlight applications related to ICM practices, such as water retention for soil improvement and the use of biofertilizers and fertilizers, even when not explicitly stated. The preceding analyses provide the contextual foundation for a more detailed examination of hydrogel applications, which are classified into four categories: bioactive products, nutritional growth inputs, pesticides, and soil conditioners (Figure 6). The Sankey diagram illustrates their distribution, with most publications addressing pesticides, followed by soil conditioners, and fewer on bioactive products, which enhance plant tissue growth and support controlled bioactive release [58]. Six studies were classified into more than one category.

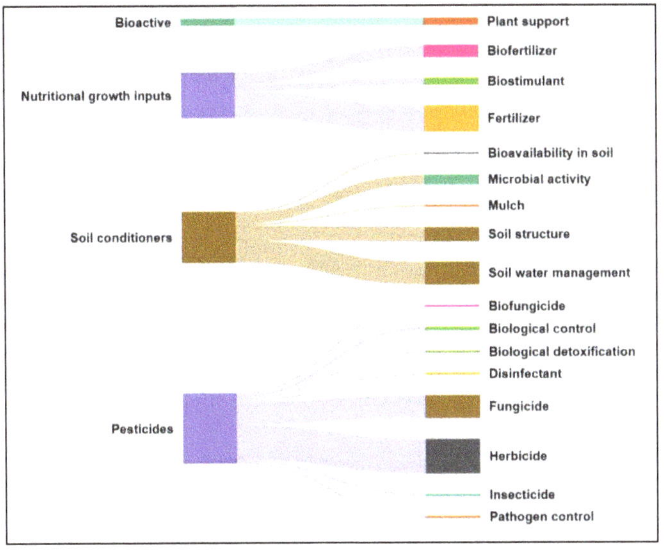


**Figure 6.** Sankey diagram of the distribution categories and subcategories identified. Image created using SankeyMATIC.

### *2.2. Analysis of Categories*

A focused analysis of these four thematic groups provides critical insight into how hydrogels are being positioned within agricultural systems. By separating the publications into categories—bioactive products, nutritional growth inputs, pesticides, and soil conditioners—it becomes possible to trace specific research trends, identify methodological approaches, and highlight persistent challenges. This categorization not only clarifies the breadth of applications but also underscores the need for targeted innovation to align hydrogel development with practical agronomic demands. The discussion proceeds according to prevalence in the dataset, beginning with pesticides and concluding with bioactive products, which, although less represented, demonstrate noteworthy potential for plant growth promotion and controlled release of active compounds.

#### 2.2.1. Pesticides

Pesticides are important in agriculture as they effectively enhance crop yields by repelling, eradicating, or mitigating the impact of pests. Their continued usage can lead to significant issues, such as environmental pollution and health hazards [59], which hydrogel-based remedies may address. This section examines various methodologies identified in the selected texts.

As previously mentioned, this category contains the highest number of documents—38 in total. Table S1 provides details on the types of hydrogels developed, the pesticides loaded, the crosslinking processes employed, and the methods used for hydrogel characterization. As shown in Figure 6, the identified subcategories include biofungicide, biological control, biological detoxification, disinfectant, fungicide, herbicide, insecticide, and pathogen control. Despite this variety, there is a clear trend toward developing environmentally friendly formulations. This tendency is reflected in the use of biodegradable materials for hydrogel production, to support the active compounds. Examples include sodium alginate (SA) and carboxymethyl cellulose (CMC) [60], gellan gum [61], and carboxymethyl chitosan (C-CS)/sodium alginate [62]. Another example is an alginate/amidated pectin hydrogel, which has been used to encapsulate the biocontrol microorganisms, including *Trichoderma koningiopsis* Th003 [63].

Figure 7 illustrates the evolution of the pesticides examined by the authors. Despite this trend, highly toxic substances such as paraquat and copper sulfate continue to be referenced in certain documents—likely due to their increasing use, particularly paraquat in 2024, given its high efficacy in weed control [64]. Novel hydrogel-based systems have been reported, including the study conducted by Dong et al. (2023) [53], which produced a metal–organic framework (MOF) nanoparticle synthesized from $FeCl_3$ and 2-aminoterephthalic acid to address challenges associated with its application. The hydrogel material contains mesoporous crystalline cages that functioned as carriers for herbicide encapsulation. Additional studies have investigated nanocomposites—comprising methylcellulose or chitosan reinforced with zeolite and further enhanced by poly(methacrylic acid)-co-polyacrylamide—to mitigate or diminish the leakage of hazardous compounds, even under fluctuating pH circumstances [56].

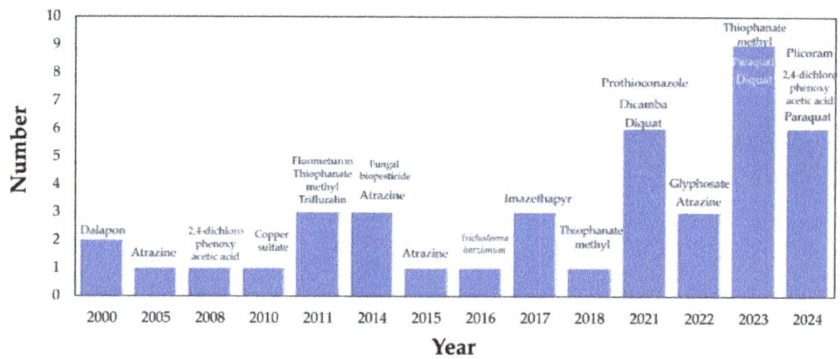


**Figure 7.** Temporal evolution of pesticide types reported in the analyzed documents.

Other commonly used chemicals, which remain toxic yet less hazardous than paraquat and copper sulfate, are also documented in Table S1 and Figure 7. These include Dalapon (sodium 2,2-dichloropropionate) [28], atrazine [51,65–67], and glyphosate [68]. In several cases, the primary objective was to achieve the controlled release of these compounds under specific conditions, particularly in aqueous environments [69].

Alongside strategies aimed at improving controlled release and reducing the adverse effects of pesticides, the adoption of eco-friendly or biologically based agents is also apparent. This includes fungal biopesticides [70], copper oxide nanoparticles that demonstrate effectiveness against various plant pathogens [71], garlic oil [46], and plant-derived compounds such as glycoalkaloids isolated from tomato and potato foliage [60].

The reviewed documents reveal a wide variety of hydrogel materials, as shown in Table S1, with 38 distinct types identified and a marked increase in diversity since 2017, as illustrated in Figure 8. Although many types were reported, a substantial proportion of formulations were based on acrylic acid, such as psyllium/acrylic acid [36], methacrylic acid [38,49,56,72], sodium alginate [60,62,73], and bio-based materials including chitosan [62,71,74] and starch [67,70,72]. The hydrogels identified include traditional copolymers such as polyacrylamide/poly(ethylene oxide), as well as semi-interpenetrating networks (semi-IPNs) based on synthetic polymers [65] and others incorporating natural polymers such as C-CS/SA [62]. Additionally, there is growing development of composites and nanocomposites that integrate inorganic nanostructures such as nanoclays [66], zeolites [37,49,56], copper oxide [71], and even nanocellulose crystals [75].

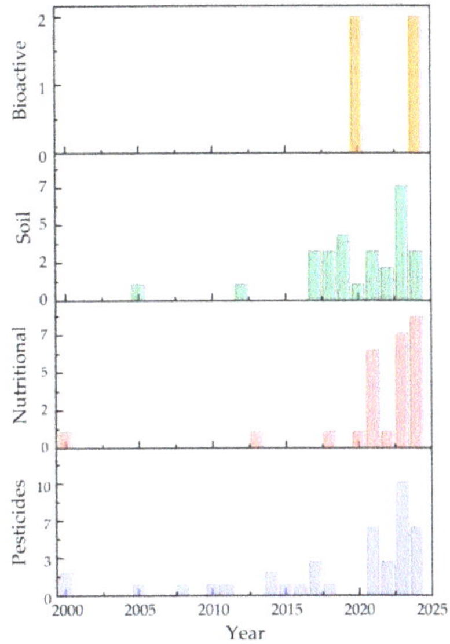


**Figure 8.** Temporal evolution of hydrogel material types reported in the analyzed documents.

A notable trend is the increasing investigation of hydrogels derived from composites and biocomposites. In summary, regarding the types of materials, as illustrated in Figures 8 and 9a, this diversity can be categorized as synthetic polymers [76,77], bio-based polymers. Some of these materials are further classified as smart hydrogels due to their ability to respond to external stimuli such as variations in pH or controlled water retention and release [57]. However, in the most recent period examined in this study, certain hydrogels have been considered as multifunctional, owing to the incorporation of diverse components and their capacity to support and adapt to multiple environmental stimuli. These functions include water absorbency at different pH, temperature, and light levels [78], as well as the controlled release of compounds from microbeads, for example, those based on Ca-alginate graft copolymer of poly[N-isopropyl acrylamide-co-N,N-diethylacrylamide] [79].

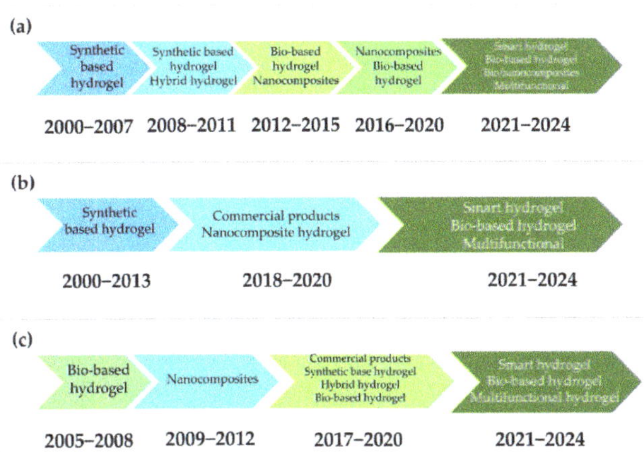


**Figure 9.** Evolution of hydrogel material types: (**a**) pesticides, (**b**) nutritional growth inputs, (**c**) soil conditioners.

Alternatively, as illustrated in Figure 10, which aggregates 21 studies featuring natural or biodegradable components, it becomes evident that there is no consistent correlation between the hydrogel type employed and the pesticides incorporated. The numbers in the figure represent the frequency of documents reporting a specific combination of hydrogel material and pesticide category. The absence of correlation arises because hydrogel selection is typically based on general properties—such as biodegradability, swelling capacity, or mechanical stability—rather than the nature of the active substance. For example, hydrogels based on chitosan (CS) have been used for both fungicides [11,62,71] and herbicides [49,56,74], while alginate (AL)-based hydrogels have been applied across all subcategories [60,68,70,73,80,81]. This finding confirms that multifunctional hydrogels are not strictly linked to a particular pesticide group.

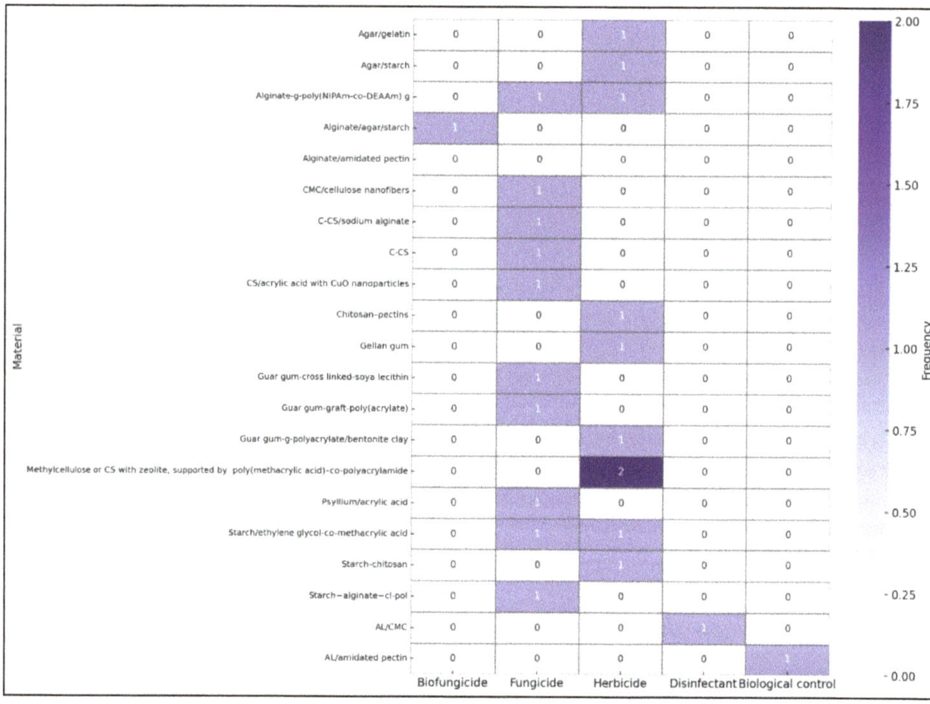


**Figure 10.** Document frequency linking bio-based hydrogels with pesticide categories. Values denote the number of studies reporting each case. Prepared with the assistance of artificial intelligence (ChatGPT, OpenAI, 2025), based on data and instructions provided by the authors.

Several authors have emphasized that the structural properties of hydrogels are essential for the regulated release of compounds [18,23,82]. Key factors include the polymer backbone, the integration of fillers—such as those utilized in composites and nanocomposites—and the density of network crosslinking [82]. The release profile can be significantly influenced by molecular interactions between the hydrogel matrix and the encapsulated substances [82]. Variations in crosslinking density during hydrogel synthesis may modify the release rate of compounds while concurrently influencing attributes such as water retention and soil-holding capacity [82]. This indicates that in developing novel hydrogel, it is essential to systematically evaluate the influence of these critical parameters on both compound absorption/release dynamics and water absorption/release characteristics.

Multiple techniques for crosslinking have been identified (see Table S1). The predominant technique involves chemical crosslinking using N,N′-methylenebisacrylamide (MBA) as a crosslinker and ammonium persulfate (APS) as an initiator. This strategy has been reported in several studies [51,69,71,83,84] and remained widely applied in works of 2024 [57]. The integration of MBA and APS is extensively utilized owing to its straightforwardness, repeatability, and tunable gel characteristics. This technique facilitates the formation of resilient hydrogel networks and offers operational simplicity, rendering it highly suitable for agricultural applications, especially in controlled pesticide release systems. In the context of APS, this initiator has been employed in conjunction with other crosslinking agents, such as hexamine [84] and additional agents like methylene bisacrylamide [82], as it is particularly effective in initiating free-radical polymerization, thereby promoting efficient and uniform network formation.

Various strategies have been employed to promote chemical or physical cross-linking, alongside increasing the number of components used to reticulate hydrogel. In certain cases, radiation approaches have been considered, originating from the early work of Saraydin et al. 2000 [27]. Nonetheless, recent studies have placed diminished emphasis on this method, while interest in procedures employing calcium agents has been increasing, likely due to the emergence of natural materials like alginate [60], pectin [74], or amidated pectin [63].

Regarding the hydrogel forms, the most commonly reported are films, beads, or granules. However, no clear correlation has been observed between the type of compound loaded and the physical form of the hydrogel. Recent innovations have focused on enhancing the spray conditions of conventional herbicides such as dicamba, particularly by achieving an optimal droplet size that minimizes drift and thereby improves efficacy. An example is the development of a hydrogel formulation based on folic acid and zinc nitrate [85], which produced droplets with effective air dispersion and confirmed the potential of this approach for improving spray performance. In the future, such strategies could be adapted to other active compounds, broadening their applicability in sustainable crop protection.

Regarding compound loading, multiple approaches have been identified (see Table S1). The predominant method, reported in 18 studies, is the swelling equilibrium method [28,37,40,49,51,53,56,65–70,72–74,82,86]. In this approach, the solid hydrogel is immersed in or comes into contact with a compound solution until entrapment occurs. This technique requires low investment and allows the evaluation of the effect of network structure, mesh size, and the affinity between hydrogel material and chemical substances. Another important advantage is its applicability for loading thermally unstable compounds [23]. Additional strategies include encapsulation techniques [11,46,60,63,75,80] or the incorporation of the molecule during hydrogel formation, utilizing an in situ method [27,57,61,62,71,76,79,84,87]. In this latter case, it is important to consider that the compound can be stabilized without compromising its integrity under the temperature, solvent, or pH conditions of fabrication. The advantage of the in situ method lies in its ability to promote stronger interactions between the hydrogel material and the compound. An example is the synthesis of a substance including the herbicide 2,4-dichlorophenoxyacetic acid (2,4-D) and poly[(1-vinyl-2-pyrrolidone)-co-(2-hydroxyethyl methacrylate)] as reported by Pizarro et al. (2008) [76]. The esterification between the hydroxyl groups of the copolymer and the acid chloride groups of the herbicide strengthens these interactions, producing a system in which chemical release is influenced by environmental pH.

Regarding the characterization techniques employed to assess the performance of the hydrogels, as illustrated in Figure 11, the most prevalent methods involve compound absorption and release studies, followed by Fourier Transform Infrared Spectroscopy

(FTIR) and swelling tests. Additional techniques such as thermogravimetry analysis (TGA), nuclear magnetic resonance alternatives (NMR), X-ray diffraction (XRD), and differential scanning calorimetry (DSC) provide structural insights into the hydrogels. This prevalence is expected, as the majority of the reviewed studies focus on hydrogel development. Ultraviolet-visible (UV-vis) spectroscopy is also frequently reported as a method to monitor compound release [37,53,63,75,84]. Swelling remains a common technique for assessing hydrogels behavior [11,27,28,38,46,49,56,63,71,79,80,82,83], and the resulting data are often compared with findings from other techniques that examine crosslinking or crystallinity, such as DSC [82].

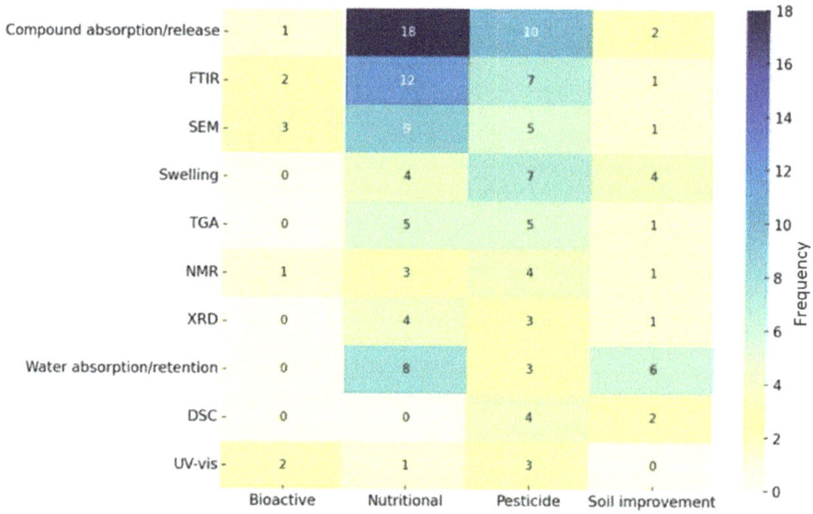


**Figure 11.** Relationship between commonly used hydrogel characterization techniques and hydrogel categories. Values denote the number of studies reporting each case. Elaborated with the assistance of artificial intelligence (ChatGPT, OpenAI, 2025), based on data and instructions provided by the authors.

Microscopic techniques such as scanning electron microscopy (SEM), transmission electron microscopy (TEM), or atomic force microscopy (AFM) are valuable for analyzing material morphology and supporting the evaluation of hydrogel behavior, for example, in studies assessing water absorption capacity [82].

The characterization of hydrogels generally includes testing of absorption, retention, and release of water, particularly in relation to soil moisture retention and the duration of water availability [82]. Some studies emphasize that the integration of specific techniques is essential for understanding the release behavior of active compounds, often using water as a model substance.

Conversely, only seven studied have reported in vivo experiment assessing hydrogel effects on plant growth or seed germination [11,46,61,69,71,79,88] (see Table S1). The plant species studied include varieties of wheat [11,46,88] and lettuce [71,79]. In addition, eight works have been investigated hydrogel-soil interaction tests, either in vivo [46,69,71,88] and ex situ [60,61,79,88], using specific soil types. These investigations focus on degradation and compound release. Despite the diversity of soils considered (see Table S1), natural derived compounds such as fungal biopesticides [70] or garlic oil [46], have been tested in agricultural soil to simulate real application conditions, whereas synthetic pesticides are typically evaluated in soils of defined texture, with attention to their mobility and leaching potential [57,69,71].

The evidence from in vivo and ex situ tests shows that research remains centered on hydrogel design and compound release, with limited attention to plant-level outcomes. Bridging this gap through plant monitoring and field validation will be essential, enabling more effective interdisciplinary collaboration and advancing hydrogel technologies toward sustainable crop protection.

#### 2.2.2. Nutritional Growth Inputs

This category, aligned with ICM strategies, comprises documents addressing compounds supplying essential macro- or micronutrients for plant metabolism, growth, and reproduction. It also encompasses biostimulants that enhance nutrient use efficiency, abiotic stress tolerance, or crop characteristics, regardless of nutrient content [7,12]. A total of 26 documents were categorized in this group, four of which were additionally classified under other categories such as pesticides [28,69], and soil improvement [2,45]. Figure 6 and Table S2 illustrate that the subcategories include fertilizers, which account for the largest number of analyzed documents, followed by biofertilizers and biostimulants.

The growing variety of fertilizer-related compounds reflects a trend similar to that seen in pesticide applications. Figure 12 shows a significant increase in the diversity of compounds reported in the reviewed documents. Urea remains the most frequently cited substance [20,32,45,50,81,89,90], primarily due to its widespread use as a fertilizer, its high nitrogen content, and low cost. However, nutrient losses and runoff in agricultural systems presents a significant challenge [50].

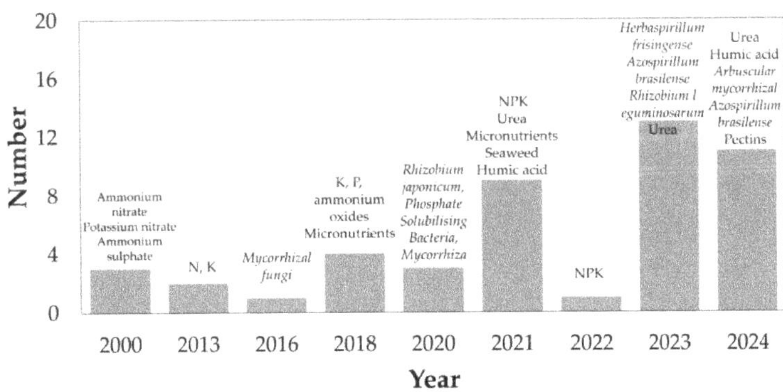


**Figure 12.** Temporal evolution of nutritional growth inputs reported in the analyzed documents.

Seven distinct hydrogel materials have been reported for mitigating urea loss. These include starch-based systems such as starch from rice-cooked wastewater with acrylamide [45], natural rubber latex–cassava starch [90], and acetylated starch–polyacrylamide [81]; natural oil-based coatings with crosslinked PAN/PAAc [50]; PVAl/humic acid hydrogels [20]; and alginate-based formulations, including composites with attapulgite, N-isopropylacrylamide–sodium alginate [89], and sodium alginate–humic acid–NIPAm–AMPS. Additionally, a hydrogel of acrylamide–methylpropanesulfonic acid–poly(ethylene glycol) with sodium alginate and humic acid as a filler has been described [32]. The objectives of these hydrogels are diverse: some aim to create hydrophilic barriers that regulate urea diffusion [50], others focus on establishing strong interactions with urea [45], and some are designed to form complexes through hydrogen bonding or ionic interactions [20]. However, the potential degradation by-products generated from these interactions and their implications for plant growth are not addressed. Overall, the trend indicates that hydrogel systems are evolving—from conventional synthetic designs to more advanced materials incorporating composite structures or smart functionalities, capable of integrating multiple responsive mechanisms, such as combined pH and temperature sensitivity [89].

Table S2 and Figure 12 indicate that humic acid (HA) is the second most frequently cited compound, with its use reported mainly in studies published between 2021 [3] and 2024 [32,34,91]. HA denotes a category of long-chain organic compounds characterized by colloidal properties and complex spatial structures, predominantly originating from plant degradation and fermentation [32]. This substance is widely recognized as an organic fertilizer capable of enhancing plant growth [91]. Researchers have utilized derivatives such as humic acid sodium salt [3] or commercial formulations [32] to develop specific hydrogels. The chemical structure of HA enables hydrogen bonding with hydrogel matrices, thereby improving network stability and while enhancing hydrophilicity, swelling behavior, and compatibility with soil [36].

The incorporation of biological agents, including *Rhizobium* [92,93], *Azospirillum* [52,93], and mycorrhizae [2], has been documented, particularly within bio-based hydrogels made from alginate [52] or pectins [93].

Figure 8 and Table S2 show that a considerable number of hydrogels have been developed, mirroring observations made in the context of pesticides. Figure 9b depicts a specific evolution in the types of polymers utilized for the incorporation of nutritional growth inputs. Between 2000 and 2013, the materials identified predominantly comprised synthetic hydrogels with limited functionality and bio-integration, including acrylamide/crotonic acid [28]. During the subsequent period, as illustrated in Figures 9b and 12, the utilization of commercial hydrogels, including Stockosorb, rose markedly. This material demonstrates benefits in alleviating drought stress in the production of olive plantlets [2]. In this timeframe, biocomposites derived from bacterial nanocellulose were also observed [22]. The latest period examined in this study, extending from 2021 to 2024, underscores the growing application of bio-based hydrogels incorporating components sourced from waste valorization, such as hydrogels derived from rice-cooked wastewater [45].

The trend in nutritional growth inputs indicates a preference for bio-based components in hydrogel production. Various natural polymers have been utilized in these systems, such as sodium alginate [9,32,34,89], starch [45,81,90,94], pectin [93,95], and cellulose [3,9,22,96]. Figure 9b illustrates that this trend has intensified in recent years and is likely to continue. Future developments may focus on multifunctional materials that enhance water retention, facilitate nutrient delivery, and respond to environmental stimuli. An example of such a multifunctional hydrogel is a pH-responsive system based on N-isopropylacrylamide (NIPAm) and 2-acrylamido-2-methyl-1-propanesulfonic acid (AMPS), which exhibits thermal sensitivity and enhanced water absorbency [32].

Further examples of multifunctional hydrogels include composites that integrate natural fillers like attapulgite [89] or biomass sources like pine resins [97]. A notable example is the leather waste–acrylic acid–maleic anhydride composite, which combines water retention, controlled fertilizer release, biodegradability, and heavy metal adsorption—specifically for Cr [III]—while also responding to various environmental parameters such as pH and coexisting ions [32]. This example highlights the incorporation of circular economy principles through the valorization of industrial by-products.

Similarly to observations in documents classified under the pesticide category, a large number of studies on nutritional growth inputs focus—either directly or indirectly—on the analysis of compound release behavior [20,22,32,34,68,81,89–91,97,98]. The structural strength of these hydrogels can be influenced by the crosslinking process [20], the incorporation of functional fillers [89], or a combination of both. An illustrative example of this integrated approach is the formulation of composites that combine pine resin polysaccharides with biochar as a filler, enhancing nutrient release and overall material performance [97].

The crosslinking processes utilized in these studies follow a pattern similar to those reported in the pesticide category. Traditional chemical crosslinking systems are widely employed, particularly those involving MBA and APS, most notably in the synthesis of hydrogels from acrylamide derivatives [81,89,97] or acrylic derivatives [35,98]. With the rising adoption of bio-based hydrogels and heightened concerns regarding environmental sustainability, alternative crosslinking strategies have emerged. Ionic crosslinking methods frequently employ divalent or trivalent metal ions, including $Ca^{2+}$ [8,34,52,93], $Cu^{2+}$ [9], and $Zn^{2+}$ [9]. Furthermore, a significant trend is observed in the development of physically crosslinked systems, where interactions such as polymer chain entanglement are crucial, particularly in formulations based on chitosan and starch [94]. Future research is likely to emphasize physical crosslinking mechanisms while reducing reliance on synthetic crosslinkers, initiators, and chemical activators.

For compounds such as humic acid and 3-indoleacetic acid, crosslinking is mainly determined by the interactions between functional groups, particularly hydroxyl and carboxylic acid moieties [3]. Strategies such as coacervation, aldehyde-mediated reactions, and citric-acid-induced crosslinking have shown significant relevance for these substances.

As observed in the case of pesticides, the most employed technique for compound loading in hydrogels for nutritional growth inputs is the swelling equilibrium method, which appears in 46% of the reviewed documents [20,28,32,34,35,45,50,81,89,91,94,96]. The subsequent prevalent technique is the in situ method (27%) [3,8,69,89,95,97,98], followed by encapsulation processes (19%) [22,52,90,93,99]. The frequent application of the swelling equilibrium method may be attributed to its appropriateness for sensitive compounds that are prone to degrade or change during hydrogel synthesis, such as biofertilizers [94]. Even for relatively stable compounds like urea, which can withstand thermal conditions, this method may enhance availability and performance [89]. The selection of a compound loading technique must carefully consider both the chemical stability and availability of the active ingredient under hydrogel fabrication conditions, as previously noted regarding the pesticide category.

Figure 11 illustrates that the main techniques employed to characterize hydrogels for nutritional growth inputs concentrate on compound absorption and release, encompassing release profiles in both water [3,35,50,91,97] and soil [3,20,89,91,97]. As previously discussed regarding pesticide-related hydrogels, structural characterization methods such as FTIR [8,22,35,50,68,81,89–91,94,95,98], SEM [9,22,35,50,68,81,89,91,93–95,98], and TGA [20,22,45,81,89,91,95,98] are also significantly utilized. Furthermore, evaluations of water absorption and retention capacity are significant for assessing the material's agricultural performance. The application of these techniques reflects a strong scholarly interest in analyzing or understanding nutrient delivery and availability. The capacity for entrapment of fertilizers, like pesticides, is frequently assessed using UV–visible spectrophotometry [32,90,91,96].

A total of 63 percent of the analyzed documents reports in vivo tests, whereas 40% include germination assays, such as the evaluation of the number of germinated seeds [9,20,45,94–96]. The analysis of plant growth encompasses root architectural parameters including length, area, and volume, as well as leaf area measurements [2,9,34,45,68,90,93,94]. Additional trails assess biomass production, including fresh plant or root biomass and dry plant biomass [2,20]. These assays are often complemented by biochemical indicators such as chlorophyll [2,9,94], protein [92], oil content [92], photosynthetic pigment [2], or phenol content [2]. In certain cases, they also provide information on plant responses under water stress conditions [2]. Recent studies have introduced advanced methods, including high-resolution digital imaging technologies, to enable more precise analyses of root architecture [99].

A notable aspect is that 79% of these documents were published between 2023 and 2024, underscoring the growing relevance of the research on hydrogel-plant interactions.

Regarding the types of plants considered, some studies have examined woody species, including olive plantlets (*Olea europaea* L. *cv.*), which are particularly valuable for long-term investigations [2]. Legumes, including soybean [92], mung beans [45,91,95], and peas [45], are commonly employed in early-stage performance testing because of their rapid growth and low experimental costs. Germination assays are frequently performed on grain crops such as maize [34,94] and sorghum [20], whereas evaluations of leafy vegetables typically emphasize traits such as leaf length, root development, and biomass accumulation.

Agricultural soils—including clay loam, sandy phosphorus-deficient, and loamy types—were among those used in the examined studies. Most studies, however, lacked systematic monitoring or characterization soil properties during the experiments. This gap highlights an important avenue for future research, especially to advance understanding of the interactions among soil, hydrogel, and the broader soil–hydrogel–plant system.

#### 2.2.3. Soil Conditioners

Soil conditioners are essential for improving agricultural productivity and mitigating environmental pollution [100,101]. Powlson et al. (2011) [100] identified the primary soil functions that support food production as follows: (1) supporting seed germination, root establishment, and root functionality, including anchorage, water uptake, and nutrient acquisition; (2) acting as a nutrient reservoir; (3) enabling the transport of water and nutrients from both native soil reserves and external inputs; (4) providing a medium for essential nutrient transformations for plant nutrition; (5) supporting diverse microbial and faunal communities that contribute to plant growth and nutrient cycling; and (6) delivering physical support for agricultural operations involving machinery, humans, and animals.

In relation to the analysis of this category, a total of 28 documents were classified. Figure 6 and Table S3 show that the subcategories identified in this review encompass soil bioavailability, microbial activity, mulching, soil structure, and soil water management. Among these, soil water management emerges as the most frequently addressed topic and is supported by the largest number of related documents.

On the other hand, hydrogels have the potential to modify and enhance the fluid dynamics of the soil liquid phase by (1) increasing water-holding capacity, (2) improving water flow characteristics, and (3) enhancing the mechanical performance of soils under diverse environmental stress conditions [102,103].

This category reveals extensive use of diverse hydrogel systems (see Figures 8 and 9c, and Table S3). Synthetic polymers, particularly polyacrylamide-based hydrogels [45,47, 104,105], include innovative examples such as poly(acrylamide-co-acrylate) derived from acrylamide processing residues [104]. Commercial products like Stockosorb 660 are also reported [2,42]. Beyond these, nanocomposite formulations, such as poly(acrylic acid-co-acrylamide)/$AlZnFe_2O_4$–potassium humate [47], and hybrid systems, including poly(acrylic acid)-graft-agar/gum Arabic [106] and chitosan/potato starch blends [10], demonstrate efforts to integrate functionality. At the same time, biopolymer-based hydrogels, often incorporating natural polymers or agricultural waste [10], highlight a clear shift toward environmentally sustainable systems consistent with circular economy principles. As further shown in Figure 9c, several hydrogels can be classified as hybrid, smart [106], or multifunctional systems within the previously defined categories. Multifunctional hydrogels play an important role in soil water management, structural improvement, and the stimulation of microbial activity [33].

On the other hand, as reported in Table S3, bio-based hydrogels are predominantly associated with soil structure [33,44,54,103,107,108] and soil water management [36,45,55,106,109–111].

Moreover, these hydrogels allow regulated water release in response to external stimuli such as temperature, pH, light, and enzymatic activity [79]. As shown in Figure 9c, since 2017 there has been a marked increase in the development of bio-based, hybrid, smart, and multifunctional hydrogels, reflecting sustained innovation directed toward sustainable agricultural applications. The application of nanocomposite hydrogels can be traced to the period between 2009 and 2012 [47].

Figure 13 highlights that soil water management consistently emerges as a central theme among authors. Soil structure has gained increasing relevance since 2012, with a particularly strong emphasis after 2021. Microbial activity has been addressed intermittently, appearing in works from 2017 [1,112], 2018 [2,10], 2020 [113], and 2023 [93]. Conversely, the subcategories of bioavailability in soil and mulch remain rarely examined, despite their potential importance for ICM strategies.

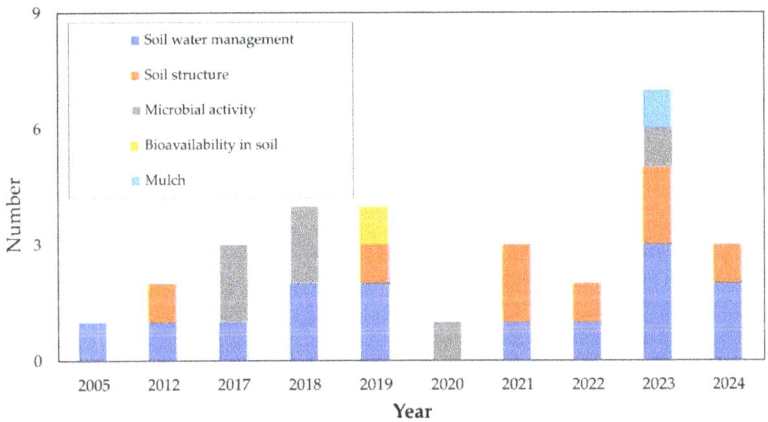


**Figure 13.** Temporal evolution of soil improvements reported in the analyzed documents.

Regarding the crosslinking process, ionic crosslinking—using divalent ions such as $Ca^{2+}$ or $Cu^{2+}$—is frequently employed, particularly in applications related to soil structure and microbial activity [93,113]. In contrast, commercial formulations or thermally crosslinked hydrogels are more common in studies focused on soil water retention. As in other categories, chemical crosslinking with MBA remains the dominant approach, especially for soil water management, due to its proven effectiveness in improving water-holding capacity [36,104,106,109].

Hydrogel presentation typically includes films and different types of granulates, such as beads, consistent with other categories. A liquid formulation was also identified, noted for its improved mixing capacity with the soil [103].

Figure 11 shows that the main hydrogel characterization methods in this category are water absorption and retention [1,10,47,48,104,107,109,110,113,114], followed by swelling analysis [10,31,55,106,107,109,112]. In addition, several studies report techniques related to thermal stability and degradation, including TGA, DSC, and related methods [10,33,45,55,107,108].

A total of 19 studies were identified that evaluated hydrogel under soil-based conditions. The soils tested were mainly characterized by low fertility, poor structure, and limited water retention, including sandy loam [1,2,36,42,47,54,103,113], saline-alkali [33], and coarse-textured soils [44,54,103,112]. This pattern underscores the importance of testing hydrogels for soil water management, a critical agricultural application. Bio-based hydrogels were also studied in cultivated soils [45,115], garden soils [110], and natural soils such as forest soil [111], reflecting attempts to design ecologically relevant testing framework across diverse conditions.

In vivo assays were documented in 13 documents, four of which evaluated germination [47,55,111,114]. Table S3 indicates that between 2012 and 2018, studies mainly focused on staple crops such as wheat [47], maize [116], and olive [2,42]. In later years, a wider range of species has been tested, including *Lactuca sativa* [115], tomatoes [48,111], potatoes [104], legumes such as *Pisum sativum* [110] and *Vigna radiata* [45,110], leafy vegetables from the *Brassica* [114] and *Capsicum* [114], and several weed species [114]. This trend demonstrates a growing interest in the broader agroecological role of hydrogel.

In relation to germination, it is also important to highlight that the incorporation of hydrogels can also alter soil porosity and permeability, thereby influencing seedling performance [47,55].

Finally, at least 30 distinct have been identified for evaluating soil-hydrogel interactions, categorized in Figure 14 and detailed in Table S3. Despite this variety, two key aspects consistently emerge as fundamental in study design: soil pH and water dynamics.

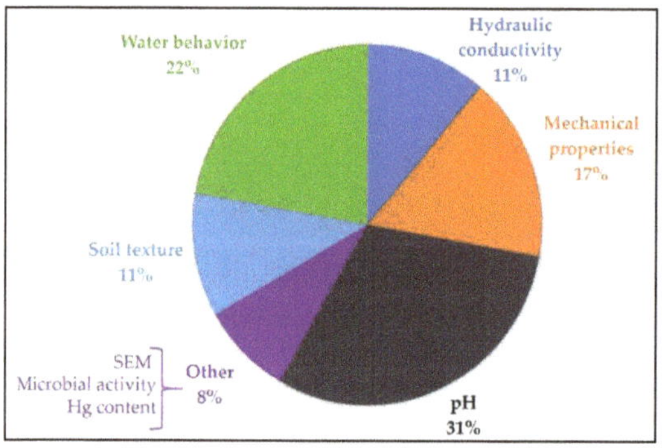


**Figure 14.** Main soil characterization parameters reported in the reviewed studies, expressed as percentage distribution.

### 2.2.4. Bioactive

Bioactive substances contribute to ICM practices by functioning as organic insecticides, growth regulators, and disease inhibitors, thereby promoting plant growth and improving soil health [117]. These compounds are typically categorized into phenolic compounds, non-phenolic compounds, and pigments [118]. In addition, bioactives can act as biochemical probes, serving as model systems for studying biochemical processes and controlled release behavior [119].

Only three documents met the criteria for this category, as reported in Table S4. These primarily examined compounds employed as model systems to track hydrogel-controlled release. Notably, phenolic molecules such as phloroglucinol (1,3,5-benzenetriol) were identified as effective probes for evaluating the performance of hydrogels synthesized from methyl-esterified pectin and humic substances derived from composted biomass [120]. Furthermore, two investigations focused on highly porous, tunable biohydrogels formed from metal ions and organic ligands, specifically amino acid-enriched collagen–starch matrices [58,121]. These systems demonstrate direct interactions with root and leaf tissues, enhancing plant metabolism, tissue development, and overall plant growth.

Although the number of documents in this category is limited (Table S4), the studies reveal a consistent reliance on bio-based materials. For instance, ionic crosslinking with $Ca^{2+}$ was applied due to the natural presence of pectins in hydrogel matrices [120], while polyurethane was incorporated into semi-interpenetrating polymer networks (semi-IPNs) [58,121]. Compound loading strategies included swelling equilibrium and encapsulation methods, as observed in earlier categories. The analytical approaches used to characterize both hydrogels and plant responses were consistent with those previously described.

Importantly, in vivo evaluations were performed with the same plant species, tomato (*Solanum lycopersicum*), in both studies [58,121]. The preference for tomato is likely linked to its widespread role as a model organism in plant science, particularly for investigations of plant defense mechanisms against pathogens affecting fruit quality [122]. This relevance is reinforced by the availability of extensive genetic and genomic resources supporting its use in experimental frameworks [123].

### 2.3. Opportunities for Future Research

This study illustrates how diverse types of hydrogels can be effectively integrated with ICM strategies to address persistent challenges in the agricultural sector. The development of hydrogels for applications such as pesticides, nutritional growth inputs, soil conditioners, and bioactive compounds represents a dynamic and expanding field of global research. Particular emphasis is observed in initiatives focused on bio-based multifunctional materials, the valorization of industrial by-products, and the reuse of post-consumer waste. These approaches enhance hydrogel performance while adhering to the principles of circular economy and green chemistry. Based on this analysis, several key opportunities for future research are identified:

- Conduct detailed investigations of hydrogel-soil interactions, with special emphasis on soil microbiota and plant–microbe dynamics.
- Expand field-based assessments in practical agricultural environments [2]. While the number of such studies has increased, further research is particularly needed in underexplored contexts, such as flooded soils.
- Examine the potential toxicity of hydrogels in soil and aquatic systems. This remains insufficiently studied, as highlighted by Kolya et al. 2023 [45], despite the likelihood of large-scale applications.
- Evaluate hydrogels performance under a wide range of environmental conditions. Research has largely prioritized drought stress, but greater attention is needed to conditions such as extreme moisture, atmospheric variability, and non-traditional systems, including soilless, urban, and peri-urban agriculture.
- Integrate regulatory aspects into hydrogel research. Future studies must address regulatory frameworks and toxicity assessments, as their absence represents a critical barrier to the safe, sustainable, and large-scale agricultural application of hydrogels.
- Despite the identification of bio-based hydrogels in this study, including those derived from agro-industrial sources, further exploration of alternatives from food waste as well as urban and industrial residues remains essential.
- None of the analyzed studies address the environmental impact of the developed hydrogels. This gap highlights a key opportunity for future research to integrate green engineering principles with environmental systems analysis, including carbon neutrality, life cycle assessment, and water footprint evaluation.
- The reviewed documents did not report the implementation of strategies such as machine learning models or simulation tools for predicting hydrogel behavior. Incorporating these approaches could improve resource utilization and leverage the expanding corpus of scientific and technical knowledge.

- Other identified opportunities involve long-term evaluation, particularly through integration of multiple ICM strategies that combine pesticides, biostimulants, biofertilizers, and soil conditioners [7] to enhance crop production, potentially complemented by mulch application [114].
- Incorporating emerging technologies, such as additive manufacturing or rapid prototyping, which rely on layer-by-layer fabrication technology [124], could support the development of complex structures. Multifunctional hydrogel may particularly benefit from these methods, which offer greater flexibility and reduced material waste [124].
- Only one study was identified that investigated the use of hydrogel as mulch. The combination of hydrogels with film-based alternatives represents a promising direction for developing multifunctional materials that integrate soil protection, water management, and compound release, particularly relevant to small-scale food production.
- This study demonstrates a growing interest in hydrogel applications; however, additional research is essential, especially for the transition from laboratory prototypes to real-scale applications. Such efforts should foster broader engagement from researchers across industrial sectors.
- Various researchers have highlighted the need to investigate alternatives to the conventional soil integration of hydrogels, such as packaging them in bags made from hygienic mask wastes [110], while also exploring techniques for reprocessing or refilling the hydrogel [23]. These methods may reduce reliance on raw materials and continuously strengthen the circular economy.
- Variations in the crosslinking process were also identified, reflecting efforts to lower the environmental impact compared with conventional methods. These trends are likely to intensify in the future, particularly for hydrogels used as chemical release devices. However, it is crucial to examine the potential by-products arising from hydrogel-compound interactions, as well as any resulting contaminants and their impacts on plants or soil.

## 3. Conclusions

This work presents a concise overview of the current development of hydrogels in agricultural practices, framed within ICM strategies. This analysis was conducted through a synthesis of bibliometric analysis and PRISMA methodology. The approach included macro- and micro-level analyses that identified the worldwide significance of the topic, identified the regions of the greatest relevance, and examined key research domains through the analysis of the author's keywords and selected publications. The trend has been increasing steadily, especially during the second decade of the twenty-first century, and appears promising for the coming years. Contributions from authors worldwide have been documented, reaffirming the topic's significance. Recent developments highlight the importance of green and sustainable agriculture, together with the expanded use of agro-waste residues in hydrogels, offering opportunities for implementing a circular economy. Guided by the ICM strategies, the selected documents were categorized into four groups: (1) pesticides, (2) nutritional growth inputs, (3) soil conditioners, and (4) bioactives. Each category was examined for patterns, with particular emphasis on identifying materials for future study in this domain.

The significance of hydrogels within the ICM framework is clear, as is the imperative to further develop advanced hydrogels from bio-based components that can function as multifunctional materials, broadening their agricultural applications. Thus, materials such as multilayer systems, nanocomposites, or biocomposites may serve as viable alternatives. Furthermore, the potential for enhancing interactions through the incorporation of the specific component within the hydrogel should be examined to adjust the crosslinking degree. However, it is essential to investigate the possible formation of by-products that may affect plants and soil during application.

A wide range of characterization methodologies were identified, most of which are based on hydrogel behavior. This underscores the need to strengthen research on hydrogel interactions with both plants and soil. Overall, this work highlights the necessity of fostering strong collaboration between industry and academia, supported by international cooperation, to advance these developments and achieve meaningful impacts in addressing global food production challenges under climate change.

## 4. Materials and Methods

This study was carried out using a rigorous process that integrated the Preferred Reporting Items for Systematic Reviews and Meta-Analyses (PRISMA) methodology and bibliometric analysis [125,126]. To minimize bias and ensure reproducibility, the PRISMA technique was selected [125,126].

Figure 15 summarizes the methodology applied in this study. The first step was to define the search-query equation. A minimum of five iterations were conducted before generating the final version. This procedure ensured a broader retrieval of relevant documents.

During this iterative process, it was possible to identify terms corresponding to different substances used as agrochemicals categories within ICM, including fertilizers, plant protection agents, and biostimulants. Given the diversity of biostimulants types and the challenges in establishing a comprehensive definition [7], this work adopted the main categories proposed by Du Jardin (2015) [7].

The final search-query was defined as [humic OR fulvic OR "amino-acid*" OR peptid* OR "protein hydrolys*" OR seaweed OR algae* OR "inorganic salt*" OR fungi* OR mycorrhiza OR bacterial* OR herbicid* OR "weed control" OR bioherbicid* OR biofertilizer* OR biostimulant*) AND [hydrogel*) AND [agricultur*) AND NOT [medicin* or scaffold* OR biosen* OR biomedical OR "solar steam" OR "stem cel*"). The Boolean operator AND NOT was used to exclude unrelated publications, particularly those addressing biological, medicine, sensors, or toxicological applications where hydrogels are also employed. To maximize retrieval, the operator asterisk (*) was used to capture variations in relevant terms.

The next step was to search for the documents in a database. In this case, the Scopus database was selected because it contains a comprehensive record set [127,128] and offers reliable access to information. However, this choice also represents a limitation, as non-database sources include all relevant publications [129]. Nonetheless, Scopus is widely used and facilitates accessible and reproducible searches.

As in previous steps, the equations were applied to the TITLE-ABS-KEY fields in Scopus. All documents published between 1999 and 31 December 2024, were retrieved. As shown Figure 15, a total of 211 records were initially obtained. Because this study focused on research articles, reviews and other document types were excluded. After applying this criterion, 158 articles remained (Figure 15). Additional exclusion criteria include language, English-only and duplicate entries, ensuring accessibility and consistency. Following these steps, the total number of articles was reduced to 146.

As indicated in Figure 15, the next step involved an extensive review of titles and abstracts. At this stage, the inclusion criterion was limited to documents explicitly addressing the use of hydrogel in soil health or quality, pest control, or crop nutrition. Documents were excluded if they focused on topics such as filtration, fruit packaging, fruit protection, sensor development, water pollution control, reviews not identified earlier, agricultural practices without hydrogel, hydrogel processing or characterization, genetics, medical applications, or biotechnology. There are fifty documents in all that have been deleted. A total of 50 documents were excluded. Ultimately, 90 articles were retained for bibliometric and in-depth analysis.

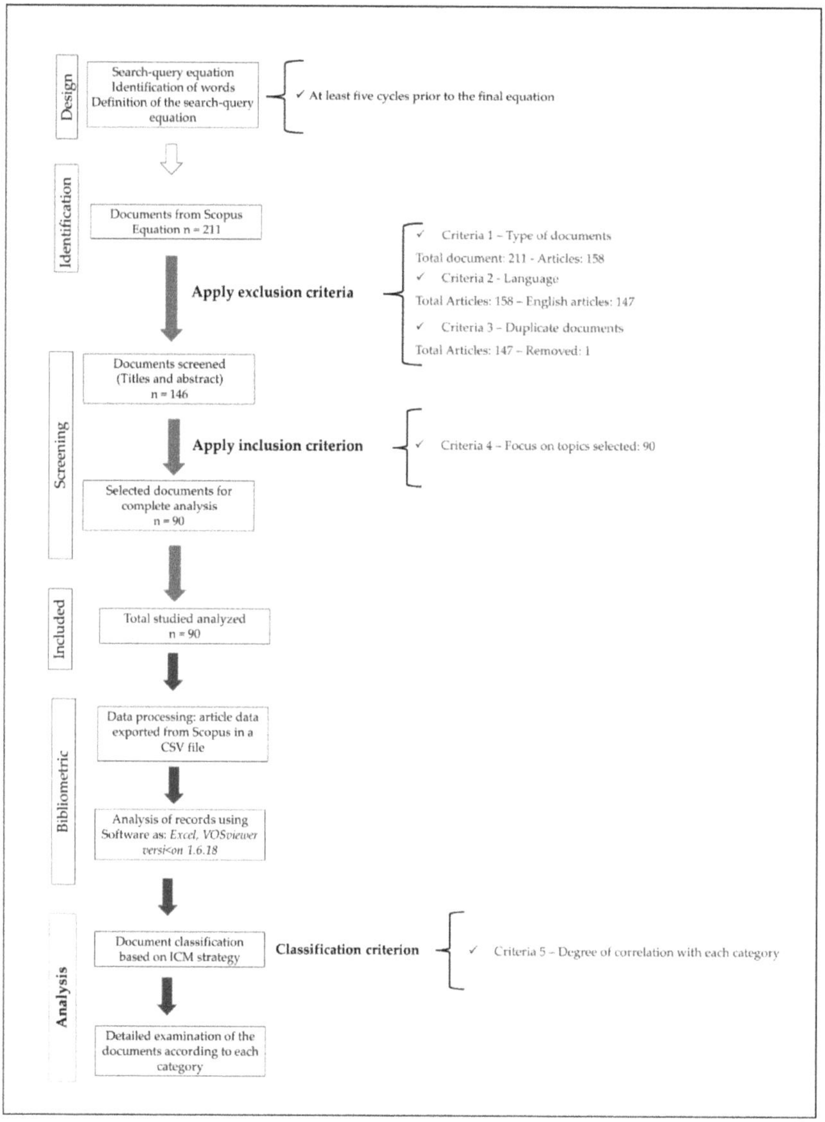


**Figure 15.** Scheme to conduct bibliometric and systematic analyses.

The Scopus data for the final set of documents was exported as comma-separated values (csv) files. Data processing and visualization were conducted using VOSviewer version 1.6.18, Microsoft Excel®, and SankeyMATIC, and were additionally supported by artificial intelligence tools (ChatGPT, OpenAI, 2025), based on data and instructions provided by the authors.

As noted in the preceding section, a macro- and micro-level bibliometric analysis was applied to identify factors such as the most active state, notable institutions, and leading authors. The visualization of data was particularly valuable for revealing collaborative networks of the co-authors, while the analysis of prevalent author keywords illustrated the evolution of research topics over time, and allowed the identification of clusters and thematic areas, as reported by other authors [26,130].

In addition, the final selected articles were thoroughly reviewed using the methodology established in prior studies [126] to determinate the most important components of the topic categories. The selected documents underwent systematic review, cataloging, and analysis across at least three work iterations. Microsoft Excel® was used to create association graphs for categories, and also it was used to create diagrams free software such as SankeyMATIC https://sankeymatic.com/ (accessed on 20 July 2025).

**Supplementary Materials:** The following supporting information can be downloaded at: https://www.mdpi.com/article/10.3390/gels11090731/s1, Table S1: Documents classified in the Pesticides category; Table S2: Documents classified in the Nutritional growth inputs category; Table S3: Documents classified in the Soil conditioners category.; Table S4: Documents classified in the Bioactive category.

**Author Contributions:** Conceptualization, C.B.-A., C.Á.-L., N.J.-Q. and G.A.H.-L.; methodology, C.B.-A., C.Á.-L., N.J.-Q., P.G.-R. and G.A.H.-L.; software, L.P.-V. and P.G.-R.; validation, C.B.-A., C.Á.-L., N.J.-Q. and G.A.H.-L.; formal analysis, C.B.-A., C.Á.-L., N.J.-Q., P.G.-R., M.T.-T. and G.A.H.-L.; investigation, C.B.-A.; resources, M.T.-T. and P.G.-R.; data curation, C.B.-A., C.Á.-L., N.J.-Q., P.G.-R., M.T.-T. and G.A.H.-L.; writing—original draft preparation, C.B.-A., L.P.-V. and P.G.-R.; writing—review and editing C.Á.-L., N.J.-Q., M.T.-T. and G.A.H.-L.; visualization, C.B.-A.; L.P.-V. and P.G.-R.; supervision, G.A.H.-L.; project administration, M.T.-T. and P.G.-R.; funding acquisition, M.T.-T. All authors have read and agreed to the published version of the manuscript.

**Funding:** This work was financed by the autonomous fund "FONDO NACIONAL DE FINANCIAMIENTO PARA LA CIENCIA, LA TECNOLOGÍA Y LA INNOVACIÓN, FRANCISCO JOSÉ DE CALDAS" grant number 112721-443-2023, by the call "Convocatoria ecosistemas en Bioeconomía, ecosistemas naturales, territorios sostenibles" del Ministerio de Ciencia, Tecnología e Innovación de Colombia" of the Ministry of Science, Technology and Innovation of Colombia, MINCIENCIAS.

**Institutional Review Board Statement:** Not applicable.

**Informed Consent Statement:** Not applicable.

**Data Availability Statement:** The data reported are available by contacting the corresponding author.

**Acknowledgments:** We thank MINCIENCIAS—Colombia for the financial support. All authors state that to enhance readability, artificial intelligence (AI) technologies such as Deepl and Quillbot were employed for language processing. ChatGPT, OpenAI, 2025 was utilized to generate images depending on data and directives supplied by the writers. The authors reviewed and edited the entire document after using these tools, and all authors assume responsibility for the publication's content.

**Conflicts of Interest:** The authors declare no conflicts of interest.

## References

1. Su, A.-Y.; Niu, S.-Q.; Liu, Y.-Z.; He, A.-L.; Zhao, Q.; Paré, P.W.; Li, M.-F.; Han, Q.-Q.; Khan, S.A.; Zhang, J.-L. Synergistic effects of *Bacillus amyloliquefaciens* (GB03) and water retaining agent on drought tolerance of perennial ryegrass. *Int. J. Mol. Sci.* **2017**, *18*, 2651. [CrossRef] [PubMed]
2. M'barki, N.; Chehab, H.; Aissaoui, F.; Dabbaghi, O.; Attia, F.; Mahjoub, Z.; Laamari, S.; Chihaoui, B.; del Giudice, T.; Jemai, A.; et al. Effects of mycorrhizal fungi inoculation and soil amendment with hydrogel on leaf anatomy, growth, and physiological performance of olive plantlets under two contrasting water regimes. *Acta Physiol. Plant.* **2018**, *40*, 113. [CrossRef]
3. Turioni, C.; Guerrini, G.; Squartini, A.; Morari, F.; Maggini, M.; Gross, S. Biodegradable hydrogels: Evaluation of degradation as a function of synthesis parameters and environmental conditions. *Soil Syst.* **2021**, *5*, 47. [CrossRef]
4. Richard, B.; Qi, A.; Fitt, B.D. Control of crop diseases through Integrated Crop Management to deliver climate-smart farming systems for low- and high-input crop production. *Plant Pathol.* **2022**, *71*, 187–206. [CrossRef]

5. Mandal, M.; Singh Lodhi, R.; Chourasia, S.; Das, S.; Das, P. A review on sustainable slow-release N, P, K fertilizer hydrogels for smart agriculture. *ChemPlusChem* **2025**, *90*, e202400643. [CrossRef]
6. Biswakarma, N.; Sharma, A.; Kumar, S.; Yadav, G.; Choudhary, A.K.; Babu, S.; Singh, R.; Rathore, S.S.; Das, A.; Singh, V.K. Integrated crop management practices for improving productivity and environmental sustainability. In *Agricultural Diversification for Sustainable Food Production*; Babu, S., Singh, R., Rathore, S.S., Das, A., Singh, V.K., Eds.; Sustainability Sciences in Asia and Africa; Springer: Singapore, 2024. [CrossRef]
7. Du Jardin, P. Plant biostimulants: Definition, concept, main categories and regulation. *Sci. Hortic.* **2015**, *196*, 3–14. [CrossRef]
8. Jaafar, A.M.; Kamarudin, N.S.; Sukor, A.S.A. Preparation and characterization of gellan gum–seaweed beads as a controlled release fertilizer: A biodegradable and environmentally friendly option. *Malays. J. Chem.* **2021**, *23*, 18–28. [CrossRef]
9. Skrzypczak, D.; Jarzembowski, Ł.; Izydorczyk, G.; Mikula, K.; Hoppe, V.; Mielko, K.A.; Pudełko-Malik, N.; Młynarz, P.; Chojnacka, K.; Witek-Krowiak, A. Hydrogel alginate seed coating as an innovative method for delivering nutrients at the early stages of plant growth. *Polymers* **2021**, *13*, 4233. [CrossRef]
10. Perez, J.J.; Francois, N.J.; Maroniche, G.A.; Borrajo, M.P.; Pereyra, M.A.; Creus, C.M. A novel, green, low-cost chitosan–starch hydrogel as potential delivery system for plant growth-promoting bacteria. *Carbohydr. Polym.* **2018**, *202*, 409–417. [CrossRef]
11. Xu, C.; Cao, L.; Bilal, M.; Cao, C.; Zhao, P.; Zhang, H.; Huang, Q. Multifunctional manganese-based carboxymethyl chitosan hydrogels for pH-triggered pesticide release and enhanced fungicidal activity. *Carbohydr. Polym.* **2021**, *262*, 117933. [CrossRef]
12. European Parliament and Council. *Regulation (EU) 2019/1009 of the European Parliament and of the Council of 5 June 2019 Laying down Rules on the Making Available on the Market of EU Fertilising Products*; L170; Publications Office of the European Union: Luxembourg, 2019; pp. 1–114. Available online: https://eur-lex.europa.eu/eli/reg/2019/1009/oj (accessed on 4 June 2025).
13. Bruce, T.J.A.; Smart, L.E.; Birch, A.N.E.; Blok, V.C.; MacKenzie, K.; Guerrieri, E.; Cascone, P.; Luna, E.; Ton, J. Prospects for plant defence activators and biocontrol in IPM—Concepts and lessons learnt so far. *Crop Prot.* **2017**, *97*, 128–134. [CrossRef]
14. Yassin, M.; Ton, J.; Rolfe, S.A.; Valentine, T.A.; Cromey, M.; Holden, N.J.; Newton, A.C. The rise, fall and resurrection of chemical-induced resistance agents. *Pest Manag. Sci.* **2021**, *77*, 3900–3909. [CrossRef]
15. Jeger, M.; Beresford, R.; Bock, C.; Brown, N.; Fox, A.; Newton, A.; Vicent, A.; Xu, X.; Yuen, J. Global challenges facing plant pathology: Multidisciplinary approaches to meet the food security and environmental challenges in the mid-twenty-first century. *CABI Agric. Biosci.* **2021**, *2*, 20. [CrossRef]
16. Campos, E.V.R.; de Oliveira, J.L.; Fraceto, L.F.; Singh, B. Polysaccharides as Safer Release Systems for Agrochemicals. *Agron. Sustain. Dev.* **2015**, *35*, 47–66. [CrossRef]
17. Pu, L.; Wang, H.; Zhao, Y.; Yuan, Z.; Zhang, Y.; Ding, J.; Qu, K.; Sun, W.; Xue, Z.; Xu, W.; et al. Skin-like hydrogels: Design strategy and mechanism, properties, and sensing applications. *J. Mater. Chem. C* **2023**, *11*, 8358–8377. [CrossRef]
18. Qin, C.; Wang, H.; Zhao, Y.; Qi, Y.; Wu, N.; Zhang, S.; Xu, W. Recent advances of hydrogel in agriculture: Synthesis, mechanism, properties and applications. *Eur. Polym. J.* **2024**, *196*, 113376. [CrossRef]
19. Priya, A.S.; Premanand, R.; Ragupathi, I.; Bhaviripudi, V.R.; Aepuru, R.; Kannan, K.; Shanmugaraj, K. Comprehensive Review of Hydrogel Synthesis, Characterization, and Emerging Applications. *J. Compos. Sci.* **2024**, *8*, 457. [CrossRef]
20. Torres-Figueroa, A.V.; de los Santos-Villalobos, S.; Rodríguez-Félix, D.E.; Moreno-Salazar, S.F.; Pérez-Martínez, C.J.; Chan-Chan, L.H.; Ochoa-Meza, A.; del Castillo-Castro, T. Physically and chemically cross-linked poly(vinyl alcohol)/humic acid hydrogels for agricultural applications. *ACS Omega* **2023**, *8*, 44784–44795. [CrossRef] [PubMed]
21. Jaramillo-Quiceno, N.; Rueda-Mira, S.; Santa Marín, J.F.; Álvarez-López, C. Development of a novel silk sericin-based hydrogel film by mixture design. *J. Polym. Res.* **2023**, *30*, 120. [CrossRef]
22. Zaharia, A.; Radu, A.-L.; Iancu, S.; Florea, A.-M.; Sandu, T.; Minca, I.; Fruth-Oprisan, V.; Teodorescu, M.; Sarbu, A.; Iordache, T.-V. Bacterial cellulose–poly(acrylic acid-co-N,N′-methylene-bis-acrylamide) interpenetrated networks for the controlled release of fertilizers. *RSC Adv.* **2018**, *8*, 17635–17644. [CrossRef] [PubMed]
23. Malka, E.; Margel, S. Engineering of PVA/PVP hydrogels for agricultural applications. *Gels* **2023**, *9*, 895. [CrossRef]
24. Fu, J.; Yap, J.X.; Leo, C.P.; Chang, C.K.; Show, P.L. Polysaccharide hydrogels for controlling the nutrient release. *Sep. Purif. Rev.* **2024**, *53*, 276–288. [CrossRef]
25. Yang, H.; Liu, L.; Yang, W.; Zhang, Y.; Zhou, Y.; Chen, L. A comprehensive overview of geopolymer composites: A bibliometric analysis and literature review. *Case Stud. Constr. Mater.* **2022**, *16*, e00830. [CrossRef]
26. Gañán, P.; Barajas, J.; Zuluaga, R.; Castro, C.; Marín, D.; Tercjak, A.; Builes, D.H. The evolution and future trends of unsaturated polyester biocomposites: A bibliometric analysis. *Polymers* **2023**, *15*, 2970. [CrossRef] [PubMed]
27. Saraydın, D.; Karadağ, E.; Güven, O. Relationship between the swelling process and the releases of water soluble agrochemicals from radiation crosslinked acrylamide/itaconic acid copolymers. *Polym. Bull.* **2000**, *45*, 287–294. [CrossRef]
28. Karadağ, E.; Saraydın, D.; Çaldiran, Y.; Güven, O. Swelling studies of copolymeric acrylamide/crotonic acid hydrogels as carriers for agricultural uses. *Polym. Adv. Technol.* **2000**, *11*, 59–68. [CrossRef]
29. Gupta, A.; Stead, T.S.; Ganti, L. Determining a meaningful R-squared value in clinical medicine. *Acad. Med. Surg.* **2024**. [CrossRef]

30. Korves, R. China, Brazil & India in Agricultural Trade Policy. Global Farmer Network. 2004. Available online: https://globalfarmernetwork.org/china-brazil-india-in-agricultural-trade-policy/ (accessed on 31 May 2025).
31. Liu, P.; Peng, J.; Li, J.; Wu, J. Radiation crosslinking of CMC-Na at low dose and its application as substitute for hydrogel. *Radiat. Phys. Chem.* **2005**, *72*, 635–638. [CrossRef]
32. Hua, B.; Wei, H.; Hu, C.; Zhang, Y.; Yang, S.; Wang, G.; Guo, T.; Li, J. Preparation of pH/Temperature-Responsive Semi-IPN Hydrogels Based on Sodium Alginate and Humic Acid as Slow-Release and Water-Retention Fertilizers. *Polym. Bull.* **2024**, *81*, 4175–4198. [CrossRef]
33. Qi, L.; Xiao, X.; Liu, T.; Ren, Z.; Ren, W.; Gao, Q.; Liu, M.; Wei, P.; Lai, Y.; Yao, W.; et al. Functionally Responsive Hydrogels with Salt-Alkali Sensitivity Effectively Target Soil Amelioration. *Sci. Total Environ.* **2024**, *918*, 170350. [CrossRef]
34. Zhang, X.; Yang, L.; Wang, W.; Xiang, Y.; Liu, J.; An, Y.; Shi, J.; Qi, H.; Huang, Z. Sodium alginate/sodium lignosulfonate hydrogel based on inert $Ca^{2+}$ activation for water conservation and growth promotion. *Environ. Res.* **2024**, *246*, 118144. [CrossRef]
35. Gao, L.; Wang, S.; Zhao, X. Synthesis and characterization of agricultural controllable humic acid superabsorbent. *J. Environ. Sci.* **2013**, *25*, S69–S76. [CrossRef]
36. Shi, C.; Lv, X.; Peng, J.; Zhu, J.; Tang, F.; Hu, L. Methylated Biochemical Fulvic Acid-Derived Hydrogels with Improved Swelling Behavior and Water Retention Capacity. *Materials* **2024**, *17*, 61448. [CrossRef]
37. Kumar, K.; Kaith, B.S. Psyllium and acrylic acid based polymeric networks synthesized under the influence of $\gamma$-radiations for sustained release of fungicide. *Fibers Polym.* **2010**, *11*, 147–152. [CrossRef]
38. Tanaka, F.C.; Junior, C.R.F.; Fernandes, R.S.; de Moura, M.R.; Aouada, F.A. Correlating pH and swelling degree parameters to understand the sorption and desorption process of diquat herbicide from nanocomposites based on polysaccharide and clinoptilolite. *J. Polym. Environ.* **2021**, *29*, 3389–3400. [CrossRef]
39. Mordor Intelligence. Russia Crop Protection Chemicals Market Size & Share Analysis (2025–2030). Available online: https://www.mordorintelligence.com/industry-reports/russia-crop-protection-chemicals-market (accessed on 19 February 2025).
40. Mordor Intelligence. Mexico Crop Chemicals Market Size & Share Analysis—Growth Trends & Forecasts (2025–2030). Available online: https://www.mordorintelligence.com/industry-reports/mexico-agrochemicals-market (accessed on 15 August 2025).
41. AgroPages; Atanor S.A. Is a Leading Manufacturer of Chemicals, Petrochemicals and Polymers in Argentina and the Largest Producer and Exporter of Agrochemicals in Latin America. *AgroPages*, 7 August 2019. Available online: https://news.agropages.com/News/Detail-31489.htm (accessed on 15 August 2025).
42. M'barki, N.; Aissaoui, F.; Chehab, H.; Dabbaghi, O.; del Giudice, T.; Boujnah, D.; Mechri, B. Cultivar dependent impact of soil amendment with water retaining polymer on olive (*Olea europaea* L.) under two water regimes. *Agric. Water Manag.* **2019**, *216*, 70–75. [CrossRef]
43. Sharma, G.; Kumar, A.; Devi, K.; Sharma, S.; Naushad, M.; Ghfar, A.A.; Ahamad, T.; Stadler, F.J. Guar gum-crosslinked-soya lecithin nanohydrogel sheets as effective adsorbent for the removal of thiophanate methyl fungicide. *Int. J. Biol. Macromol.* **2018**, *114*, 295–305. [CrossRef] [PubMed]
44. Barrientos-Sanhueza, C.; Mondaca, P.; Tamayo, M.; Álvaro, J.E.; Díaz-Barrera, A.; Cuneo, I.F. Enhancing the mechanical and hydraulic properties of coarse quartz sand using a water-soluble hydrogel based on bacterial alginate for novel application in agricultural contexts. *Soil Sci. Soc. Am. J.* **2021**, *85*, 1880–1893. [CrossRef]
45. Kolya, H.; Kang, C.-W. Synthesis of starch-based smart hydrogel derived from rice-cooked wastewater for agricultural use. *Int. J. Biol. Macromol.* **2023**, *226*, 1477–1489. [CrossRef]
46. Oliveira Frühauf, G.; Rossa Beltrami, L.V.; Francisquetti, E.; Borsoi, C. Fungicidal potential of polymeric hydrogels based on carboxymethylcellulose and yerba mate residues with the incorporation of garlic oil. *Starch/Stärke* **2022**, *75*, 2200182. [CrossRef]
47. Shahid, S.A.; Qidwai, A.A.; Anwar, F.; Ullah, I.; Rashid, U. Effects of a novel poly(AA-co-AAm)/$AlZnFe_2O_4$/potassium humate superabsorbent hydrogel nanocomposite on water retention of sandy loam soil and wheat seedling growth. *Molecules* **2012**, *17*, 12587–12602. [CrossRef] [PubMed]
48. Kim, S.M.; Rhie, Y.H.; Kong, S.M.; Kim, Y.S.; Na, Y.H. Synthesis of nanocomposite hydrogels for improved water retention in horticultural soil. *ACS Agric. Sci. Technol.* **2022**, *2*, 1206–1217. [CrossRef]
49. Tanaka, F.C.; Yonezawa, U.G.; de Moura, M.R.; Aouada, F.A. Obtention, characterization, and herbicide diquat carrier/release properties by nanocomposite hydrogels based on the polysaccharides and zeolite for future use in agriculture. *Environ. Nanotechnol. Monit. Manag.* **2023**, *20*, 100880. [CrossRef]
50. Ghobashy, M.M.; Mousaa, I.M.; El-Sayyad, G.S. Radiation synthesis of urea/hydrogel core shells coated with three different natural oils via a layer-by-layer approach: An investigation of their slow release and effects on plant growth-promoting rhizobacteria. *Prog. Org. Coat.* **2021**, *151*, 106022. [CrossRef]
51. Singh, B.; Sharma, D.K.; Negi, S.; Dhiman, A. Synthesis and characterization of agar–starch based hydrogels for slow herbicide delivery applications. *Int. J. Plast. Technol.* **2015**, *19*, 263–274. [CrossRef]

52. Mafune, K.K.; Kasson, M.T.; Winkler, M.-K.H. Building blocks toward sustainable biofertilizers: Variation in arbuscular mycorrhizal spore germination when immobilized with diazotrophic bacteria in biodegradable hydrogel beads. *J. Appl. Microbiol.* **2024**, *135*, lxae167. [CrossRef] [PubMed]
53. Dong, J.; Han, A.; Zhao, Y.; Li, H.; Yang, Y.; Yuan, B.; Wang, Y.; Liu, R.; Yin, X.; Du, X. Smart, degradable, and eco-friendly carboxymethyl cellulose-CaII hydrogel-like networks gated MIL-101(FeIII) nanoherbicides for paraquat delivery. *Sci. Total Environ.* **2023**, *903*, 166424. [CrossRef]
54. Dorochesi, F.; Barrientos-Sanhueza, C.; Díaz-Barrera, Á.; Cuneo, I.F. Enhancing soil resilience: Bacterial alginate hydrogel vs. algal alginate in mitigating agricultural challenges. *Gels* **2023**, *9*, 988. [CrossRef]
55. Tang, X.; Zhang, Z.; Shi, H.; Wang, Y.; Yan, A.; Yang, F.; Xie, K. Synthesis and characterization of agarose-bacterial cellulose hydrogel for promoting seed germination by improving soil water retention. *Polym. Bull.* **2024**, *81*, 8215–8227. [CrossRef]
56. Tanaka, F.C.; Yonezawa, U.G.; de Moura, M.R.; Aouada, F.A. Studies of the sorption-desorption of pesticides from cellulose-based derivative nanocomposite hydrogels. *Molecules* **2024**, *29*, 204932. [CrossRef] [PubMed]
57. Singh, J.; Singh, B.; Vishavnath. Agar-gelatin-derived hydrogel-based controlled delivery devices for linuron herbicide to prevent environmental hazards. *Environ. Chem. Ecotoxicol.* **2024**, *6*, 153–163. [CrossRef]
58. Flores-Urquieta, M.P.; Claudio-Rizo, J.A.; Cabrera-Munguía, D.A.; Cano-Salazar, L.F.; Flores-Guía, T.E.; Espinoza-Neira, R.; Becerra-Rodriguez, J.J. Studying the impact of selenium complex chemical structure on collagen-starch hydrogels for enhanced tissue growth. *Soft Mater.* **2024**, *22*, 325–341. [CrossRef]
59. Barzman, M.; Bàrberi, P.; Birch, A.N.E.; Boonekamp, P.; Dachbrodt-Saaydeh, S.; Graf, B.; Hommel, B.; Jensen, J.E.; Kiss, J.; Kudsk, P.; et al. Eight Principles of Integrated Pest Management. *Agron. Sustain. Dev.* **2015**, *35*, 1199–1215. [CrossRef]
60. Clemente, I.; Baglioni, M.; Bonechi, C.; Bisozzi, F.; Rossi, C.; Tamasi, G. Green hydrogels loaded with extracts from *Solanaceae* for the controlled disinfection of agricultural soils. *Polymers* **2023**, *15*, 4455. [CrossRef]
61. Azahari, N.A.; Jaafar, A.M.; Rayung, M.; Asib, N.; Mamat, M. Evaluation of gellan gum-glufosinate ammonium beads and its release study performance. *Malays. J. Chem.* **2023**, *25*, 36–45. [CrossRef]
62. Xu, C.; Cao, L.; Cao, C.; Chen, H.; Zhang, H.; Li, Y.; Huang, Q. Fungicide itself as a trigger to facilely construct Hymexazol-Encapsulated polysaccharide supramolecular hydrogels with controllable rheological properties and reduced environmental risks. *Chem. Eng. J.* **2023**, *452*, 139195. [CrossRef]
63. Cruz-Barrera, M.; Izquierdo-García, L.F.; Gómez-Marroquín, M.; Santos-Díaz, A.; Uribe-Gutiérrez, L.; Moreno-Velandia, C.A. Hydrogel Capsules as New Delivery System for *Trichoderma koningiopsis* Th003 to Control *Rhizoctonia solani* in Rice (*Oryza sativa*). *World J. Microbiol. Biotechnol.* **2024**, *40*, 108. [CrossRef]
64. Market Research Future. Paraquat Market Research Report—Forecast till 2032. Available online: https://www.marketresearchfuture.com/reports/paraquat-market-25859 (accessed on 12 June 2025).
65. Abd El-Rehim, H.A.; Hegazy, E.-S.A.; Abd El-Mohdy, H.L. Properties of polyacrylamide-based hydrogels prepared by electron beam irradiation for possible use as bioactive controlled delivery matrices. *J. Appl. Polym. Sci.* **2005**, *98*, 1262–1270. [CrossRef]
66. Liao, R.; Ren, S.; Yang, P. Quantitative fractal evaluation of herbicide effects on the water-absorbing capacity of superabsorbent polymers. *J. Nanomater.* **2014**, *2014*, 905630. [CrossRef]
67. Supare, K.; Mahanwar, P. Starch-chitosan hydrogels for the controlled-release of herbicide in agricultural applications: A study on the effect of the concentration of raw materials and crosslinkers. *J. Polym. Environ.* **2022**, *30*, 2448–2461. [CrossRef]
68. Zheng, D.; Wang, K.; Bai, B.; Hu, N.; Wang, H. Swelling and glyphosate-controlled release behavior of multi-responsive alginate-g-P(NIPAm-co-NDEAm)-based hydrogel. *Carbohydr. Polym.* **2022**, *282*, 119113. [CrossRef]
69. Kumar, V.; Singh, A.; Das, T.K.; Sarkar, D.J.; Singh, S.B.; Dhaka, R.; Kumar, A. Release behavior and bioefficacy of imazethapyr formulations based on biopolymeric hydrogels. *J. Environ. Sci. Health B* **2017**, *52*, 402–409. [CrossRef]
70. Namasivayam, S.K.R.; Vidyasankar, A. Biocompatible formulation of potential fungal biopesticide *Nomuraea rileyi* (F.) Samson for the improved post treatment persistence and biocontrol potential. *Nat. Environ. Pollut. Technol.* **2014**, *13*, 835–838.
71. Shang, H.; Ma, C.; Li, C.; Zhao, J.; Elmer, W.; White, J.C.; Xing, B. Copper oxide nanoparticle-embedded hydrogels enhance nutrient supply and growth of lettuce (*Lactuca sativa*) infected with *Fusarium oxysporum* f. sp. *lactucae*. *Environ. Sci. Technol.* **2021**, *55*, 13432–13442. [CrossRef]
72. Abd El-Mohdy, H.L.; Hegazy, E.A.; El-Nesr, E.M.; El-Wahab, M.A. Control release of some pesticides from starch/(ethylene glycol-co-methacrylic acid) copolymers prepared by $\gamma$-irradiation. *J. Polym. Res.* **2011**, *18*, 1605–1615. [CrossRef]
73. Naboulsi, A.; Haydari, I.; Bouzid, T.; Grich, A.; Aziz, F.; Regti, A.; Himri, M.E.; Haddad, M.E. Fixed-bed adsorption of pesticide agricultural waste using cross-linked adsorptive hydrogel composite beads. *Environ. Sci. Pollut. Res.* **2024**, *31*, 32320–32338. [CrossRef] [PubMed]
74. Adi Sulianto, A.; Adiyaksa, I.P.; Wibisono, Y.; Khan, E.; Ivanov, A.; Drannikov, A.; Ozaltin, K.; Di Martino, A. From fruit waste to hydrogels for agricultural applications. *Clean Technol.* **2024**, *6*, 1–17. [CrossRef]
75. You, C.; Ji, X.; Lin, H.; Ma, N.; Wei, W.; Long, L.; Ning, L.; Wang, F. Preparation of UV-responsive hydrogels based on nanocellulose and their utilization in fungicide delivery. *Environ. Sci. Nano* **2024**, *11*, 1442–1451. [CrossRef]

76. Pizzaro, G.d.C.; Marambio, O.G.; Jeria-Orell, M.; Avellán, M.C.; Rivas, B.L. Synthesis, characterization and application of poly[(1-vinyl-2-pyrrolidone)-co-(2-hydroxyethyl methacrylate)] as controlled-release polymeric system for 2,4-dichlorophenoxyacetic chloride using an ultrafiltration technique. *Polym. Int.* **2008**, *57*, 897–904. [CrossRef]
77. Smagin, A.V.; Smagina, M.V.; Kolganihina, G.B.; Sadovnikova, N.B.; Gulbe, A.Y.; Bashina, A.S. Fungicidal and antibacterial activity of the hydrogel compositions with silver. *Int. J. Eng. Technol.* **2018**, *7*, 14–20. [CrossRef]
78. Singh, R.K.; Mishra, S.; Pandey, A.; Singh, A.; Tripathi, D.K. Multifunctional polymeric hydrogels for agricultural sustainability. *Polym. Test.* **2023**, *126*, 107579.
79. Zheng, L.; Seidi, F.; Wu, W.; Pan, Y.; Xiao, H. Dual-functional lignin-based hydrogels for sustained release of agrochemicals and heavy metal ion complexation. *Int. J. Biol. Macromol.* **2023**, *235*, 123701. [CrossRef] [PubMed]
80. Singh, J.; Singh, B.; Vishavnath. Designing starch–alginate hydrogels for controlled delivery of fungicide for the alleviation of environmental pollution. *ACS Agric. Sci. Technol.* **2022**, *2*, 1239–1250. [CrossRef]
81. Singh, J.; Singh, B.; Vishavnath. Design of innovative agrochemical delivery system for sustainable agriculture and to alleviate environmental and health hazards. *Sustain. Chem. Environ.* **2023**, *4*, 100046. [CrossRef]
82. Singh, A.; Kaith, B.S.; Sud, D.; Bhatti, M.S. Fabrication of smart hydrogel based on functionally modified *Colocasia esculenta* as potential nutrient carrier and soil conditioner. *Water Air Soil Pollut.* **2023**, *234*, 402. [CrossRef]
83. Chandrika, K.P.; Singh, A.; Rathore, A.; Kumar, A. Novel cross-linked guar gum-g-poly(acrylate) porous superabsorbent hydrogels: Characterization and swelling behaviour in different environments. *Carbohydr. Polym.* **2016**, *149*, 175–185. [CrossRef]
84. Sharma, R.; Kalia, S.; Kaith, B.S.; Kumar, A.; Thakur, P.; Pathania, D.; Srivastava, M.K. Guar gum-poly(itaconic acid) based superabsorbents via two-step free-radical aqueous polymerization for environmental and antibacterial applications. *J. Polym. Environ.* **2017**, *25*, 176–191. [CrossRef]
85. Song, Y.; Zhu, F.; Cao, C.; Cao, L.; Li, F.; Zhao, P.; Huang, Q. Reducing pesticide spraying drift by folate/$Zn^{2+}$ supramolecular hydrogels. *Pest Manag. Sci.* **2021**, *77*, 5278–5285. [CrossRef]
86. Wang, Y.-X.; Shen, X.-F.; Zhang, J.-Q.; Pang, Y.-H. Fabrication of β-cyclodextrin-polyacrylamide/covalent organic framework hydrogel at room temperature for the efficient removal of triazole fungicides from environmental water. *Environ. Pollut.* **2023**, *333*, 121971. [CrossRef]
87. Fu, X.; Xu, M.; Li, T.; Li, Y.; Zhang, H.; Zhang, C. The improved expression and stability of zearalenone lactonohydrolase from *Escherichia coli* BL21 (DE3). *Appl. Biochem. Microbiol.* **2021**, *57*, 79–85. [CrossRef]
88. Kolesnikov, L.E.; Uspenskaya, M.V.; Kremenevskaya, M.I.; Orlova, A.G.; Razumova, I.E.; Kolesnikova, Y.R. The biological basis for the use of acrylic hydrogel and protein growth stimulant in the soft wheat and triticale cultivation. *Agron. Res.* **2021**, *19*, 1545–1561. [CrossRef]
89. Wu, J.; Yang, H.; Hou, C.; Guan, H.; Gu, S.; Yin, Y.; Jing, H.; Wang, Y.; Wang, M. Preparation of an attapulgite-modified composite hydrogel and application in an environmentally responsive green fertilizer. *ACS Appl. Polym. Mater.* **2023**, *5*, 10217–10225. [CrossRef]
90. Phansroy, N.; Boonyod, S.; Mulasake, O.; Uttha, A.; Songkram, C.; Somboon, T.; Kongon, J.; Lersuwannapong, N.; Saengsuwan, S.; Khawdas, W.; et al. Innovative environment-friendly liquid fertilizer bead from sodium alginate coating with IPN membrane derived from natural rubber and cassava starch. *J. Polym. Res.* **2024**, *31*, 67. [CrossRef]
91. Cui, S.; Li, P.; Ji, L.; Wang, T.; Liu, Y.; Lan, Y.; Jiang, J. Superabsorbent quaternary ammonium guar gum hydrogel with controlled release of humic acid for soil improvement and plant growth. *Carbohydr. Polym.* **2024**, *337*, 122188. [CrossRef]
92. Kumar, A.; Meena, R.S.; Nirmal, D.E.; Gurjar, D.S.; Singh, A.; Yadav, G.S.; Pradhan, G. Response of polymers and biofertilizers on soybean (*Glycine max*) yield under rainfed condition. *Indian J. Agric. Sci.* **2020**, *90*, 767–770. [CrossRef]
93. Chaparro-Rodríguez, M.; Estrada-Bonilla, G.; Rosas-Pérez, J.; Gómez-Álvarez, M.; Cruz-Barrera, M. Hydrogel capsules as new approach for increasing drying survival of plant biostimulant gram-negative consortium. *Appl. Microbiol. Biotechnol.* **2023**, *107*, 6671–6682. [CrossRef] [PubMed]
94. Lima-Tenório, M.K.; Furmam-Cherobim, F.; Karas, P.R.; Hyeda, D.; Takahashi, W.Y.; Pinto Junior, A.S.; Galvão, C.W.; Tenório-Neto, E.T.; Etto, R.M. *Azospirillum brasilense* AbV5/6 encapsulation in dual-crosslinked beads based on cationic starch. *Carbohydr. Polym.* **2023**, *308*, 120631. [CrossRef]
95. Kanagalakshmi, M.; Devi, S.G.; Subasini, S.; Amalan, A.J.; Pius, A. Experimental assessment of biostimulants on mung bean growth on a soilless culture system using superabsorbent pectin based hydrogel. *Int. J. Biol. Macromol.* **2024**, *273*, 133058. [CrossRef]
96. Rodríguez-Quesada, J.; Rodríguez Mora, K.; Bernal-Samaniego, C.A.; Jirón-García, E.G.; Rojas-Alvarado, C. Development of nanocellulose hydrogels from *Sargassum* seaweed as controlled nutrient release systems and their application in germination. *Ecol. Eng. Environ. Technol.* **2024**, *25*, 308–320. [CrossRef]
97. Das, S.K.; Ghosh, G.K. Hydrogel-biochar composite for agricultural applications and controlled release fertilizer: A step towards pollution free environment. *Energy* **2022**, *242*, 122977. [CrossRef]

98. Hu, Z.-Y.; Chen, G.; Yi, S.-H.; Wang, Y.; Liu, Q.; Wang, R. Multifunctional porous hydrogel with nutrient controlled-release and excellent biodegradation. *J. Environ. Chem. Eng.* **2021**, *9*, 106146. [CrossRef]
99. Lima-Tenório, M.K.; Karas, L.P.; Furmam-Cherobim, F.; Guerlinguer, E.; Rubira, A.F.; Steffens, M.B.R.; Galvão, C.W.; Tenório-Neto, E.T.; Etto, R.M. Encapsulation of plant growth-promoting bacteria with gum arabic hydrogels: A potential system for sustainable agriculture. *J. Polym. Environ.* **2024**, *32*, 5702–5712. [CrossRef]
100. Powlson, D.S.; Gregory, P.J.; Whalley, W.R.; Quinton, J.N.; Hopkins, D.W.; Whitmore, A.P.; Hirsch, P.R.; Goulding, K.W.T. Soil management in relation to sustainable agriculture and ecosystem services. *Food Policy* **2011**, *36*, 72–87. [CrossRef]
101. Shah, F.; Wu, W. Soil and crop management strategies to ensure higher crop productivity within sustainable environments. *Sustainability* **2019**, *11*, 1485. [CrossRef]
102. Machalík, R.; Wandzik, I. A mini-review on chitosan-based hydrogels with potential for sustainable agriculture applications. *Polymers* **2020**, *12*, 2425. [CrossRef]
103. Barrientos-Sanhueza, C.; Cargnino-Cisternas, D.; Díaz-Barrera, A.; Cuneo, I.F. Bacterial alginate-based hydrogel reduces hydro-mechanical soil-related problems in agriculture facing climate change. *Polymers* **2022**, *14*, 922. [CrossRef]
104. Smagin, A.; Sadovnikova, N.; Smagina, M. Synthetic gel structures in soils for sustainable potato farming. *Sci. Rep.* **2019**, *9*, 55205. [CrossRef]
105. Turull, M.; Fontàs, C.; Díez, S. Conventional and novel techniques for the determination of Hg uptake by lettuce in amended agricultural peri-urban soils. *Sci. Total Environ.* **2019**, *668*, 40–46. [CrossRef]
106. Hasija, V.; Sharma, K.; Kumar, V.; Sharma, S.; Sharma, V. Green synthesis of agar/gum arabic based superabsorbent as an alternative for irrigation in agriculture. *Vacuum* **2018**, *157*, 458–464. [CrossRef]
107. Baidakova, M.; Sitnikova, V.; Uspenskaya, M.; Olekhnovich, R.; Kremenevskaya, M. Polymer acrylic hydrogels with protein filler: Synthesis and characterization. *Agron. Res.* **2019**, *17*, 913–922. [CrossRef]
108. Ding, C.; Zhang, S.; Fu, X.; Liu, T.; Shao, L.; Fei, M.; Hao, C.; Liu, Y.; Zhong, W.-H. Robust supramolecular composite hydrogels for sustainable and "visible" agriculture irrigation. *J. Mater. Chem. A* **2021**, *9*, 24613–24621. [CrossRef]
109. Saha, A.; Gupt, C.B.; Sekharan, S. Recycling natural fibre to superabsorbent hydrogel composite for conservation of irrigation water in semi-arid regions. *Waste Biomass Valoriz.* **2021**, *12*, 6433–6448. [CrossRef]
110. Kolya, H.; Kang, C.-W. A New approach for agricultural water management using pillows made from covid-19waste face masks and filled with a hydrogel polymer: Preliminary studies. *Agriculture* **2023**, *13*, 152. [CrossRef]
111. Sorze, A.; Valentini, F.; Dorigato, A.; Pegoretti, A. Development of a xanthan gum based superabsorbent and water retaining composites for agricultural and forestry applications. *Molecules* **2023**, *28*, 1952. [CrossRef]
112. Pham, V.T.H.; Murugaraj, P.; Mathes, F.; Tan, B.K.; Truong, V.K.; Murphy, D.V.; Mainwaring, D.E. Copolymers enhance selective bacterial community colonization for potential root zone applications. *Sci. Rep.* **2017**, *7*, 16253. [CrossRef]
113. Taha, T.H.; Elnouby, M.S.; Abu-Saied, M.A.; Alamri, S. Tested functionalization of alginate-immobilized ureolytic bacteria for improvement of soil biocementation and maximizing water retention. *RSC Adv.* **2020**, *10*, 21350–21359. [CrossRef]
114. Li, S.-S.; Wang, S.-B.; Chen, Y.; Zhu, Q.-S.; Lan, L.-M.; Bu, H.; Hu, T.; Jiang, G.-B. Biodegradable, anti-freezing and self-healable hydrogel mulch film for weed control. *Chem. Eng. J.* **2023**, *462*, 142211. [CrossRef]
115. Mayerová, M.; Šimon, T.; Stehlík, M.; Madaras, M. Improving the stability of soil aggregates using soil additives and revegetation by grassland. *Plant Soil Environ.* **2023**, *69*, 282–290. [CrossRef]
116. Sandoval, A.P.; Gerardo Yáñez-Chávez, L.; Sánchez-Cohen, I.; Samaniego-Gaxiola, J.A.; Trejo-Calzada, R. Hydrogel, biocompost and its effect on photosynthetic activity and production of forage maize plants (*Zea mays* L.). *Acta Agron.* **2017**, *66*, 63–68.
117. Varma, J.; Kulshrestha, A.K.; Pankaj, P.P.; Upadhye, V.J.; Shrivastav, A. Bioactive compounds: Industrial and agricultural applications. In *Biotechnological Intervention in Production of Bioactive Compounds*; Devi, J., Ed.; Sustainable Landscape Planning and Natural Resources Management; Springer: Cham, Switzerland, 2025. [CrossRef]
118. Aruwa, C.E.; Amoo, S.O.; Kudanga, T. Opuntia (Cactaceae) plant compounds, biological activities and prospects—A comprehensive review. *Food Res. Int.* **2018**, *112*, 328–344. [CrossRef] [PubMed]
119. Amanzholkyzy, A.; Zhumagaliyeva, S.; Sultanova, N.; Abilov, Z.; Ongalbek, D.; Donbayeva, E.; Niyazbekova, A.; Mukazhanova, Z. Hydrogel delivery systems for biological active substances: Properties and the role of HPMC as a carrier. *Molecules* **2025**, *30*, 1354. [CrossRef]
120. Nuzzo, A.; Mazzei, P.; Drosos, M.; Piccolo, A. Novel humo-pectic hydrogels for controlled release of agroproducts. *ACS Sustain. Chem. Eng.* **2020**, *8*, 10079–10088. [CrossRef]
121. Valdés-Lozano, C.I.; Claudio-Rizo, J.A.; Cabrera-Munguía, D.A.; León-Campos, M.I.; Mendoza-Villafaña, J.J.; Becerra-Rodriguez, J.J. Modulation of animal and plant tissue growth with collagen-starch-organic molybdenum networks hydrogel biomatrices. *Polym. Adv. Technol.* **2024**, *35*, e6568. [CrossRef]
122. Liu, W.; Liu, K.; Chen, D.; Zhang, Z.; Li, B.; El-Mogy, M.M.; Tian, S.; Chen, T. *Solanum lycopersicum*, a model plant for the studies in developmental biology, stress biology and food science. *Foods* **2022**, *11*, 2402. [CrossRef] [PubMed]

123. Rosa-Martínez, E.; Bovy, A.; Plazas, M.; Tikunov, Y.; Prohens, J.; Pereira-Dias, L. Genetics and breeding of phenolic content in tomato, eggplant and pepper fruits. *Front. Plant Sci.* **2023**, *14*, 1135237. [CrossRef] [PubMed]
124. Li, J.; Wu, C.; Chu, P.K.; Gelinsky, M. 3D Printing of Hydrogels: Rational Design Strategies and Emerging Biomedical Applications. *Mater. Sci. Eng. R Rep.* **2020**, *140*, 100543. [CrossRef]
125. Rethlefsen, M.L.; Kirtley, S.; Waffenschmidt, S.; Ayala, A.P.; Moher, D.; Page, M.J.; Koffel, J.B.; Blunt, H.; PRISMA-S Group. An extension to the PRISMA Statement for Reporting Literature Searches in Systematic Reviews. *Syst. Rev.* **2021**, *10*, 39. [CrossRef]
126. López, L.C.; Hincapié-Llanos, G.A. Comparison of Mango (*Mangifera indica*) dehydration technologies: A systematic review. *Agriengineering* **2024**, *6*, 2694–2717. [CrossRef]
127. Aghaei Chadegani, A.; Salehi, H.; Yunus, M.; Farhadi, H.; Farhadi, M.; Ale Ebrahim, N. A comparison between two main academic literature collections: Web of science and Scopus databases. *Asian Soc. Sci.* **2013**, *9*, 18–26. [CrossRef]
128. Bergman, E.M.L. Finding citations to social work literature: The relative benefits of using Web of Science, Scopus, or Google Scholar. *J. Acad. Librariansh.* **2012**, *38*, 370–379. [CrossRef]
129. Aria, M.; Cuccurullo, C. bibliometrix: An R-tool for comprehensive science mapping analysis. *J. Informetr.* **2017**, *11*, 959–975. [CrossRef]
130. Cobo, M.J.; López-Herrera, A.G.; Herrera-Viedma, E.; Herrera, F. SciMAT: A new science mapping analysis software tool. *J. Am. Soc. Inf. Sci.* **2012**, *63*, 1609–1630. [CrossRef]

*gels*

*Review*

# From Agro-Industrial Waste to Natural Hydrogels: A Sustainable Alternative to Reduce Water Use in Agriculture

**César F. Alonso-Cuevas [1], Nathiely Ramírez-Guzmán [2,3,*], Liliana Serna-Cock [4], Marcelo Guancha-Chalapud [5], Jorge A. Aguirre-Joya [6], David R. Aguillón-Gutiérrez [6], Alejandro Claudio-Rizo [7] and Cristian Torres-León [3,6,*]**

1 Center for Interdisciplinary Studies and Research (CEII), Universidad Autonoma de Coahuila, Arteaga 25280, Coahuila, Mexico; calonso@uadec.edu.mx
2 School of Biological Sciences, Universidad Autonoma de Coahuila, Torreón 27276, Coahuila, Mexico
3 Agri-Food and Agro-Industrial Bioeconomy Research Group, Universidad Autonoma de Coahuila, Torreón 27276, Coahuila, Mexico
4 Faculty of Engineering and Administration, Universidad Nacional de Colombia, Palmira 763533, Colombia; lserna@unal.edu.co
5 Centro Nacional de Asistencia Técnica a la Industria (ASTIN), Servicio Nacional de Aprendizaje (SENA), Cali 760004, Colombia; mguanchac@sena.edu.co
6 Research Center and Ethnobiological Garden (CIJE), Autonomous University of Coahuila, Viesca 27480, Coahuila, Mexico; jorge_aguirre@uadec.edu.mx (J.A.A.-J.); david_aguillon@uadec.edu.mx (D.R.A.-G.)
7 Faculty of Chemical Sciences, Universidad Autonoma de Coahuila, Saltillo 25280, Coahuila, Mexico; jclaudio@uadec.edu.mx
* Correspondence: nathiely.ramirez@uadec.edu.mx (N.R.-G.); ctorresleon@uadec.edu.mx (C.T.-L.)

## Abstract

The increasing demand for food necessitates that agri-food systems adopt innovative techniques to enhance food production while optimizing the use of limited resources, such as water. In agriculture, hydrogels are being increasingly used to enhance water retention and reduce irrigation requirements. However, most of these materials are based on synthetic polymers that are not biodegradable. This raises serious environmental and health concerns, highlighting the urgent need for sustainable, biodegradable alternatives. Biomass-derived from agro-industrial waste presents a substantial potential for producing hydrogels, which can effectively function as water collectors and suppliers for crops. This review article provides a comprehensive overview of recent advancements in the application of agro-industrial waste for the formulation of hydrogels. Additionally, it offers a critical analysis of the development of hydrogels utilizing natural and compostable materials. Agro-industrial and food waste, which are rich in hemicellulose and cellulose, have been utilized to enhance the mechanical properties and water absorption capacity of hydrogels. These biomaterials hold significant potential for the development of effective hydrogels in agricultural applications; they can be either hybrid or natural materials that exhibit efficacy in enhancing seed germination, improving water retention capabilities, and facilitating the controlled release of fertilizers. Natural hydrogels derived from agro-industrial waste present a sustainable technological alternative that is environmentally benign.

**Keywords:** agro-industrial waste; agricultural hydrogels; water capability; bio compostable hydrogels; food sustainability; bioeconomy



https://doi.org/10.3390/gels11080616

## 1. Introduction

The growth of the world's population has increased the demand for food, raising concerns about the control of water use for agricultural purposes [1]. The scarcity of this resource has led to a search for new technologies that help improve agricultural performance. Hydrogels have demonstrated promising results in capturing water and enhancing water availability for plants [1]. Hydrogels are superabsorbent polymeric materials that play a vital role in addressing critical challenges within food systems, particularly in mitigating water scarcity in agricultural practices [2], nutrient management in plants [2], crop yield in the field [2], and the removal of contaminants in the soil [3].

However, commercial hydrogels are currently manufactured with materials that do not fully degrade in the environment [2]. Numerous hydrogels used in the present day exhibit prolonged degradation rates, resulting in potential accumulation in soils and aquatic environments [4]. This characteristic raises concerns regarding their long-term impact on ecosystems. The most common hydrogels are synthesized with acrylamides and acrylates that are not biodegradable [5]. Hydrogel degradation can be carried out by chemical, biological, thermal, and mechanical processes, but the degradation condition could vary by the way in which it is synthesized, the monomer used, as well as the cross-linkers and initiators [6].

In recent years, the production of hydrogels from agricultural waste has been studied [7]. The primary use of these hydrogels is to capture water from the soil [8]. Cellulose-based hydrogels can be used in various settings due to their excellent swelling capacity, water retention, and biodegradability [9].

Since hydrogels for agricultural use are a technology under constant research [6,8], key aspects of their performance may vary depending on certain conditions [10], the incorporation of agro-industrial waste gives way to sustainability and the circular bioeconomy, also to the development of systems that use natural resources for subsistence, as set out in the Sustainable Development Goals (SDG) in objective number 12 [11]. Figure 1 shows the contribution to the circular bioeconomy by utilizing disposable resources and obtaining new components for incorporation into hydrogels. Selected plants are initially processed to extract primary fibers, which are used in the production of industrial goods. The residual biomass, which is traditionally discarded, is further valorized by extracting key biopolymers, including cellulose, lignin, and hemicellulose. These biopolymers, when combined with natural monomers, serve as raw materials for the synthesis of hydrogels—materials that can retain significant amounts of water. These hydrogels can then be reintegrated into agricultural systems to improve soil moisture retention, especially in arid regions. This closed-loop system exemplifies how crop by-products can be transformed into high-value functional materials, contributing to a more sustainable and circular bioeconomy.

Tariq et al. [4] previously authored an insightful and comprehensive review on the innovative development of hydrogels derived from pure biopolymers, including starch, chitosan, rubber, gelatin, lignin, and alginate. In a notable recent review, Zhu et al. [12] undertook an exhaustive examination of cutting-edge methods for producing hydrogels from natural polymers sourced from agricultural waste. However, despite the depth of these analyses, they fall short in addressing the current landscape of research focused on leveraging agroindustrial waste as a sustainable raw material for creating hydrogels for agricultural applications. Furthermore, they do not explore the remarkable potential these materials hold within a circular bioeconomy framework, a critical aspect for promoting sustainability in agriculture.

Based on the above, this review article aims to analyze the current trend in utilizing hydrogels derived from agro-industrial waste and their application as a viable alternative for water capture in agricultural areas.

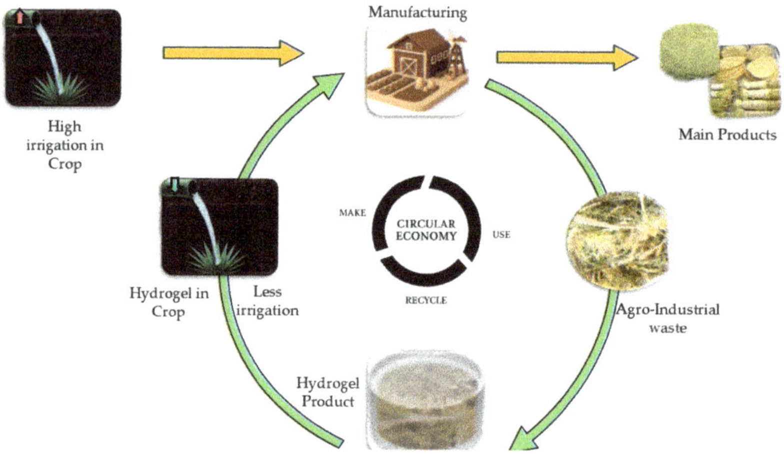


**Figure 1.** Leveraging agro-industrial waste as a model for promoting a circular bioeconomy in the processes of fiber extraction and hydrogel synthesis.

## 2. Technologies and Trends in Sustainable Food Production

For decades, the way food is produced has undergone significant changes, including the improved use of fertilizers to nourish crops, the use of specialized substrates, and the conservation of water resources. The low availability of water due to droughts, resource misuse, soil erosion, and deforestation is the main problem worldwide in food cultivation [13]. Additionally, the problem is exacerbated by climate change [5], which affects security and sustainability in agri-food systems [13]. In addition, the high demand for food due to population growth has led to an increase in agricultural waste, a polluting resource [8].

Over the years, various technologies and techniques have been implemented in food production, including irrigation control, the use of fertilizers and pesticides, and soil treatment, to achieve crops with high yields [14]. The way the soil is used is also important due to its role in providing nutrients for food production. Factors such as erosion and pollution have led to the adoption of crop rotation, intercropping, and even organic farming, the latter of which is often inefficient [14].

Irrigation is a fundamental factor because water must be used efficiently. Therefore, its use varies according to soil, growing season, and crop types. Drip irrigation is a method for harnessing water, particularly in greenhouses [15], which supplies water and nutrients in a more controlled manner with minimal filtration into the soil. Food is also produced through hydroponics; this technology was developed in response to urbanization, the lack of suitable soils for cultivation, and the contamination of existing soils. Hydroponics can supply plants with essential nutrients for growth; this technique offers high yields and serves as a viable alternative to traditional agriculture [16].

Hydrogels have also been used to enhance water retention capacity and reduce irrigation frequency while also serving to aerate substrates [17]. This article focuses on an analysis of the development of hydrogel technologies that utilize agro-industrial residues in recent years.

## 3. Hydrogels in Agriculture

A hydrogel is a network of cross-linked polymer chains with water-retaining capacity produced by a reaction of one or more monomers. Figure 2 shows the schematic representation of a hydrogel for agricultural applications, indicating the formation of a three-dimensional polymeric network through the interaction of monomers and a crosslinking agent (a), which enables the structure to retain significant amounts of water molecules (b), a key feature for improving soil moisture retention in agricultural systems.

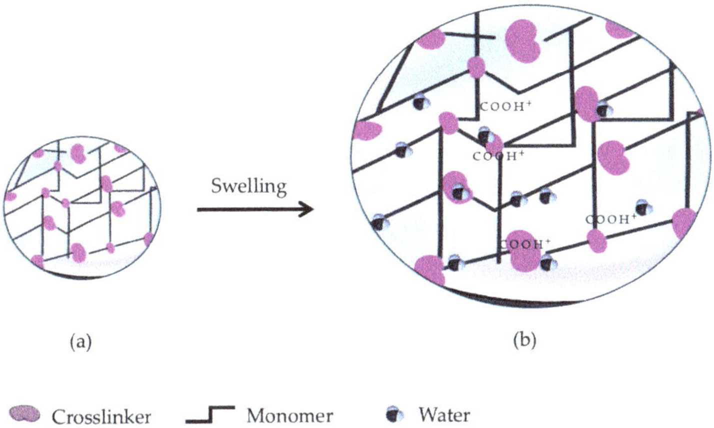


**Figure 2.** Schematic representation of a hydrogel for agricultural applications: Connections between monomers and a crosslinker (**a**) create a three-dimensional polymeric network that absorbs water molecules (**b**).

The hydrogels are hydrophilic and resistant to dissolution. Hydrogels can be synthesized with natural and synthetic components [18]. The use of hydrogels as substrates in horticulture has been highlighted in recent years [18]. Water retention, controlled release, and conservation are properties that are sought to be improved, in addition to the use of biopolymers that do not promote contamination [19].

The use of polymeric materials in agriculture is driven by the challenges posed by increasing food demand, but primarily by the need to utilize water more efficiently. These materials can retain and release water, promoting crop growth and soil conditioning [20]. Although hydrogels have advantages, they have also recently generated many questions: Can they biodegrade? Can they affect the soil? Can they release nano- or microplastics? Can this nano- or microplastic reach food or be deposited in the human body, causing disease? Do they have a good cost-benefit ratio? [21]. As a proposal to answer these questions, there are natural biopolymers derived from agro-industrial waste, which have a lower environmental impact. Generally, determining the environmental impact of hydrogels is based on their type, which refers to their composition, including petroleum-derived hydrogels, synthetic hydrogels, hydrogels derived from natural compounds, natural hydrogels, or hybrid hydrogels. Consequently, synthetic hydrogels generate more pollution than natural compounds.

### *3.1. Synthetic Hydrogels*

Synthetic hydrogels are derived from petroleum sources and exhibit good performance in terms of mechanical and chemical properties, although they have lower biological activity [22]. Most superabsorbent hydrogels on the market are based on polyacrylamide or polyacrylate [8,10], or they are mixed with cellulose. However, concerns exist regarding their toxicity for use in agriculture or applications related to human consumption, such as their low biological activity and poor biodegradability [23].

Implications of Hydrogel Nanoplastics in Food Safety

In recent years, concern has grown over commercial agricultural hydrogels. Commercial agricultural hydrogels are made from synthetic polymers such as polyacrylamide and polyacrylates, which are non-biodegradable and harmful to the environment [2]. Synthetic polymers play a crucial role in enhancing soil moisture retention and minimizing irrigation requirements. However, the durability of these non-biodegradable materials poses a significant challenge, as they can break down in the environment, releasing micro- and nanoplastic particles that persist in the soil [4].

Acrylamide monomers can be absorbed through the skin and lungs, posing significant health risks, recognized neurotoxin and a potential carcinogen [24]. The absorption and accumulation of these nanoplastics in crops introduce potential contamination into the food chain, raising unforeseen risks to human health. Pinzón-Moreno et al. [25], demonstrated that synthetic hydrogels formulated from polyacrylate can generate polymeric nanoparticles, which are capable of being released into agricultural soils when exposed to water. Commercial polyacrylate-based hydrogels are not biodegradable and decompose slowly in the soil. During this degradation process, polymer fragments or nanoparticles can form, which may persist in the environment [26].

Currently, no long-term studies are proving the accumulation of commercial hydrogel nanoparticles in crops or their migration to consumers. However, existing research indicates a potential risk of soil contamination. For example, polystyrene nanoplastics have been found to accumulate in soil and be absorbed by edible crops, such as rice and peanuts, negatively impacting their nutritional quality [27]. Research has demonstrated that the small size of nanoplastics enables them to cross biological membranes and potentially cause adverse health effects [28]. Addressing the increasing concerns regarding nanoplastics in agricultural systems and their potential implications for human health presents an opportunity to advance measurement technologies and foster further research in this critical area.

### 3.2. *Natural Hydrogels*

Natural hydrogels are composed of natural sources, including polysaccharides and proteins. These natural resources can be derived from the agricultural sector, waste, or organisms belonging to the animal kingdom [29]. Natural biopolymers based on chitosan could be a considerable alternative to promote agricultural crop growth, as they are biodegradable and environmentally friendly [30]. One of the primary challenges for hydrogels composed of natural compounds is to identify materials with mechanical properties comparable to those of synthetic materials while maintaining biodegradability and biocompatibility [31].

### 3.3. *Natural Hydrogels with Agro-Industrial Waste (Hybrids)*

The global accumulation of crop residues is particularly important, as it can affect fields and contribute to pollution [32]. However, these residues contain significant amounts of cellulose and hemicellulose, which allows for compatibility with hydrogel synthesis [8]. Studies indicate that hydrogels based on various agricultural residues possess good swelling and water retention capacities, thereby improving resource availability [33].

Table 1 presents the advantages and disadvantages of various types of hydrogels based on their composition. In addition, the study objectives in hydrogels are generalized, as mentioned in the previous sections, as mechanical resistance, absorption capacity, and biocompatibility. Synthetic hydrogels are more capable of swollen water but have a negative environmental impact. Natural polymers have low mechanical properties; nevertheless, they are biodegradable and exhibit good water retention. Hybrid hydrogels share similar characteristics.

**Table 1.** Main advantages and disadvantages of hydrogel classification.

| Hydrogel Classification | Advantage | Disadvantage |
|---|---|---|
| Synthetics | • Great water retention.<br>• High wear resistance.<br>• Its manufacturing can be more controlled. | • Low biodegradability.<br>• It has a negative environmental impact.<br>• Its synthesis often contaminates |
| Natural | • Good water retention<br>• They are biodegradable<br>• Monomers are from residual natural matter | • Low mechanical properties<br>• High manufacturing costs |
| Hybrids | • Good water retention<br>• Synthetic and natural monomers are used<br>• Recyclable waste material can be used<br>• They are biocompatible | • The synthesis is complex<br>• Their mechanical properties vary<br>• High manufacturing costs |

## 4. Natural Components in the Manufacture of Hydrogels

Given that natural and hybrid hydrogels exhibit superior biocompatibility, understanding the key components commonly employed in their synthesis is crucial. These components typically include polysaccharides such as starch, chitosan, alginate, and cellulose, as well as proteins like gelatin and collagen. Comprehensive information about each of these components is provided below.

### *4.1. Polysaccharides*

#### 4.1.1. Starch

Starch is a natural polysaccharide composed of glucose monomers linked primarily by α-1,4-glycosidic bonds, with occasional α-1,6-glycosidic branches. Its general chemical formula is $(C_6H_{10}O_5)n$, where *n* represents the number of glucose units. The molecular weight of starch varies widely depending on the source and degree of polymerization, typically ranging from 300,000 to several million Daltons. This polysaccharide possesses functional groups that enable it to form effective hydrogels, and it is both renewable and economically accessible. Starch copolymer hydrogels can be utilized in agriculture, serving as controlled-release media for pesticides and fertilizers, as well as binders for seed germination [34]. Additionally, a starch copolymer has been investigated for dye adsorption, where the adsorption capacity varies with temperature and dye concentration [35]. Starch is a useful and important component in the food industry, due to its great physicochemical properties, which allow to production of starch nanoparticles with high biocompatibility, minimal toxicity, and water dispersibility, including digestible resistant starch [36].

#### 4.1.2. Chitosan

Chitosan is a linear polysaccharide derived from the deacetylation of chitin [37]; consisting mainly of β-(1→4)-linked D-glucosamine units with varying amounts of N-acetyl-D-glucosamine. Its general chemical formula is $(C_6H_{11}NO_4)n$, where *n* indicates the degree of polymerization; the molecular weight can range from 50,000 to over 1,000,000 Daltons.

Chitosan is a polysaccharide derived from chitin, one of the most abundant natural polymers after cellulose. Chitosan comes from the deacetylation of chitin through three processes: a homogeneous reaction, a heterogeneous reaction, and an enzymatic method [38]. Chitosan has a wide range of applications, including the agro-industry, where it is used in crops and post-harvest processes, as well as in the medical field, where it is utilized as a carrier for agents and dressings due to its physicochemical properties [38].

#### 4.1.3. Cellulose

Cellulose can be found in the cell walls of plants; it is considered a biopolymer of great abundance and an unlimited source of raw material. Its general chemical formula is $(C_6H_{10}O_5)n$ where *n* represents the number of repeating glucose units. The molecular weight of cellulose varies depending on its source and degree of polymerization, typically ranging from 100,000 to over 1,000,000 Daltons. The capacity for polymer synthesis can vary according to the source from which the cellulose is obtained and the physical and chemical treatments applied [39]. Cellulose is difficult to dissolve, and it presents complications during cross-linking. Techniques such as physical and chemical polymerization facilitate this process [40]. Hydrogels synthesized from cellulose derivatives exhibit excellent swelling, retention, and water control properties in crops [23].

### *4.2. Proteins*

#### Gelatin

Gelatin, derived from collagen, is well-suited for various branches of science. This polymer is derived from the remains of mammals, including pigs and cattle [41]. Unlike polysaccharides, gelatin does not have a fixed chemical formula due to its complex protein structure. Still, it can be generally represented by the empirical formula $C_{102}H_{151}O_{39}N_{31}$ for an average gelatin polypeptide unit. Its molecular weight varies significantly depending on the extraction method and degree of hydrolysis, typically ranging from 20,000 to 300,000 Daltons.

In some processes, such as capsule production, there is a considerable accumulation of gelatin waste; this residue could be used for hydrogel synthesis [42]. Furthermore, it has been studied that the gelatin market is growing due to its versatility in applications, particularly in biomedical and agricultural areas [43].

Gelatin copolymer hydrogels with good mechanical properties have been synthesized, and their biodegradability has been proven [44].

### *4.3. Cross-Linkers*

The use of cross-linkers is crucial in the development of hydrogels derived from natural polymers, such as gelatin, as it significantly enhances both the structural integrity and functional performance of these materials. Cross-linking agents stabilize the polymer network by forming covalent or ionic bonds between polymer chains. Among the noteworthy natural cross-linking agents are Tannic Acid, Genipin, and Citric Acid.

#### 4.3.1. Tannic Acid

Tannic acid is a natural cross-linker found in plants. Tannic acid is one of the most abundant reserve materials in plants and represents a significant source of tannins. It is available commercially, with its chemical composition primarily denoted as $C_{76}H_{52}O_{46}$, corresponding to decagalloyl glucose. However, it is important to note that commercial tannic acid is generally a mixture of galloyl glucose molecules [45]. The approximate molecular weight of tannic acid is around 1701.19 g/mol. This acid is readily accessible, non-toxic, and has demonstrated its innovation in the manufacturing of biopolymers [46]. Like other natural cross-linkers, it has the advantage of being applied in different research areas. Additionally, it can enhance the physical and mechanical properties of the substances with which it is used due to the presence of hydroxyl groups in its structure [47].

#### 4.3.2. Genipin

Genipin is a naturally occurring compound extracted from the fruits *Gardenia jasminoides* and *Genipa americana*; it is used as a natural crosslinking agent in biomaterials. The chemical formula of genipin is $C_{11}H_{14}O_5$, and its molecular weight is 226.23 g/mol. It has been used more frequently in recent studies, such as in the evaluation of the performance of genipin in the synthesis of hydrogels with chitosan, as well as evaluating its functional groups [29]. It has been reported that using genipin as a cross-linker improves the absorption capacity in media of different pH levels [48].

#### 4.3.3. Citric Acid

Citric acid is a tricarboxylic acid that occurs naturally in citrus fruits. Additionally, it can be synthesized through the fermentation of carbohydrates, including starch and glucose [49]. This compound is commonly used as a pH regulator and crosslinker in polymeric systems. Its chemical formula is $C_6H_8O_7$, and its molecular weight is 192.12 g/mol.

In a copolymer, citric acid can increase the swelling ratio as the concentration increases [44]. Additionally, combinations of cellulose compounds have been studied and compared for their absorption behavior in the presence of citric acid, concluding that they are viable for agricultural use [50].

### 4.4. *Agro-Industrial Waste as a Source of Polysaccharides*

Plants are a source of polymeric compounds, and agricultural waste has been utilized to develop new technologies in recent years. These wastes are a source of cellulose and hemicellulose [9]. Once extracted, the components are used to manufacture hydrogels, which function in seed germination [51]. The following examples illustrate the utilization of agro-industrial waste as a source of natural polysaccharides to produce hydrogels.

#### 4.4.1. *Furcraea bedinghausii*

Residues from the fique plant, which belongs to the Agave family, have proven to be compatible with obtaining polymeric materials [47] and can improve the mechanical properties of hydrogels due to their cellulose content [52]. This plant is used in Colombia and is a high-production resource, producing approximately thirty thousand tons per year [52].

#### 4.4.2. *Agave tequilana* Weber

One of the plants with the highest waste production, which contains hemicellulose, cellulose, and lignin, has been utilized to develop hydrogel films. Additionally, hybrid hydrogels have been synthesized from this waste. This waste is important due to its high production and utilization [53]. *Agave tequilana* Weber, commonly known as blue agave, is primarily cultivated to produce tequila, a traditional Mexican alcoholic beverage made through fermentation and distillation of its sugars.

#### 4.4.3. *Agave lechuguilla* Torr.

This plant is a resource used to obtain the fiber called "ixtle", which is utilized in the manufacture of brushes, construction, as well as in wickerwork and basketry [54]. Lechuguilla has been found to have potential for biotechnological applications such as the production of biofuels and chemicals with high added value, in addition to having phytochemical properties such as the content of saponins, sapogenins, phenolic compounds, and fructans, as well as lignins, cellulose, hemicellulose, and antioxidant capacity [55].

## 5. Waste-Based Hydrogels and Their Impact on Reducing Water Usage in Agriculture

The incorporation of agro-industrial residues into hydrogel formulations has demonstrated significant potential to enhance water absorption and retention.

Jong et al. [51], developed a hydrogel utilizing waste from the agro-paper industry, achieving a remarkable maximum water absorption capacity of 465.5%. This impressive result is attributed to the inclusion of 3% cellulose in the formulation. The authors also reported positive outcomes in rice germination. Hydrogels have been successfully synthesized from black liquor, a byproduct of the paper industry [56], utilizing a graft copolymerization process initiated with free radicals. Research indicates that these hydrogels exhibit a significant increase in water retention capacity, measuring at 45.25%. This enhancement is attributed to the presence of lignin and polysaccharides within the black liquor, which contributes to the formation of hydrophilic groups, such as carboxyl and hydroxyl groups, on the surface of the hydrogel. Additionally, the modification of the surface plays a critical role in enhancing the mechanical properties of hydrogels produced through bulk polymerization [18].

Guancha-Chalapud et al. [52] used nanofibers (3% $w/w$) from the agro-industrial waste of *F. bedinghausii* to form hydrogels using the solution polymerization method, The authors report that this hydrogel allows reducing the irrigation frequency by up to 90%. Greenhouse experiments have showcased the remarkable benefits of hydrogels crafted from agro-industrial waste abundant in polysaccharides. In a study by Madramootoo et al. [57], the use of cellulose-rich hydrogels in greenhouse tomato cultivation resulted in a 20% reduction in irrigation water—translating to a savings of 225 mm—when compared to traditional control treatments.

The agro-industrial waste of coconut fiber has also been effectively utilized in the production of hydrogels through the graft polymerization method [58]. These hydrogels possess a remarkable water absorption capacity of 342 g of water per gram of dry gel when tested in distilled water. This can potentially enhance water availability in agricultural soils by as much as 125%.

Recently, Gayen et al. [59] conducted research in which they utilized rice straw and tamarind seed residues to create hydrogels. Their findings indicated that soil incorporated with these hydrogels demonstrated a 33% enhancement in maximum water holding capacity. Furthermore, the residue from the date palm (*Phoenix dactylifera* L.) has shown significant potential in hydrogel formulation through a carboxymethylation process, followed by crosslinking with citric acid. This approach resulted in a remarkable equilibrium swelling capacity of 700% [60].

The methods of delivering water to plants vary significantly. By assessing soil moisture [51] and measuring soil electrical conductivity [56] over time, we can evaluate the availability of water for plants. A study by Sulianto et al. [61] demonstrates that a pectin-starch hydrogel can retain 62% of water in the soil after five days, whereas soil without the hydrogel shows no water availability. Furthermore, increasing the concentration of hydrogels in the soil can lead to a 125% increase in water availability [58].

These findings highlight the remarkable potential of biopolymer-rich agro-industrial waste as sustainable precursors for hydrogel synthesis. By leveraging advanced polymerization techniques, such as graft copolymerization and solution or bulk polymerization, we can unlock innovative pathways that promise both environmental sustainability and cutting-edge applications.

## 6. Trend in the Formulation of Hydrogels with Agro-Industrial Waste

As shown in Figure 3, there has been a shift in the study trend regarding the use of agro-industrial waste for hydrogel production over the past 16 years. The study of hydrogels based on agro-industrial waste for agricultural and food technology applications reveals a clear upward trend over the past decade. Between 2009 and 2015, the volume of relevant publications in the field remained low, suggesting that it was still in the early stages of development and had not yet garnered substantial academic interest during that timeframe.

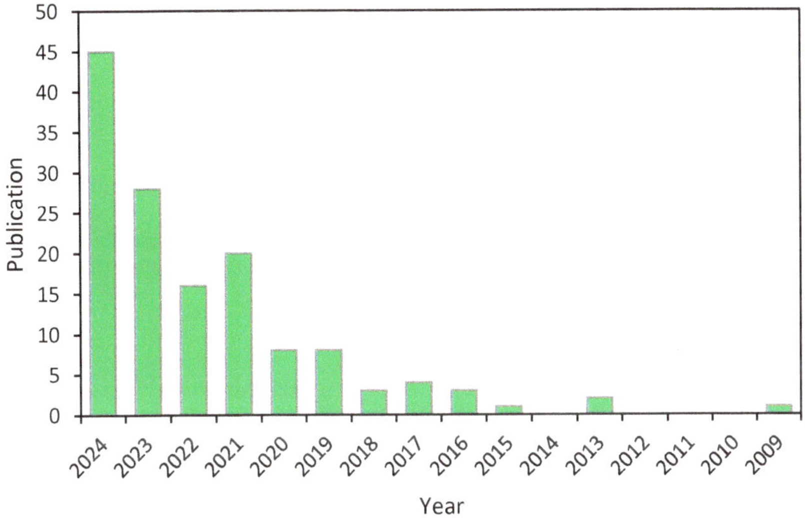


**Figure 3.** Scientific publications on hydrogels from agro-industrial and food waste sources over time: results from an advanced search in Web of Science.

Since 2016, the number of publications has steadily increased, particularly in 2019 and 2020, with eight articles published each year. From 2021 onward, this trend intensified, reaching a peak in 2024 with 45 published articles (Figure 3). This growth demonstrates that the topic has transitioned from a niche area to a recognized field of study. Continued growth is anticipated as global policies increasingly prioritize the valorization of waste materials within sustainable development initiatives.

Figure 4 illustrates the correlation between authorship and the number of articles on the research topic, highlighting the trend in publishing articles on hydrogels derived from natural waste. The topics of "hydrogel," "waste," and "water" remain relevant from 2022 to 2024. Additionally, it can be established that the keyword "hydrogel" predominates, with the highest number of correlation links. Additionally, it is worth noting how the words "cellulose" and "chitosan" are related to the topic "hydrogel" from 2022, which is relevant due to the use of natural components for obtaining biopolymers. The words "cross-linking" and "kinetics" reflect topics addressed in obtaining said biopolymers, using appropriate cross-linkers, carrying out extractions of chemical compounds, and testing their versatility. Also, the word "nanocomposite" and "sodium alginate," according to the software, is the incidence of authors who worked on those topics, but the word "nanocomposite" has greater relevance according to the graph, showing a greater number of links, due to the use of nanocomposites in the synthesis of hydrogels.

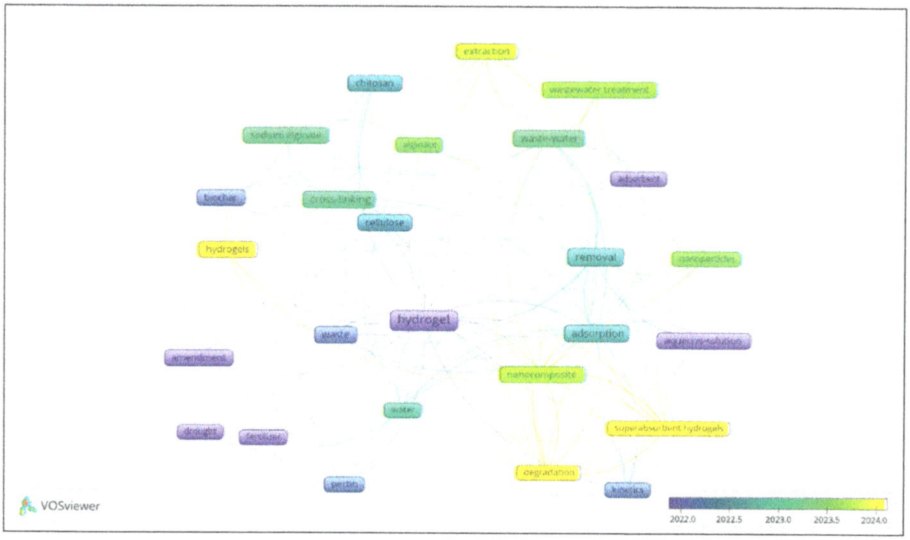


**Figure 4.** Related research trend by keywords "Based waste hydrogel" AND "Plants" from 2021 to 2024, excluding review articles, using "VOSviewer".

Additionally, Figure 5 illustrates the correlation with hydrogels, analyzing the most frequently used keywords in published articles and highlighting some possible research objectives, particularly in publications focused on Agricultural Sciences, Food Sciences, and Technology. This analysis reaffirms that the primary focus is on studying hydrogels based on natural plant residues. The word "hydrogel" is linked to three blocks denoted in different colors, representing the keywords used together for the publication of scientific articles. Figure 5 shows that the words "available water" are the second most relevant, indicating that the issue of water availability is directly related to the study of hydrogels based on agro-industrial waste.

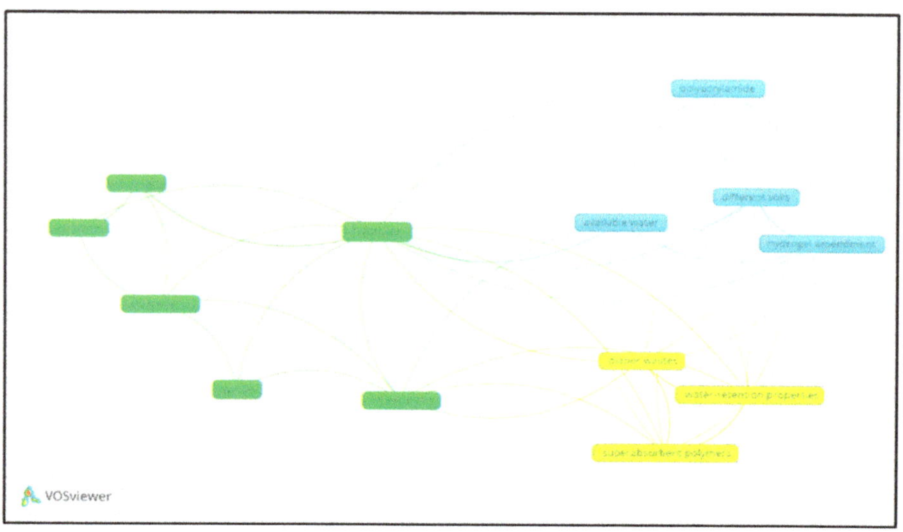


**Figure 5.** Incidence of keywords searching for "Based wasted hydrogel" AND "Plants" excluding review articles, in research areas "Agriculture and Food Science Technology".

Table 2 presents the study trend, as determined by the publication of articles, based on a systematic search using the "Web of Science" database, which defines the type of natural component used, the type of hydrogel, its application, the results found, and the authorship. Cellulose can be used as an agent in the manufacture of hydrogels. Additionally, waste from the paper agro-industry has been utilized to produce cellulose and to germinate rice seeds [51]. Guancha-Chalapud et al. report that obtaining fibers from Fique plant waste through the delignification process can be used as reinforcements of the mechanical properties of hydrogels, resulting in the improvement of hydrogels, which are capable of having greater water absorption, reducing the frequency of irrigation [52].

**Table 2.** Trends in the Use of Agro-Industrial Waste for the Development of Polysaccharide-Based Hydrogels.

| Residue | Application | Results | Polysaccharide | Swelling Capacity (%) | Authorship |
|---|---|---|---|---|---|
| Paper waste | Seed germination | Increases percentage of germination compared to soil without hydrogel | Cellulose | 465 | [51] |
| Paper waste | Tomato cultivation | They had a better response to water reabsorption | Cellulose | 415 | [57] |
| Waste from the paper industry | Water retention and controlled release of fertilizers | The values recorded for absorption capacity were better compared to the control. | Lignin | 386 | [56] |
| Fique plant residue | Reinforcing the mechanical properties of hydrogels | Increased absorption capacity | Cellulose | 474 | [52] |
| Coconut fiber | Reinforcing the mechanical properties of hydrogels | Good reswelling capacity | Cellulose and lignin | 342 | [58] |
| Rice straw and Tamarind seeds | Reinforcing mechanical properties and nutrient releaser | Great swelling capacity, long release nutrient | Cellulose | 7722 | [59] |
| Date Palm rachis | Seed germination, polymer component | Increases swelling capacity as germination seed | Lignin and Cellulose | 777.8 | [60] |
| Orange, apple, and banana peels | water content in sandy soils | High swelling capacity and increase in water content in sandy soils up to 12 days | Pectin and Starch | 400 | [61] |

These residues shown in Table 2 are usable thanks to their lignin, hemicellulose and cellulose content, from which they can be extracted or not. Each residue has favored the swelling capacity of the hydrogels that have been synthesized using it. The residue of the Fique plant [52] and the residue of the coconut plant [58] have also been useful in reinforcing mechanical properties. In the case of black liquor, without performing component extraction, it helps to retain water, releasing it over a longer time. Cellulose can also be extracted from paper waste, which, when incorporated into hydrogel polymeric networks, can increase the percentage of seed germination, in addition to its performance during crop development. The importance of these residues is that they can be environmentally friendly, and this biomass can be used to generate technology for use in agriculture.

## 7. Conclusions

Various types of hydrogels have proven effective in agricultural practices, for water harvesting, fertilizer release, and as soil substrates. However, these can be improved to make them easier to manufacture. Natural hydrogels composed of components from agro-industrial waste represent a sustainable alternative with a low environmental impact. They can improve water use, reduce production costs, and are biologically friendly. The use of waste promotes a circular economy and aligns with the United Nations Sustainable Development Goals (SDGs): SDG 2, Zero Hunger, and SDG 12, Responsible Production and Consumption. Based on a review of the literature, it was found that agricultural waste has potential for the development of hydrogels. The number of publications each year demonstrates that this topic continues to gain momentum as a growing scientific research trend. Further research is needed to obtain relevant information on the use of agro-industrial waste in developing hydrogels and formulating a completely natural hydrogel with high water absorption.

**Author Contributions:** Writing—original draft preparation, visualization, and methodology, C.F.A.-C. and C.T.-L.; methodology, validation, formal analysis, investigation, resources, writing—review and editing, and supervision, L.S.-C., A.C.-R., M.G.-C. and J.A.A.-J.; investigation, resources, writing—review and editing, D.R.A.-G.; supervision, project administration.; funding acquisition, N.R.-G. and C.T.-L. All authors have read and agreed to the published version of the manuscript.

**Funding:** This research was funded by the Secretariat of Science, Humanities, Technology, and Innovation (Secihti) (Proyecto RENAJEB-2023-17).

**Institutional Review Board Statement:** Not applicable.

**Informed Consent Statement:** Not applicable.

**Data Availability Statement:** Data are contained within the article.

**Acknowledgments:** The authors acknowledge the support from the Center for Interdisciplinary Studies and Research CEII, Autonomous University of Coahuila, Arteaga, Coahuila, Mexico.

**Conflicts of Interest:** The authors declare no conflicts of interest.

## References

1. Abdallah, A.M. The effect of hydrogel particle size on water retention properties and availability under water stress. *Int. Soil Water Conserv. Res.* **2019**, *7*, 275–285. [CrossRef]
2. Anuar, W.A.N.W.; Ramli, R.A.; El-Sayed, M.M.; Warkar, S.G. Recent study on biodegradable hydrogels for agriculture application: A review. *J. Environ. Chem. Eng.* **2025**, *13*, 115679. [CrossRef]
3. Yudaev, P.; Semenova, A.; Chistyakov, E. Gel based on modified chitosan for oil spill cleanup. *J. Appl. Polym. Sci.* **2024**, *141*, e54838. [CrossRef]
4. Tariq, Z.; Iqbal, D.N.; Rizwan, M.; Ahmad, M.; Faheem, M.; Ahmed, M. Significance of biopolymer-based hydrogels and their applications in agriculture: A review in perspective of synthesis and their degree of swelling for water holding. *RSC Adv.* **2023**, *13*, 24731–24754. [CrossRef] [PubMed]
5. Saha, A.; Sekharan, S.; Manna, U. Superabsorbent hydrogel (SAH) as a soil amendment for drought management: A review. *Soil Tillage Res.* **2020**, *204*, 104736. [CrossRef]
6. Caulfield, M.J.; Qiao, G.G.; Solomon, D.H. Some aspects of the properties and degradation of poly-acrylamides. *Chem. Rev.* **2002**, *102*, 3067–3083. [CrossRef]
7. Sahmat, S.S.; Rafii, M.Y.; Oladosu, Y.; Jusoh, M.; Hakiman, M.; Mohidin, H. A systematic review of the potential of a dynamic hydrogel as a substrate for sustainable agriculture. *Horticulturae* **2022**, *8*, 1026. [CrossRef]
8. Li, S.; Chen, G. Agricultural waste-derived superabsorbent hydrogels: Preparation, performance, and socioeconomic impacts. *J. Clean. Prod.* **2019**, *251*, 119669. [CrossRef]
9. Kabir, S.M.F.; Sikdar, P.P.; Haque, B.; Bhuiyan, M.A.R.; Ali, A.; Islam, M.N. Cellulose-based hydrogel materials: Chemistry, properties and their prospective applications. *Prog. Biomater.* **2018**, *7*, 153–174. [CrossRef]
10. Adjuik, T.A.; Nokes, S.E.; Montross, M.D.; Wendroth, O. The impacts of bio-based and synthetic hydrogels on soil hydraulic properties: A review. *Polymers* **2022**, *14*, 4721. [CrossRef]
11. Torres-León, C.; Ramírez-Guzmán, N.; Londoño-Hernández, L.; Martínez-Medina, G.A.; Díaz-Herrera, R.; Navarro-Macías, V.; Aguilar, C.N. Food waste and byproducts: An opportunity to minimize malnutrition and hunger in developing countries. *Front. Sustain. Food Syst.* **2018**, *2*, 52. [CrossRef]
12. Zhu, J.; Zhang, Z.; Wen, Y.; Song, X.; Tan, W.K.; Ong, C.N.; Li, J. Recent advances in superabsorbent hydrogels derived from agro waste materials for sustainable agriculture: A review. *J. Agric. Food Chem.* **2024**, *72*, 22399–22419. [CrossRef] [PubMed]
13. Geng, S.M.; Yan, D.H.; Zhang, T.X.; Weng, B.S.; Zhang, Z.B.; Qin, T.L. Effects of drought stress on agriculture soil. *Nat. Hazards* **2015**, *75*, 1997–2011. [CrossRef]
14. Richard, B.; Qi, A.; Fitt, B.D.L. Control of crop diseases through Integrated Crop Management to deliver climate-smart farming systems for low- and high-input crop production. *Plant Pathol.* **2022**, *71*, 1–15. [CrossRef]
15. De Pascale, S.; Dalla Costa, L.; Vallone, S.; Barbieri, G.; Maggio, A. Increasing water use efficiency in vegetable crop production: From plant to irrigation systems efficiency. *HortTechnology* **2022**, *21*, 301–308. [CrossRef]

16. Velazquez-Gonzalez, R.S.; Garcia-Garcia, A.L.; Ventura-Zapata, E.; Barceinas-Sanchez, J.D.O.; Sosa-Savedra, J.C. A review on hydroponics and the technologies associated for medium- and small-scale operations. *Agriculture* **2022**, *12*, 646. [CrossRef]
17. Fonteno, W.C.; Bilderback, T.E. Impact of hydrogel on physical properties of coarse-structured horticultural substrates. *J. Am. Soc. Hortic. Sci.* **2019**, *118*, 217–222. [CrossRef]
18. Ahmed, E.M. Hydrogel: Preparation, characterization, and applications: A review. *J. Adv. Res.* **2015**, *6*, 105–121. [CrossRef]
19. Paradelo, R.; Basanta, R.; Barral, M.T. Water-holding capacity and plant growth in compost-based substrates modified with polyacrylamide, guar gum or bentonite. *Sci. Hortic.* **2019**, *243*, 344–349. [CrossRef]
20. Zhu, J.; Tan, W.K.; Song, X.; Gao, Z.; Wen, Y.; Ong, C.N.; Loh, C.S.; Swarup, S.; Li, J. Converting Okara to superabsorbent hydrogels as soil supplements for enhancing the growth of Choy Sum (*Brassica* sp.) under water-limited conditions. *ACS Sustain. Chem. Eng.* **2020**, *8*, 9425–9433. [CrossRef]
21. Vedovello, P.; Sanches, L.V.; da Silva Teodoro, G.; Majaron, V.F.; Bortoletto-Santos, R.; Ribeiro, C.; Putti, F.F. An overview of polymeric hydrogel applications for sustainable agriculture. *Agriculture* **2024**, *14*, 840. [CrossRef]
22. Bashir, S.; Hina, M.; Iqbal, J.; Rajpar, A.H.; Mujtaba, M.A.; Alghamdi, N.A.; Wageh, S.; Ramesh, K.; Ramesh, S. Fundamental concepts of hydrogels: Synthesis, properties, and their applications. *Polymers* **2020**, *12*, 2702. [CrossRef]
23. Demitri, C.; Scalera, F.; Madaghiele, M.; Sannino, A.; Maffezzoli, A. Potential of cellulose-based superabsorbent hydrogels as water reservoir in agriculture. *Int. J. Polym. Sci.* **2013**, *2013*, 435073. [CrossRef]
24. Xiong, B.; Loss, R.D.; Shields, D.; Pawlik, T.; Hochreiter, R.; Zydney, A.L.; Kumar, M. Polyacrylamide degradation and its implications in environmental systems. *NPJ Clean Water* **2018**, *1*, 17. [CrossRef]
25. Pinzon-Moreno, D.D.; Maurate-Fernandez, I.R.; Flores-Valdeon, Y.; Neciosup-Puican, A.A.; Carranza-Oropeza, M.V. Degradation of hydrogels based on potassium and sodium polyacrylate by ionic interaction and its influence on water. *Polymers* **2022**, *14*, 2656. [CrossRef]
26. Liang, D.; Du, C.; Ma, F.; Shen, Y.; Wu, K.; Zhou, J. Degradation of polyacrylate in the outdoor agricultural soil measured by FTIR-PAS and LIBS. *Polymers* **2018**, *10*, 1296. [CrossRef]
27. Jiang, M.; Wang, B.; Ye, R.; Yu, N.; Xie, Z.; Hua, Y.; Dai, S. Evidence and impacts of nanoplastic accumulation on crop grains. *Adv. Sci.* **2022**, *9*, 2202336. [CrossRef]
28. Vogel, A.; Tentschert, J.; Pieters, R.; Bennet, F.; Dirven, H.; van den Berg, A.; Haase, A. Towards a risk assessment framework for micro- and nanoplastic particles for human health. *Part. Fibre Toxicol.* **2024**, *21*, 48. [CrossRef] [PubMed]
29. Klein, M.; Poverenov, E. Natural biopolymer-based hydrogels for use in food and agriculture. *J. Sci. Food Agric.* **2020**, *100*, 3255–3262. [CrossRef] [PubMed]
30. Chakraborty, M.; Hasanuzzaman, M.; Rahman, M.; Khan, M.A.R.; Bhowmik, P.; Bhowmik, P.; Mahmud, N.U.; Tanveer, M.; Islam, T. Mechanism of plant growth promotion and disease suppression by chitosan biopolymer. *Agriculture* **2020**, *10*, 624. [CrossRef]
31. Mortier, C.; Costa, D.C.S.; Oliveira, M.B.; Haugen, H.J.; Lyngstadaas, S.P.; Blaker, J.J.; Mano, J.F. Advanced hydrogels based on natural macromolecules: Chemical routes to achieve mechanical versatility. *Mater. Today Chem.* **2022**, *26*, 101222. [CrossRef]
32. Krishna, V.V.; Mkondiwa, M. Economics of crop residue management. *Ann. Rev. Resour. Econ.* **2025**, *15*, 19–39. [CrossRef]
33. Heise, K.; Kirsten, M.; Schne-ider, Y.; Jaros, D.; Keller, H.; Rohm, H.; Kalbitz, K.; Fischer, S. From agricultural byproducts to value-added materials: Wheat straw-based hydrogels as soil conditioners? *ACS Sustain. Chem. Eng.* **2019**, *7*, 8604–8612. [CrossRef]
34. Ismail, H.; Irani, M.; Ahmad, Z. Starch-based hydrogels: Present status and applications. *Int. J. Polym. Mater. Polym. Biomater.* **2013**, *62*, 411–420. [CrossRef]
35. Mahmoud, G.A.; Abdel-Aal, S.E.; Badway, N.A.; Abo Farha, S.A.; Alshafei, E.A. Radiation synthesis and characterization of starch-based hydrogels for removal of acid dye. *Starch/Stärke* **2014**, *66*, 400–408. [CrossRef]
36. Tian, S.; Chen, Y.; Chen, Z.; Yang, Y.; Wang, Y. Preparation and characteristics of starch esters and its effects on dough physicochemical properties. *Int. J. Food Sci.* **2018**, *1*, 1395978. [CrossRef]
37. Kou, S.; Peters, L.; Mucalo, M. Chitosan: A review of sources and preparation methods. *Int. J. Biol. Macromol.* **2021**, *169*, 85–94. [CrossRef]
38. Giraldo Pedraza, J.D.; Universidad Austral de Chile, Valdivia, Chile. Propiedades, Obtención, Caracterización y Aplicaciones del Quitosano. 2015. Available online: https://www.researchgate.net/publication/277302110_PROPIEDADES_OBTENCION_CARACTERIZACION_Y_APLICACIONES_DEL_QUITOSANO (accessed on 25 June 2025). [CrossRef]
39. Bhaladhare, S.; Das, D. Cellulose: A fascinating biopolymer for hydrogel synthesis. *RSC Adv.* **2022**, *10*, 1923–1945. [CrossRef] [PubMed]
40. Zainal, S.H.; Mohd, N.H.; Suhaili, N.; Anuar, F.H.; Lazim, A.M.; Othaman, R. Preparation of cellulose-based hydrogel: A review. *J. Mater. Res. Technol.* **2021**, *10*, 1–15. [CrossRef]
41. Haug, I.J.; Draget, K.I. *Gelatin. Handbook of Hydrocolloids*, 2nd ed.; Elsevier Inc.: Amsterdam, The Netherlands, 2009; pp. 142–163. [CrossRef]

42. Charoenchaitrakool, M.; Tulathon, P.; Meesangnil, W.; Niamnuy, C.; Seubsai, A.; Sudsakorn, K. Carboxymethyl cellulose and gelatin composite hydrogel for environmentally friendly urea delivery. *Colloids Surf. A Physicochem. Eng. Asp.* **2024**, *690*, 133774. [CrossRef]
43. Andreazza, R.; Morales, A.; Pieniz, S.; Labidi, J. Gelatin-based hydrogels: Potential biomaterials for remediation. *Polymers* **2023**, *15*, 1026. [CrossRef] [PubMed]
44. Tang, Q.; Wu, J.; Lin, J.; Fan, S.; Hu, D. A multifunctional poly(acrylic acid)/gelatin hydrogel. *J. Mater. Res.* **2009**, *24*, 1653–1661. [CrossRef]
45. Torres-León, C.; Ventura-Sobrevilla, J.; Serna-Cock, L.; Ascacio-Valdés, J.A.; Contreras-Esquivel, J.; Aguilar, C.N. Pentagalloylglucose (PGG): A valuable phenolic compound with functional properties. *J. Funct. Foods* **2017**, *37*, 1–10. [CrossRef]
46. Zheng, L.Y.; Shi, J.M.; Chi, Y.H. Tannic acid physically cross-linked responsive hydrogel. *Macromol. Chem. Phys.* **2018**, *219*, 1900234. [CrossRef]
47. Moghaddam, S.Y.Z.; Biazar, E.; Esmaeili, J.; Kheilnezhad, B.; Goleij, F.; Heidari, S. Tannic acid as a green cross-linker for biomaterial applications. *Mini-Rev. Med. Chem.* **2022**, *23*, 1320–1340. [CrossRef]
48. Meena, R.; Prasad, K.; Siddhanta, A.K. Development of a stable hydrogel network based on agar-kappa-carrageenan blend cross-linked with genipin. *Food Hydrocoll.* **2009**, *23*, 497–509. [CrossRef]
49. Franklin, D.S.; Guhanathan, S. Synthesis and characterization of citric acid-based pH-sensitive biopolymeric hydrogels. *Polym. Bull.* **2014**, *71*, 93–110. [CrossRef]
50. Demitri, C.; Del Sole, R.; Scalera, F.; Sannino, A.; Vasapollo, G.; Maffezzoli, A.; Ambrosio, L.; Nicolais, L. Novel superabsorbent cellulose-based hydrogels crosslinked with citric acid. *J. Appl. Polym. Sci.* **2008**, *110*, 2453–2460. [CrossRef]
51. Jong, S.J.; KarunaKaran, K.; Wasli, M.E.; Musa, Z.; Chin, S.F. Cellulose-based hydrogel as a natural medium for paddy seed germination. *Starch/Stärke* **2023**, *76*, 2200234. [CrossRef]
52. Guancha-Chalapud, M.A.; Serna-Cock, L.; Tirado, D.F. Hydrogels are reinforced with Colombian fique nanofibers to improve techno-functional properties for agricultural purposes. *Agriculture* **2022**, *12*, 117. [CrossRef]
53. Tovar-Carrillo, K.L.; Nakasone, K.; Sugita, S.; Tagaya, M.; Kobayashi, T. Effects of sodium hypochlorite on Agave tequilana Weber bagasse fibers used to elaborate cyto and biocompatible hydrogel films. *Mater. Sci. Eng. C* **2014**, *42*, 808–815. [CrossRef]
54. Hernández, R.; Lugo, E.C.; Díaz, L.; Villanueva, S. Extracción y cuantificación indirecta de las saponinas de agave lechuguilla Torrey. *e-Gnosis* **2005**, *3*, 3–12.
55. Carmona, J.E.; Morales-Marti, T.K.; Mussatto, S.I.; Castillo-Quiroz, D.; Ri, L.J. Propiedades químicas, estructurales y funcionales de la lechuguilla (*Agave lechuguilla* Torr.). *Rev. Mex. De Cienc. For.* **2017**, *8*, 100–122.
56. Liu, X.; Li, Y.; Meng, Y.; Lu, J.; Cheng, Y.; Tao, Y.; Wang, H. Pulping black liquor-based polymer hydrogel as water retention material and slow-release fertilizer. *Ind. Crops Prod.* **2021**, *165*, 113445. [CrossRef]
57. Madramootoo, C.A.; Jain, A.; Oliva, C.; Wang, Y.; Abbasi, N.A. Growth and yield of tomato on soil amended with waste paper based hydrogels. *Sci. Hortic.* **2023**, *310*, 111752. [CrossRef]
58. Saha, A.; Gupt, C.B.; Sekharan, S. Recycling natural fibre to superabsorbent hydrogel composite for conservation of irrigation water in semi-arid regions. *Waste Biomass Valorization* **2021**, *12*, 6433–6448. [CrossRef]
59. Gayen, T.K.; Ali, M.A.; Warkar, S.G. Cellulose microfibers-embedded carboxymethyl tamarind kernel gum hydrogels as soil conditioners and plant nutrients carriers. *ACS Agric. Sci. Technol.* **2025**, *5*, 1132–1144. [CrossRef]
60. Alsubaie, F.S.; Srdar, M.; Fayraa, O.; Alsulami, F.M.; Omran, F.; Alamry, K.A. Development of eco-friendly date palm biomass-based hydrogels for enhanced water retention in soil. *Gels* **2025**, *11*, 349. [CrossRef] [PubMed]
61. Sulianto, A.A.; Adiyaksa, I.P.; Wibisono, Y.; Khan, E.; Ivanov, A.; Drannikov, A.; Di Martino, A. From fruit waste to hydrogels for agricultural applications. *Clean Technol.* **2023**, *6*, 1–17. [CrossRef]

MDPI AG
Grosspeteranlage 5
4052 Basel
Switzerland
Tel.: +41 61 683 77 34

*Gels* Editorial Office
E-mail: gels@mdpi.com
www.mdpi.com/journal/gels

www.ingramcontent.com/pod-product-compliance
Lightning Source LLC
Chambersburg PA
CBHW061521210926
51850CB00008B/1994
*9783725862825*